EINSTEIN'S JURY

EINSTEIN'S
JURY

The Race to Test Relativity

Jeffrey Crelinsten

PRINCETON UNIVERSITY PRESS

PRINCETON AND OXFORD

Copyright © 2006 by Princeton University Press
Published by Princeton University Press, 41 William Street,
Princeton, New Jersey 08540
In the United Kingdom: Princeton University Press,
3 Market Place, Woodstock, Oxfordshire OX20 1SY

Library of Congress Cataloging-in-Publication Data

Crelinsten, Jeffrey.
Einstein's jury : the race to test relativity / Jeffrey Crelinsten.
p. cm.
Includes bibliographical references and index.
ISBN-13: 978-0-691-12310-3 (hardcover : acid-free paper)
ISBN-10: 0-691-12310-1 (hardcover : acid-free paper)
1. Relativity (Physics) 2. Einstein, Albert, 1879–1955. 3. Astrophysics—
History—20th century. 4. Physics—History—20th century.
5. Astronomy—History—20th century. I. Title.
QC173.585.C74   2006
530.11—dc22   2005032681

British Library Cataloging-in-Publication Data is available

This book has been composed in Sabon with Gill Sans display

Printed on acid-free paper. ∞

pup.princeton.edu

Printed in the United States of America

1   3   5   7   9   10   8   6   4   2

To my parents,

Dorothy and Abe Crelinsten,

in loving memory

CONTENTS

ILLUSTRATIONS

TABLES

PREFACE

THIS BOOK GREW out of a long-standing interest in the world of science and public attitudes toward scientists and their work. After obtaining two science degrees (B.Sc. in physics and M.Sc. in astronomy), I taught undergraduates for six years at Concordia University in Montreal in an innovative program called "Science and Human Affairs." The chairman of the department in those days, David Wade Chambers, now at Deakin University in Victoria, Australia, introduced me to the formal discipline of the history of science. I left teaching to take a Ph.D. at l'Institut d'histoire et de sociopolitique des sciences at the University of Montreal. My dissertation was on the reception of Einstein's general theory of relativity by the American astronomy community.[1] This book is based on the research I did for my thesis and afterwards.

I was fortunate in finding a remarkable community of scholars at the institute, all of whom, in one way or another, impressed upon me the exciting possibilities for research in this field. In particular, Lewis Pyenson, my thesis supervisor, first introduced me to the history of the physical sciences, and above all, to the literature on Einstein and the genesis and reception of his theories of relativity. This latter subject captivated me. With my background in astronomy, it was a small step to conceive the idea for the research that ultimately led to this book.

After a year of postdoctoral research, I published a major article on the early part of my study and then left academic pursuits to engage in a science-writing career. For a number of years I wrote documentaries for radio and television, before I cofounded The Impact Group, a consulting and publishing firm that specializes in science and technology communications, education, and policy.

In the intervening years since completing my dissertation and postdoctoral work, the "Einstein industry" has flourished. Before my dissertation was finished, I was fortunate to participate in some of the celebrations of the centenary of Einstein's birth. I wrote a two-hour radio biography on Einstein for the Canadian Broadcasting Corporation and a film on relativity for the National Film Board of Canada. Since then, I have watched the remarkable progress of the Einstein Papers Project, as it moved through the early years and is now preparing to publish correspondence and papers from the 1920s. I have enjoyed reading the various biographies that have come out during this period and have marveled at the increasing richness of the narratives as more historical material on Einstein becomes available. Nonetheless, I have also been struck by the fact that the story that I uncovered in my dissertation research is still largely untold.

The story of how general relativity was received among astronomers in its early years and through the 1920s is of interest because it addresses a number of important historical issues. It documents how scientists came to accept general relativity as a valid theory for serious consideration. It elucidates how attitudes toward Einstein and his theory were formed and became entrenched in the general culture, even to this day. It also covers a period of revolutionary change in astronomy and physics, both intellectually and institutionally. The shift of scientific dominance from Europe to America started in earnest in the astronomy community during this period, a decade earlier than the well-documented migration of physicists in the 1930s and after.

I have rewritten parts of my dissertation and reorganized some sections. I have added new material from my postdoctoral work and from work done by others in the intervening years. In the first part of the book, I borrow liberally from the article I published during my postdoctoral year on William Wallace Campbell and relativity in *Historical Studies of the Physical Sciences*. I thank John Heilbron, editor of *HSPS*, for granting me permission to do so.[2]

One of the more gratifying aspects of publishing a work that has been dormant for two decades is the opportunity to thank publicly the many people who were helpful to me while I was conducting archival research during my doctoral and postdoctoral years. Spencer Weart and Joan Warnow of the Center for the History of Physics, then at the American Institute of Physics in New York, showed an active interest in my project and opened the center's archives to me. I am grateful for their assistance as well as that offered to me by the staff of the Niels Bohr Library. Art Hoag, then director of Lowell Observatory in Flagstaff, Arizona, gave me access to the observatory archives. When I was at Mars Hill, he and his staff treated me to a warm hospitality that I shall never forget. While there, I also had the pleasure of meeting Bill Hoyt, who generously shared with me some of his vast knowledge regarding Vesto Melvin Slipher and the Lowell Observatory. Chick Capen kindly took time from his observing to show me Mars through the Lowell refractor. Judith Goodstein, then the archivist, and now director, at the Millikan Memorial Library of the California Institute of Technology, was most friendly and helpful during my visit there, as was her assistant Susan Trauger. The late Mary Shane, then curator of the Lick archives at Santa Cruz, made me feel as if I could stay in that lovely part of the world forever. She generously gave much of her valuable time to assist me with the files. I benefited immeasurably from her close familiarity with the archives and her long experience with the Lick community. Her late husband, Donald Shane, a one-time director of

Lick, also gave me much to think about during my visit to the Shane household. The then director of Lick Observatory, Donald Osterbrock, showed a keen interest in my project. He encouraged me greatly. He and his wife, Irene, helped to make my long stay in Santa Cruz a most pleasant one. Dorothy Schaumberg also aided my research while I was in Santa Cruz and by correspondence after I had returned to Montreal. Rem Stone gave me an illuminating tour of the Lick instruments during my visit to Mount Hamilton. While I was working in the Lick archives, I had the good fortune to meet the late Helen Wright, who was also conducting research there. Her interest and support were a source of inspiration for me. She also aided me immensely by sharing some documents that she obtained from Walter Sydney Adams that would otherwise have been inaccessible to me. Helen and I became good friends, and we continued to share insight and information through my postdoctoral work and afterwards. I also wish to acknowledge with thanks the late Allie Vibert Douglas of Kingston, Ontario, for the fascinating conversations we had in Montreal, and at her home in Kingston, about Eddington and astronomical life in the 1920s and 1930s. Her generosity in making available some manuscript material that she had collected while writing her biography of Eddington enriched my own study. At her request, I subsequently donated this material on her behalf to Cambridge University in England through David Dewhirst. George Gatewood, then director of Allegheny Observatory in Pittsburgh, welcomed me hospitably during my visit there. Wallace Beardsley of Pittsburgh enlightened me by correspondence concerning his personal acquaintance with Keivin Burns. I gratefully acknowledge the help of the archivists at the Hillman Library, University of Pittsburgh; the Seeley Mudd Library at Princeton University; the Harvard University archives; the Smithsonian Institution archives, the National Academy of Sciences, and the National Archives in Washington, D.C.; the Royal Astronomical Society, the Royal Society, and University College in London; Royal Greenwich Observatory archives in Herstmonceux; the Robarts Library, University of Toronto; and the McGill University Archives. Zane Sterns, former librarian at the David Dunlap Observatory in Richmond Hill, Ontario, gave me valuable assistance when I was doing periodical research there.

Conversations with the following during my doctoral and postdoctoral years were extremely valuable in refining various aspects of my thinking on this subject: Stanley Goldberg, John Heilbron, Karl Hufbauer, Lewis Pyenson, Mark Rothenberg, David DeVorkin, Allan Sandage, and Robert Kargon. My understanding of relativity matured during interviews I conducted while working for the Canadian Broadcasting Corporation, writing radio documentaries on aspects of science. Banesh Hoffmann, John Archibald Wheeler, Richard Feynman, Morris Kline, and Martin Klein come to mind as most helpful in developing my appreciation of the theory. I was further able to talk about related matters with many experts in the field while preparing my radio biography of Albert Einstein for the centenary year of his birth. These discussions contributed to my general appreciation of Einstein's ideas and historical circumstances. Without listing individuals here, I wish to acknowledge their influence and record my appreciation for the time they all gave to me, and for their contributions to the radio project. Thanks also to Bernie Lucht, who produced the show, and Digby Peers, executive producer.

I gratefully acknowledge financial assistance during my doctoral and postdoctoral years from the Canada Council, Social Sciences and Humanities Research Council of Canada, and Ministère de l'Education du Québec.

Several people were especially helpful to me in finding photos and archival material and granting permissions for publication. Thanks especially to Dorothy Schaumberg and Cheryl Danbridge, Mary Lea Shane archives of the Lick Observatory; Heather Lindsay, American Institute of Physics; John Strom and Tina McDowell, Carnegie Institution of Washington; Kurt Arlt, Astrophysikalisches Institut, Potsdam; Christine Bärtsch, ETH-Bibliothek, Zurich; Antoinette Beiser, Lowell Observatory; Barbara Wolff, Einstein Archives, Jerusalem; Marianne Kasica, University of Pittsburgh; Mark Hurn, University of Cambridge Observatories, England; Susanne Uebele, Archiv zur Geschichte der Max-Planck-Gesellschaft, Berlin; Norbert Ludwig, Bildarchiv Preußischer Kulturbesitz, Berlin; Jennifer Downes and Gloria Clifton, Royal Observatory, Greenwich; Peter Hingley, Royal Astronomical Society; Adam Perkins, University Library, Cambridge.

Katie Easterling and Catherine Spence provided invaluable assistance with photo research and preparing the final manuscript for submission.

Thanks to Ingrid Gnerlich and her assistant Daniel Ranbom at Princeton University Press for guiding the manuscript through the approval and submission process, to Terri O'Prey for shepherding the manuscript through production, to Alice Calaprice for her careful copyediting and helpful comments, and to Ann Truesdale for preparing the index.

There are a few friends and colleagues who inspired me to write this book. I want to thank Robert Fripp and Mark Bernstein for their encouragement and gentle persistence and for leading by example; Robert Fripp for invaluable editorial advice; Rita Fundner for comments on an early version of the manuscript; Rudy Lindner, Henry Green, and Diana Kormos Buchwald for their advice and support; and my wife, Paula, for being there for me all the way.

INTRODUCTION

THE EINSTEIN YEAR, 2005, marked the centenary of Einstein's first publication on his theory of relativity and the other remarkable scientific contributions he made in 1905. This book takes a detailed historical look at the subsequent development of relativity, one of the most dramatic scientific revolutions of the twentieth century. Through yet unpublished correspondence and manuscripts, and published papers in scientific journals and newspapers, we follow Einstein and a small community of astronomers as they first hear of Einstein's theory and become involved in testing some of its astronomical predictions. We watch as they come to realize that Einstein has opened up an entirely new field of research on the nature of the universe.

During his lifetime, Einstein was hailed as the "new Copernicus." Just as the sixteenth-century astronomer ripped us from the center of the universe to its periphery, so Einstein revolutionized our concepts of space and time. It took more than a hundred years for Copernicus's ideas of a Sun-centered universe to percolate to general culture. So it should be no surprise to us that Einstein's notions of space and time are not yet part of our everyday thinking. We still live in a Newtonian world. We still sense space as a container in which we live and move. We feel the passage of time, inexorably taking us into the future toward our mortal end. Many specialists—physicists, astronomers, mathematicians—use Einstein's equations to explore the natural world. Do they teach their children to think in terms of Einstein's space-time? Likely not. Their homes are firmly rooted in Newton's absolute space and time, just as ours are.

The fundamental concept of Einstein's relativity was that moving observers can consider themselves at rest. He first explored this notion with uniform motion. Observers moving in a straight line at constant velocity will have the same laws of physics. No experiment could be performed by any of them to determine whether they are moving or not. Einstein's innovation was to extend this principle of relativity to all physics, including electricity, magnetism, and optics. The results were startling. Observers in relative motion will not agree on simultaneity of events. They disagree about lengths of objects in the direction of motion. They disagree about rates of clocks. They disagree about mass. Velocities add in such a way that nothing can move faster than the speed of light. Energy and mass are equivalent. Matter is energy made manifest.

Einstein extended his principle of relativity to accelerated motions and ended up with a radically different theory of gravitation from Newton's. Mass-energy distorts space-time. Planets move in orbits along curved paths under no influence of any force. They "coast" along curves in the geometry of space-time. These startling ideas have revolutionized physics and astronomy. Yet they have not yet been assimilated into our thinking or our culture.

This book tells the story of how Einstein's ideas first attracted attention in a community that had a unique stake in whether or not it should take Einstein seriously. Astronomers make a living studying the solar system, stars, and nebulae and the overall structure of our universe. Einstein's theory touched upon their area of expertise and promised to change it radically. How they debated and judged the merits of Einstein reflect how new ideas are assimilated into culture and society.

Einstein had reason to care what astronomers thought about his theory. Five years after he wrote his first paper on relativity in 1905, astronomers began discussing its implications regarding gravitation and Newtonian mechanics. In 1911 Einstein published new calculations, based on his equivalence principle, predicting certain measurable astronomical effects. Tests of these predictions depended on newly developed techniques in astronomical photography and spectroscopy. Einstein tried to convince astronomers to tackle the observational problem. It was not long before a few began adapting research programs to look for these effects.

Einstein published his general theory of relativity in 1915, in the dark days of the First World War. It contained three predictions, all astronomical in nature. One of them neatly explained a long-standing anomaly in Mercury's measured orbit that had stirred up heated debates about competing theories, none of them satisfactory. The other two were a shift toward the red of spectral lines emitted by large gravitating bodies; and the bending of the path of light in a gravitational field, observable as an outward displacement of stars in the vicinity of the eclipsed Sun. After the war ended, British astronomers verified the light-bending prediction. Their announcement in 1919 catapulted Einstein and his theory to world fame. Within a few years astronomers showed that the third effect exists in the Sun; and in 1925 astronomers invoked the effect to prove the existence of incredibly dense stars called "white dwarfs."

A direct consequence of general relativity was the possibility of an expanding universe. In 1929, astronomers in California presented dramatic results concerning motions of spiral nebulae that meant the universe is indeed expanding. This research opened up a new field of relativistic cosmology in the 1930s.

Work on the astronomical consequences of relativity greatly aided its acceptance by the scientific community. Yet the vast majority of astronomers had a poor understanding of Einstein's theory. Vociferous debates occurred and international reputations were at stake. Unparalleled public interest in relativity and its originator, Einstein, created added pressures. Issues outside the realm of science came to the fore.

During this period, the center of gravity of important areas of astronomical research was shifting from Europe to the United States. While testing of astronomical predictions of general relativity was an international enterprise, Americans were acknowledged leaders in several lines of research relevant to relativity. Major observational and technological developments at two leading California observatories were instrumental in deciding the observational case for relativity. The emergence of observational cosmology was solely based on the observing programs of a few observatories in the American West.

Americans led in observational fields, particularly in the West with its great mountain observatories. Europeans were stronger in theory. The most influential theoretical work on relativity within the international astronomical community came out of England and Holland. British publications were the primary source of information about relativity for American astronomers. The fact of common language and of being allies in the war fostered particularly close ties between British and American astronomers. Yet clearly distinct national styles existed in each country. Theoretical strength in England and observational prowess in America led to a mutual interdependence, resulting in an interesting blend of cooperation and competition.

Part One of this volume sets the context for the story. Chapter 1 places Einstein within the global physics community, describes the nature of the astronomy community, and introduces several of its key players.

Chapter 2 discusses the first articles published on relativity in astronomical journals in Britain and the United States. General readers can skim or omit this chapter without losing the main story. Astronomers were an "educated public" looking at the relativity development emerging from the physics community. How they assimilated the new theory into their own discipline reflects how society later received relativity.

Part Two deals with astronomers' involvement with relativity when Einstein was still developing the general theory and after he published it during the early years of the First World War. This period covers how several astronomers adopted relativity testing into their research before the war and their continuing work through the war, to just after British astronomers verified the relativistic bending of light, announcing it to the world in November 1919.

Part Three follows the debate that raged over relativity in the postwar decade. Two leading American observatories, Lick and Mount Wilson, emerged as arbiters of the debate, based on their superior technology and impeccable reputation as leaders in astrophysical research.

Part Four relates how astronomers came to a gradual if reluctant acceptance of relativity in the latter half of the 1920s. Astronomers at Lick and Mount Wilson were able to overwhelm critics by publishing new results verifying relativity and systematically countering their specific attacks. Relativity's gravitational redshift also helped verify a dramatic consequence of Eddington's latest theory of stellar interiors—the existence of super-high-density white dwarf stars.

The Epilogue briefly describes astronomers' transition from Einstein's jury to witnesses on his behalf. In 1929, Mount Wilson astronomer Edwin Hubble discovered that spiral galaxies are receding from us, confirming general relativity's usefulness in studying the large-scale structure of the universe.

The history of how astronomers received general relativity in the first part of the twentieth century helps us understand popular attitudes held today. Many reactions, pro and con, foreshadowed attitudes expressed later in general culture after Einstein and his theory became famous. Perhaps the story of how professional astronomers tried to assimilate Einstein's theory will explain why Einstein's revolutionary ideas about space and time have not yet become part of our everyday culture.

NOTATION CONVENTION
FOR ANGULAR MEASURE

In the astronomical literature, the symbols for angular units are placed over the decimal point.

e.g.   1.″in75 (1.75 seconds of arc)
       0.″80 (0.80 minutes of arc)

ABBREVIATIONS

ORGANIZATIONS

AAAS	American Association for the Advancement of Science
AAS	American Astronomical Society
AIP	American Institute of Physics
ASP	Astronomical Society of the Pacific
CIW	Carnegie Institution of Washington
IAU	International Astronomical Union
JPEC	Joint Permanent Eclipse Committee
NAS	National Academy of Sciences
RAS	Royal Astronomical Society
RASC	Royal Astronomical Society of Canada

ASTRONOMICAL MEASURES

A	angstrom
dec.	declination
H.A.	hour angle
km	kilometer
p.e.	probable error
R.A.	right ascension
μ	micron
″	seconds of arc
′	minutes of arc

ARCHIVES

AAS	Papers of the American Astronomical Society. Located in the archives of the Center for the History of Physics, AIP.
CA	Archives of the California Institute of Technology, R. A. Millikan Memorial Library, Pasadena, California.
CP	Papers of Heber Doust Curtis. Located at the University of Pittsburgh Libraries, Pittsburgh, Pennsylvania. Call number of collection: 64:22.
CPAE	Collected Papers of Albert Einstein (see Bibliography).
HA	Harvard University Library, Pickering Collection.

HM   Hale Microfilm. Correspondence of George Ellery Hale, located
       in the archives of the Center for the History of Physics, AIP.
LC   Willem de Sitter Collection, Leiden Observatory.
LO   Mary Lea Shane archives of the Lick Observatory. Located in
       the University Archives, University of California—Santa Cruz,
       California.
LP   Papers of Carl Otto Lampland. Located in the archives of the
       Lowell Observatory, Flagstaff, Arizona.
MA   McGill University Archives. Correspondence of Louis King,
       Accession Number 454–705.
MC   Papers of William F. Meggers. Located in the archives of the
       Center for the History of Physics, AIP.
MW   Walter Adams correspondence, formerly in the possession of the
       late Helen Wright, and currently located in the archives of
       the Carnegie Institution of Washington.
NA   National Academy of Sciences, Washington, D.C.
RGO   Royal Observatory, Herstmoncaux, England (now at Cambridge
       University).
SM   Schwarzschild Microfilm. Correspondence of Karl Schwarz-
       schild, located in the archives of the Center for the History of
       Physics, AIP.
SP   Papers of Vesto Melvin Slipher. Located in the archives of the
       Lowell Observatory. Flagstaff, Arizona.
UM   Heber Doust Curtis Collection of the Michigan Historical Collec-
       tions, Bentley Historical Library, University of Michigan.
US   United States National Archives, Washington, D.C.
USNO   United States Naval Observatory.

JOURNALS

Abhandl. Preuss. Ak. W. Math.-Phys. Kl.	Abhandlungen der mathematisch-physika-lischen Klasse der königlichen Preussischen Akademie der Wissenschaften
AJ	The Astronomical Journal
Amer. J. Sci.	American Journal of Science
AN	Astronomische Nachrichten
Ann. der Phys.	Annalen der Physik
Ann. Report NAS	Annual Report of the NAS
Ann. Soc. Sci. Bruxelles	Annales de la Société scientifique de Bruxelles
Ap. J.	The Astrophysical Journal
BMNAS	Biographical Memoirs of the NAS
BMRAS	Biographical Memoirs of the RAS

BMRS	Biographical Memoirs of the RS
DSB	Dictionary of Scientific Biography
HSPS	Historical Studies in the Physical Sciences
JBAA	Journal of the British Astronomical Association
JHA	Journal for History of Astronomy
JOSA	Journal of the Optical Society of America
J. de Phys.	Journal de Physique
JRASC	Journal of the RASC
Kod. Bull.	Kodaikanal Observatory Bulletin
Kokaikanal Report	Annual Report of the Kodaikanal Observatory
LOB	Lick Observatory Bulletin
Lowell Obs. Bull.	Lowell Observatory Bulletin
MNRAS	Monthly Notices of the RAS
NW	Naturwissenschaft
PAAS	Publications of the AAS
PASP	Publications of the ASP
Phil. Mag.	The London, Edinburgh, and Dublin Philosophical Magazine and Journal of Science
Phil. Rev.	The Philosophical Review
Phys. Perspect.	Physics in Perspective
Phys. Rev.	The Physical Review
Phys. Rev. Lett.	Physical Review Letters
Phys. Z.	Physikalische Zeitung
Phys. Zs.	Physikalische Zeitschrift
Pop. Astr.	Popular Astronomy
Proc. BAAS	Proceedings of the the BAAS
Proc. NAS	Proceedings of the NAS
Pub. Allegh. Obs.	Publications of the Allegheny Observatory
Rev. Mod. Phys.	Reviews of Modern Physics
Sci. Monthly	Scientific Monthly
Sitzgsb. Ak. W. München	Sitzungsberichte der mathematisch-physikalischen Klasse der Königlichen Bayerischen Akademie der Wissenschaften zu München
Trans. IAU	Transactions of the IAU
Verh. D. Deutsch. Phys. Ges.	Verhandlungen der Deutschen physikalischen Gesellschaft
Zs. Astrophys.	Zeitschrift für Astrophysik

PART ONE

1905–1911
Early Encounters with Relativity

Chapter One

EINSTEIN AND THE WORLD COMMUNITY

OF PHYSICISTS AND ASTRONOMERS

Einstein Enters the World Stage

Einstein introduced his theory of relativity into a world that was changing dramatically. Scientific research and technological development were increasingly seen as valuable resources for nations. While applied research went on in industry, most basic physics research around the world was done in academic institutions. The normal path for professional advancement was to find a job as a professor. In Germany, a student who succeeded in obtaining an academic post had "arrived": "The professor had reason to be proud of himself. He had outdistanced most of his fellow graduate students. His 10,000 marks a year placed him in the upper bourgeoisie: in Prussia, for example, less than 1 percent of the population had incomes in excess of 9,500 marks in 1900." German-speaking Europe was the place to be for physics, and Berlin was the center of the physics universe. The leading countries worldwide were the United States, Germany, the United Kingdom, and France. These "big four" had the largest number of academic physicists, the highest total expenditure on academic physics, and produced the largest number of physics papers in leading journals. Though America had the greatest quantity, Germany led the world in quality and prestige. Being published in a German-language physics journal meant that your work would be read widely and by the best physicists in the world. German-speaking universities were spread all over Europe, from Switzerland to eastern Europe as far as Russia, and also in South America. Theoretical physics, which was Einstein's specialty, was a relatively new subdiscipline within physics. The center of gravity for theoretical physics was in German-speaking Europe. Germany led the world's nations in number of university chairs in theoretical physics. The Netherlands had the highest concentration of theoretical chairs per physics post. In the United Kingdom, and even more so in the United States, experimental physics was more common. In the United States, theoretical physicists were rare.[1]

Einstein was completely unknown when he came up with his theory of relativity.[2] He was a junior patent clerk in German-speaking Switzerland when he published his first article on relativity in 1905 at the age of twenty-six. His theory was revolutionary. Since the age of sixteen, he had been wondering about the nature of light and what the world would look like from the vantage point of a light beam. His intuition told him that it was not possible to travel at the speed of light, because no one had ever observed what he expected a person riding a light beam would see. Nor did theory predict it. Einstein knew that uniform motion (constant speed in a straight line) should not affect the laws of mechanics. We experience this principle when we sit in a moving car—it feels the same as if we were stationary. Without looking out the window, we can't tell the difference between moving at uniform speed and sitting at rest. No experiment that we perform in the car will tell us whether we are moving or are at rest. Observers in different states of uniform motion can assume they are at rest and the laws of mechanics will still hold true. This is the principle of relativity. Einstein elevated this principle for all physics, including electrodynamics and optics. All observers in uniform motion can assume they are at rest. No optical or electrodynamics experiment that we perform will tell us whether we are moving or not, including measuring the speed of light. Everyone gets the same answer.

The consequences of this simple statement are startling. Einstein showed that all observers agree on a certain combination of space and time measurements, but not on specific lengths and time intervals. Someone moving past you at high speed will tell you your meter stick is shorter than theirs, and you will measure their meter stick as shorter than yours by the same amount. The extent of the contraction depends on your relative speed. Their clock will run slower than yours, but they will tell you your clock is running slower than theirs. Yet you will both agree when you each measure the speed of light. If you measure the speed of light emitted by the headlight on a speeding train from the ground, you will get the same value as a person standing on the train. The reason you agree is that your meter sticks and clocks measure different lengths and times.

Einstein's theory yielded the same equations that Dutch physicist Hendrik Antoon Lorentz had derived using different concepts. Physicists had hypothesized the existence of a luminiferous ether to account for how light waves propagate. They believed that light comprises undulations in an otherwise invisible ether, which permeates all of space. Lorentz had developed an electron theory to explain why it is impossible to measure Earth's motion through the ether. His idea was that objects compress as they move through the resisting medium. He came up with the same formula for length contraction that Einstein got. Moving lengths contract in the direction they move, the amount of contraction being greater the faster the speed. Einstein's theory captured the fact that each observer sees the other length as contracted. His theory eliminated the need to interpret the contraction physically. The ether became superfluous. It also predicted other consequences that Lorentz's electron theory did not. Only Einstein's theory derived the time effect: moving clocks run slow. He also showed that observers do not agree on simultaneity of events. Two simultaneous events for one observer might not be simultaneous for another. Einstein also showed that if two clocks are synchronous at the same place, and one moves at high speed on a closed path, ending up at the same starting point, the two clocks will no longer be synchronous. The traveling clock will have slowed down relative to the stationary one during the journey.

Einstein submitted his relativity article to the prestigious journal *Annalen der Physik*, published in Berlin. The coeditor, Max Planck, who was responsible for theoretical articles, saw its merit and published it.[3] At first, Einstein's paper elicited discussion only in Germany. The French ignored it as if it had never appeared. The British took some time to react to it, and then they misunderstood and reinterpreted it in terms of a mechanical, luminiferous ether. The Americans also ignored it for awhile, but then most physicists reacted strongly against it for not being practical.[4] The Göttingen mathematician Hermann Minkowski caught physicists' attention with his four-dimensional space-time formulation of Einstein's theory. At a lecture in 1908, Minkowski coined the term "space-time" and made the grand statement: "Henceforth space on its own and time on its own will decline into mere shadows, and only a kind of union between the two will preserve its independence." His formalism involved a four-dimensional space-time in which he treated time as the fourth dimension. This geometrical interpretation made it easy to calculate consequences of Einstein's theory. Yet it also encouraged physicists and mathematicians to conceptualize an absolute four-dimensional world. This notion runs counter to Einstein's entire approach, which was to eliminate the concept of absolute space, a "container" in which objects in the world reside. Einstein did not like Minkowski's formalism, thinking that the elegant mathematics confused the physics. He would later change his mind as he grappled to generalize his theory to accelerated motion.[5]

Einstein's 1905 relativity paper was third in a series of four outstanding papers he published that year. The first was a "revolutionary" paper on light quanta and the second a paper on Brownian motion. His fourth contribution was a supplement to his relativity paper, in which he derived the equivalence of matter and energy as expressed in the now-famous formula $E = mc^2$.[6] Within a couple of years, Einstein was corresponding with the leading physicists in Germany. They were shocked when they discovered that he was a lowly patent clerk and not an eminent professor. In September 1907, Einstein received a letter from the publisher of the prestigious firm of Teubner in Leipzig stating "my presses will always be at your disposal in case you have any literary plans." Several publishers approached him to write a popular account of relativity. Einstein replied: "I cannot imagine how this topic could be made accessible to broad circles. Comprehension of the subject demands a certain schooling in abstract thought, which most people do not acquire because they have no need of it."[7] This reticence to popularize his theory would haunt him, and others, for decades. Together with its unfamiliar notions of space and time, relativity's lack of accessibility to the layperson would contribute to the myth that Einstein's theory is incomprehensible.

In 1907, Einstein agreed to write a comprehensive review article on relativity for Johannes Stark's *Yearbook of Radioactivity and Electronics*. Entitled "On the Relativity Principle and the Conclusions Drawn from It," Einstein's paper went beyond his 1905 papers and introduced for the first time his attempt to incorporate gravitation into his relativity framework. "Now I am concerned with another relativity-theory reflection on the law of gravitation, by which I hope to explain the still unexplained secular changes in the perihelion distance of Mercury . . . but so far it doesn't seem to work out."[8]

In 1909 Einstein finally left the Patent Office to become extraordinary professor of theoretical physics at the University of Zurich. He did not get the post easily. First, there was no position for a theoretical physicist. Second, the physics professor, his former teacher Alfred Kleiner, had another candidate in mind, Swiss-born Friedrich Adler. These two obstacles were cleared when Adler, who had been a close friend of Einstein's at school years earlier, himself suggested Einstein, and when Kleiner was elected rector of the university and promptly created the new post. Einstein then had to demonstrate to Kleiner that he could teach. He failed at first, but then redeemed himself. The Zurich faculty voted on the matter and picked Einstein. Luckily, the faculty recognized that he was a rising star and recommended him despite his Semitic origins. Kleiner noted that "about the personal character of Dr. Einstein nothing but the best reports are made by all who know him." Personally, he was "unhesitatingly prepared to have him as a colleague in my immediate proximity." The dean added to the faculty's recommendation:

> The above remarks by our colleague Kleiner, based as they are on many years of personal contact, were the more valuable to the commission, and indeed to the department as a whole, as Herr Dr. Einstein is an Israelite, and as the Israelites are credited among scholars with a variety of disagreeable character traits, such as importunateness, impertinence, a shopkeeper's mind in their understanding of their academic position, etc., and in numerous cases with some justification. On the other hand, it may be said that among the Israelites, too, there are men without even a trace of these unpleasant characteristics and that it would therefore not be appropriate to disqualify a man merely because he happens to be a Jew. After all, even among non-Jewish scientists

there are occasionally people who, with regard to a mercantile understanding of their academic profession, display attitudes which one is otherwise accustomed to regard as specifically "Jewish." Neither the commission, nor the department as a whole, therefore thought it compatible with its dignity to write "anti-Semitism" as a principle on its banner, and the information which our colleague Herr Kleiner was able to furnish on Herr Dr. Einstein has put our minds completely at rest.[9]

The Directorate of Education still wanted Adler. By good fortune, Adler gallantly took himself out of the running and Einstein got the job.

Einstein's reputation in German-speaking Europe grew quickly. In 1910 news from Czechoslovakia that he had been nominated for a full professorship at the German university in Prague prompted Zurich University to raise his salary. An attractive offer from Prague eventually came, however, and Einstein moved his family in April 1911. It was during his stay in Prague that Einstein would return to his deliberations about relativity and gravitation. This work would bring the young genius into contact with the world of astronomy.

THE ASTRONOMY COMMUNITY

The greatest number of astronomers were in the same "big four" countries as for physics. The United States and Germany had the largest communities, followed by France and the United Kingdom, closely followed by Russia, with Italy not far behind. A distinctive feature of the world astronomical community was that it had a larger institutional base than physics. The basic home of the observational astronomer was the observatory. In each country one found many men and women who were not attached to universities and colleges. State-supported institutions mandated to provide time service and astronomical data useful for civilian needs were also important centers of basic research. They often housed leading astronomers. Theoreticians, too, were often outside of academia. Nautical-almanac departments of the navy and geodetic institutes employed computers and higher-grade celestial mechanics specialists. In addition to these state-supported institutions, private observatories funded by individuals with an interest in astronomy were common. Philanthropic foundations dedicated to financing scientific research also initiated and financed observatories. The most notable in this period were the Carnegie Institution and the Rockefeller Foundation in the United States. In addition to astronomers working in these research observatories and institutions, there were those at colleges and universities. For them, the job structure was similar to that of their physicist colleagues, with the observatory being analogous to the physics research institute or laboratory.[10]

The top position in any observatory was the directorship. In academically affiliated places, the director usually held the astronomy chair. Centers weak in astronomy often had one of the mathematics or physics professors run the observatory. The director decided the research program, allowing his staff lesser or greater freedom in following their own interests. Instrumental facilities played a key role in determining what a director might or might not do. One needed a good refractor for double-star work.[11] Large reflectors were necessary for nebular photography. Climate also played a role. Spectroscopy required less atmospheric transparency than photometry. Observatories located in poorer climates, or near large cities, concentrated on stellar radial velocities and other spectroscopic work.[12] Within these constraints, the director could do what he wished, but he usually had to make compromises along the way. For example, Heber D. Curtis, who played an important role in the relativity story, took the director's position at Allegheny Observatory in Pittsburgh after spending years at Lick Observatory on Mount Hamilton in California. The mountain observatory had offered a clear sky perfect for nebular photography. Less than a month after taking up his new post, he wrote to his former chief at Lick, William Wallace Campbell: "This place for several years has been just a parallax machine, without much chance for individual work. Am planning to reduce the program, very gradually, and without destroying the value of unfinished work, to about six-tenths of its present scope, so that everyone can have some time for himself. But it will long remain, I think, one of the things we can do here."[13] A year later Curtis had changed his tune:

> My first year here seems mainly to have been spent in finding out the things that I cant [sic] do here, and there are quite a lot of them . . . the California combination of instruments *plus* climate is a hard one to beat. Parallax and photometry we can do here to great advantage, however. But not photoelectric photometry, I fear. I have naturally had "in the back of my head" various plans for changing the character of the work here, but am gradually coming round to the conviction that what we are doing is not only the thing we can do best, but also the field most needed today.[14]

The parallax program was a very heavy one: "Like the old yarn of the man who 'caught' the bear, it is something which we cant let go even in part as yet without losing the value of much which has been done here in the past." Curtis estimated that it would be "a year or so yet" before he could reduce the work to the 60 percent level that was his goal.

Observatory directors varied in how much autonomy they gave to their staff. Depending on the size of the institution, there could be several different levels of seniority below the top job. In the United States the most senior were called "astronomer" or "associate astronomer," depending on how sophisticated a pay scale was needed. In Germany, these senior staff were called "Hauptobservator," in the United Kingdom "chief assistant," and in France "astronome" or "astronome adjoint." In large observatories, senior staff were in charge of one instrument and/or one of the main research programs. Within guidelines set down by the director, they had free reign on what research they chose to carry out. In smaller institutions, these ranks were rarely filled, as the director usually required an assistant before he could afford a research colleague. Middle ranks were filled by "assistant astronomer" (U.S.), "observator" (Germany), "second assistant" (U.K.) and "aide-astronome" (France). In most observatories, particularly the thriving research centers, seniority and pay scale were the only practical differences between senior and middle ranks. At the lowest rung of the observatory hierarchy one finds the "assistants" ("Hilfsarbeiter" in German and sometimes "third assistant" in larger U.K. establishments), "computers," aids, mechanics, and secretaries. The assistant might perform a host of tasks such as routine night observing, preparing lists of stars or other objects to include in one of the research programs, developing photographic plates, and measuring plates. The computer might determine orbits based on series of photographs of comets, asteroids or planets, measure and tabulate positions of spectral lines, or calculate stellar parallaxes. In major observatories, a large staff of computers might be hired with a chief computer in charge. In small observatories, a director and one or two assistants might comprise the entire staff. Assistants might in reality do the work of an astronomer at larger places, depending on the director's disposition.

Even at major observatories primarily dedicated to research, assistants might have a great deal of latitude in their work. Heber Curtis, referring almost two decades later to his first years as an assistant on the staff of Lick Observatory, claimed "about the only difference between an Assistant and an Astronomer at that hot-bed of research is that the latter is older and draws a bigger salary."[15] Not all assistants were as lucky. As we shall see, Erwin Freundlich, a young assistant at the Royal Observatory in Berlin, ran into difficulties with his director, Hermann Struve. Freundlich wanted to conduct observational tests of Einstein's theory, whereas Struve wanted to keep him occupied on routine observational and computational tasks.[16]

THE ASTROPHYSICS REVOLUTION

The latter part of the nineteenth century ushered in two new technologies—photography and spectroscopy. For the first time since humans began to observe the heavens, astronomers could study the motions and physics

Figure 1.1. Lick Observatory on the summit of Mount Hamilton, California, near San Jose, ca. 1923. The large dome houses the 36-in refractor. (Courtesy Mary Lea Shane Archives of the Lick Observatory, University Library, University of California-Santa Cruz.)

of stars. The spectroscope allowed astronomers to study the stars' chemical composition and their velocities in the line of sight. Photography provided permanent recording of images that astronomers could measure and analyze in detail. These dramatic changes marked the beginnings of a shift away from the older tradition of positional astronomy toward study of the physics of celestial objects. A new field of research—astrophysics—was born.[17]

As the fledgling discipline began to flourish about the turn of the century, the American astronomical community took a leadership role. American astronomical journals began to appear, stemming the tide of American papers flowing across the Atlantic for publication in European journals. The Americans designed and built new research observatories and applied advanced technology to astrophysical problems with great vigor. Four major observatories devoted to astrophysical research went into operation around this time—Lick Observatory in northern California (fig. 1.1); Lowell Observatory near Flagstaff, Arizona; Yerkes Observatory in Williams Bay near Chicago; and Mount Wilson Observatory in southern California (fig. 1.2). All of them were privately funded.

Figure 1.2. Mount Wilson Observatory in the San Gabriel Mountains outside Pasadena, California, 1931. *Left to right*: horizontal Snow telescope, 60-foot solar tower, 150-foot solar tower, 60-inch telescope dome, 50-foot interferometer building, and dome of the Hooker 100-in telescope. The first telescope put into operation was the Snow telescope completed in 1905 and used by George Hale. The 60-ft solar tower was erected in 1908; the 60-in telescope was completed in 1909; the 150-ft solar tower was completed in 1912; and the Hooker 100-in telescope was tested successfully in 1917. (Courtesy Carnegie Institution of Washington.)

The combination of advanced equipment and excellent observing conditions thrust the Pacific observatories into the forefront of astrophysical research. Percival Lowell founded his observatory in Flagstaff because of the excellent "seeing" in the desert climate (figs. 1.3a,b). The astronomer Vesto Melvin Slipher utilized the favorable observing conditions, advanced technology, and his ingenuity to pioneer spectroscopy of faint nebula with the Lowell 24-inch refractor (fig. 1.4). Lowell was primarily interested in planetary work, especially observations of Mars, though he allowed Slipher to spend half his time on his own interests. In 1909, Lowell assigned Slipher the task of photographing spectra of spiral nebulae in the hope of learning more about the origin of our solar system. At the time, he and many other astronomers believed that nebulae were in our own Milky Way stellar system and were birthplaces of stars. By 1912, Slipher was able to photograph the spectrum of the Andromeda

Figure 1.3a. Lowell Observatory near Flagstaff, Arizona. Dome housing the 24-inch Clark refractor. (Courtesy Lowell Observatory.)

nebula, an object so faint that no one had succeeded before him. The spectrum showed a remarkable displacement of lines toward the red. The redshift was larger than for any other object, indicating an enormous velocity of recession. Slipher got the first good plate of the spectrum, showing displacement of spectral lines, in August 1912, three years after he first took up the problem. The exposure took nine hours on a single night.[18] Slipher shifted his work to other spirals, dominating the field for years until a 100-inch reflecting telescope went into operation at Mount Wilson. Slipher's work revolutionized nebular spectroscopy and opened an avenue of research that led directly to the discovery of the expanding universe. He took over the observatory directorship after Lowell's death.

William Wallace Campbell had been appointed to the Lick Observatory staff in 1891, where he had use of the Lick 36-inch refracting telescope (fig. 1.5a,b). The Lick instrument was then the largest of its kind in the world and later second only to the 40-inch refractor of the Yerkes Observatory.[19] Campbell designed a new and powerful spectrograph for use with the 36-inch. With it he set new standards of precision in stellar spectroscopy.[20] When Lick director James Keeler died suddenly and unexpectedly at the turn of the century, the Lick trustees consulted twelve leading astronomers for advice about a successor. They all chose Campbell.[21] As director of Lick Observatory, Campbell embarked on a systematic pro-

Figure 1.3b. Percival Lowell observing at the 24-inch Clark telescope. (Courtesy Lowell Observatory.)

gram to determine radial velocities of stars brighter than a specified value. His goal was to determine statistically the structure of the stellar system and the Sun's motion through it. With financial assistance from Darius Ogden Mills, a friend of the observatory who had financed Campbell's spectrograph, the new Lick director had a second spectrograph built. He set it up at a station in Chile, so that the southern part of the sky could be included in his massive observing program.[22]

In 1911, the year that astronomers first learned of Einstein's gravitational light-bending prediction, Campbell published preliminary results

Figure 1.4. Vesto Melvin Slipher observing with the Brashear spectrograph mounted on the 24-inch telescope. (Courtesy Lowell Observatory.)

of the Lick radial velocity program. His paper caused a stir. Campbell showed that there was a systematic shift in spectra of certain types of stars. If interpreted as a velocity, these shifts meant that the stars were receding from the Sun by as much as 4 km per second. The cause of this so-called K-term became an important research topic. More than a decade later, Campbell's K-term would be interpreted and misinterpreted in terms of Einstein's general theory of relativity.[23] Campbell's accurate and systematic program on radial velocities inspired many other observatory directors to put similar programs on their research agendas. Campbell's technical and organizational experience in the area put him in demand as a consultant, and his reputation and that of his observatory spread rapidly.

Campbell was also a champion for science and a strong American scientific community. He was an experienced fund-raiser and organizer, continually looking for ways to fund science. Consider the circular letter he sent in 1915 to Western scientists as president of the American Association for the Advancement of Science:

Figure 1.5a. William Wallace Campbell in 1893 at the eye end of the Lick 36-inch refractor. (Courtesy of the Mary Lea Shane Archives of the Lick Observatory, University Library, University of California–Santa Cruz.)

The Pacific Division of the American Association for the Advancement of Science desires to form a card catalogue of the men and women in the Pacific region who would assist, or at least take a friendly interest, in the advancement and dissemination of knowledge. . . . Our present want is especially a list of those who are trustees or supporters of educational institutions, art galleries, museums, libraries, etc.; of physicians, attorneys, merchants, and

Figure 1.5b. Campbell thirty years later, at the same telescope, several months before he announced results of eclipse observations made in Australia in 1922. (Courtesy of the Mary Lea Shane Archives of the Lick Observatory, University Library, University of California–Santa Cruz.)

other leading citizens who take a personal interest in the intellectual advancement of their communities; and of persons who have a special interest in
some branch of science. Would you be willing to form such a list for your
own region, but letting it include addresses in any part of your state? We
need to know full names and post office addresses, titles, occupations or
professions, and we should be glad to have a line describing their leading
interest or service.[24]

This combination of organizational ability and attention to the wider support of science led the University of California to ask Campbell to be
president in 1922. The National Academy of Sciences made the same
request after he had retired some eight years later.[25] For the first two decades of the twentieth century and more, Campbell directed research at
Lick Observatory with characteristic acumen.

Campbell's opposite number at Mount Wilson, George Ellery Hale,
was scientific entrepreneur par excellence (fig. 1.6). Hale, more than any
one person, contributed to the advancement and institutionalization of
astrophysics research in the United States. He was the driving force behind
Yerkes and Mount Wilson. After graduation from M.I.T. in 1890, he built
his own 12-inch telescope at home and set up Kenwood Observatory,
which he ran for six years. In August 1894, while associate professor of
astrophysics at the University of Chicago, he started the *Astrophysical
Journal*. He was also one of the founding members of the American Astronomical and Astrophysical Society.[26] Hale was among the first American
astronomers to dedicate himself single-mindedly to astrophysical research
as opposed to the older positional astronomy. As a boy he had the opportunity to help George Washington Hough, director of the Dearborn Observatory in Evanston, Illinois, with time determination. Though he loved
doing it at the time, he decided then that such work could never satisfy
him. "The reason lay in the fact that I was born an experimentalist, and
I was bound to find the way for combining physics and chemistry with
astronomy."[27]

After founding Yerkes Observatory and making it one of the world's
leading astrophysics observatories, Hale was drawn to the clear skies out
West. He established the Mount Wilson Solar Observatory at Mount Wilson in southern California and became its first director. As soon as research began at Mount Wilson, the world's astronomers were amazed at
the technical advances. In 1908 Walter Sydney Adams published remarkable results from a spectroscopic investigation of the Sun's rotation. Cape
of Good Hope astronomer Jacob Halm congratulated Hale on Adams's
paper; but, he admitted, "while reading it I could not suppress a feeling
of sadness at the astounding fact that he could do on one plate what took
me a whole year's troublesome visual observations."[28] Hale's vision was

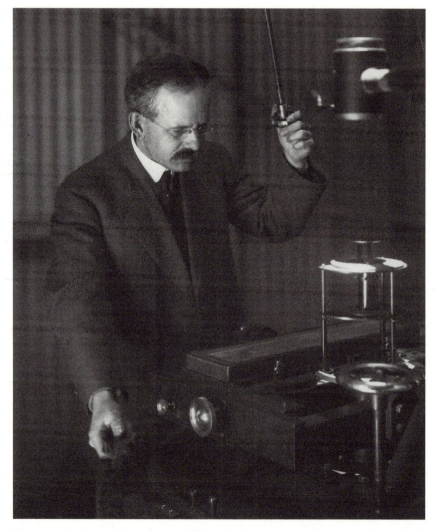

Figure 1.6. Astronomer George Ellery Hale operating his spectroheliograph, which he used to study magnetic fields in sunspots. (Courtesy Carnegie Institution of Washington.)

to establish the world's leading center for research, including the observatory, a physical laboratory, and leading theoreticians nearby. He was instrumental in creating the California Institute of Technology to provide a first-rate education and research institution that would complement the work of the observatory.[29]

While the Americans were developing a strong capability in astrophysics research, the leading European countries were by no means insignificant in comparison. Some of the most influential early pioneers in astrophysics were British. During the first decade of the twentieth century, the largest proportion of observatories engaged in astrophysical work was in Britain. Solar physics particularly interested British astronomers. More solar research occurred in Britain than in the other leading European countries and the United States. Germany, too, had a strong commitment to astrophysics and to research in general. On the eve of the First World War, Edward Charles Pickering, director of Harvard College Observatory, could boast that "the United States has attained an enviable position in the newer departments of astronomy," meaning astrophysics. Yet he wondered whether the American lead could be maintained. He noted, "In Europe, especially in Germany, observatories and instruments of the highest grade are now being constructed, the government furnishing appliances with the most liberal hand."[30]

The outbreak of World War I and subsequent years of international chaos drastically changed the situation. After the war, Germany was thrown into a financial crisis that severely affected resources for scientific research. William F. Meggers of the Bureau of Standards in Washington traveled in Europe for four months during the summer of 1921. He visited scientists in England, Holland, France, Germany, Austria, Switzerland, and Italy. He observed that German and Austrian scientists were in a bad situation. The Germans could not afford subscriptions to foreign scientific journals. In many cases they were reticent to try, because of resentments due to the war. Inflation was crippling Austria, reducing the University of Vienna's annual appropriation for its physics institute to one-sixtieth of its 1914 level, or "less than 15 shillings in British money!"[31] In the fall of 1923 a concerned Meggers wrote Heinrich Kayser in Bonn asking incredulously whether what he had been reading in U.S. newspapers about the German mark was true or whether it was a joke. Meggers had been "so positive that your country would survive that I bet on marks when they were 3 for 1 cent and now they are 1,000,000 for 2 cents."[32]

With the economic situation as bad as it was, Germany could not compete on the international scene. Scientific isolation from allied countries exacerbated the situation. French astronomers adamantly opposed having any association with German colleagues. Though the British were less single-minded about the issue, the general feeling among European astronomers was to exclude Germany from the International Astronomical Union, which the allied countries created after the war.[33] The American position was equally anti-German. Unlike some of their transatlantic colleagues, the Americans did not want to extend the ban to neutrals and urged that they be admitted to the union as early as possible. At the Brussels meeting of the International Council in July 1919, the neutrals were invited to membership.[34] Germany was another story. At Campbell's initiation,[35] the board of directors of the Astronomical Society of the Pacific removed the Berlin Observatory from a list of six nominating observatories for the society's Bruce Medal. They substituted the National Observatory of the Argentine Republic at Cordoba.[36] In spring 1922 Joel Stebbins, secretary of the American Astronomical Society, canvassed members on the issue of renewed relations with the Germans. Out of twenty-nine respondents, nine were opposed and twenty were for renewal, "but more or less with reservations to the extent that we should not do anything to offend the French and Belgians."[37] The British, as winners, did not suffer as much as the Central Powers; and, due to their physical isolation from the continent, not as much as their French allies. The United States came out ahead of the Europeans in general. In astronomy their strong position before the war turned to world leadership after it.

EUROPEAN BRAINS AND AMERICAN MONEY

In 1905, the same year that Einstein published his first article on relativity, Hale successfully launched the International Union for Solar Research. Its fourth conference took place at Mount Wilson in August and September of 1910. About one hundred astronomers attended the event, many of whom were from Europe. One of the principal decisions that emerged from the sessions was to extend the union to incorporate stellar astrophysics.[38] Among the festivities was a tour of the observatory, where Hale had established a sophisticated laboratory for spectroscopic research as part of the installation. The guests included Karl Schwarzschild, the dynamic director of the Astrophysical Observatory in Potsdam; Heinrich Kayser of Bonn; and a young German spectroscopist, Heinrich Konen. Konen was on an extended visit to U.S. observatories and laboratories as part of a traditional *Studienreise*, or study travel year. He submitted a report of his trip, which included the following assessment of American astrophysics and physics: "One ought not to be deceived by well meaning articles in the newspapers which repeat the old view that Americans have money and institutes but no researchers or ideas. Perhaps this is still true in other fields, but in astrophysics and physics it has been out of date for a long time, and implies a fatal error. The Americans possess both, men and ideas, money and instruments; and they apply them with reckless energy."[39] The European perception of America as the nouveau riche of science persisted in Britain as well as Germany. In February 1911, the president of the Royal Astronomical Society used the phrase "British brains and American money" when referring to lunar tables calculated by Ernest W. Brown, a British-born theoretical astronomer who had crossed the Atlantic to take a position at Yale. Ironically, he was arguing for more support for the older astronomical discipline of astrodynamics, which he felt was being neglected in favor of the newer fields of astrometry and astrophysics.[40]

On his *Studienreise*, Konen had perceived the strength of the American astrophysics community. Men like Hale had organized it into an extremely interdisciplinary group. Spectroscopists and other physicists interacted with the astronomers. Everyone tried to solve problems that bore directly or indirectly on astrophysical questions. The physicists Konen met in this milieu were largely practical men whose contact with practitioners from other disciplines generated exciting new lines of research. Most of them were not theoretical physicists. In fact, the Europeans were not far off in their assessment of the Americans' theoretical abilities. Hale, Campbell, and others trying to build up the research community in America were keenly aware of their deficiencies in theory.

Theoretical developments such as relativity and quantum mechanics coming out of Germany increasingly motivated these leaders to beef up their strength in modern theoretical physics. They relied on European institutions to give the younger astronomers more physics training. Paul Merrill got his Ph.D. in astronomy from the University of California at Berkeley in 1913. After completing his thesis at Lick Observatory, he landed a job as astronomy instructor at the University of Michigan. Before he left, Campbell had advised him that "a knowledge of the modern developments of Physics" would be of "value to an astronomer." Two years later, Merrill asked Campbell if he should go abroad after the war to study in Europe. Campbell replied:

> My opinion on the importance to an astronomer of a knowledge of the modern developments of physics is stronger now than when you were here. Next academic year might be the time to go abroad to Cambridge, Manchester, Paris, or Germany, provided the war ends within the next few months, but this is not probable; and to plan for residing in Europe while the war is in progress would be folly. Professor Millikan, of the University of Chicago, is a good man in certain phases of modern physics, but you would get so many other advantages through European experience I would advise waiting for conditions there to improve.[41]

Robert Andrews Millikan had been a student of Albert Abraham Michelson, winner of the Nobel Prize for physics in 1907. Michelson's pioneering work in optical interferometry had earned him an international reputation as an experimental physicist of the highest caliber. Millikan, trained in the same tradition, became equally renowned for measuring the elementary electronic charge, experimentally verifying Einstein's quantum formula for the photoelectric effect and measuring Planck's constant.[42] Millikan was a fine experimenter, but he was no theoretician. His friend, physicist Frank B. Jewett, president of the American Institute of Electrical Engineers, presented the Institute's Edison medal to Millikan in 1923, the same year that Millikan received the Nobel Prize in physics. Jewett had sent him a copy of the speech he would be giving at the presentation. At Millikan's request, he deleted the following sentence: "I do not think that Millikan is a great physicist in the sense that we look upon Newton, Kelvin, Helmholtz or JJ Thomson, that is, as a man who has produced or will produce revolutionary ideas." Jewett obviously thought highly of Millikan, but he understood that he was a specific brand of experimentalist. The great theoreticians he had mentioned had come from the United Kingdom and Germany.[43]

In the United Kingdom, mathematicians trained at Cambridge University had a virtual monopoly on physics posts as well as astronomy. The term "wrangler" was given to graduates who obtained first-class degrees in mathematics at Cambridge. "Senior wranglers" came first in the infamous tripos examination. In 1914, the observational astronomer Arthur Hinks complained "the whole trend of policy in Cambridge & England generally . . . is to take astronomical posts as sustenance for mathematicians." Hinks was bitter because he had been passed over in favor of a younger man, Arthur Stanley Eddington (fig. 1.7), to succeed Sir Robert Ball as director of the observatory at Cambridge. Hinks had been chief assistant since 1903. Eddington was an up-and-coming theoretician of exceptional ability. In 1904 he was the youngest senior wrangler in the history of Cambridge and became one of the world's leading theoretical astrophysicists. Hinks was a positional astronomer of the old school and did not like the shift in orientation toward astrophysics. He felt that he rightfully deserved the Cambridge directorship and resigned rather than stay as chief assistant under Eddington. "They must have been mad to imagine that a man who had had the ambition to do what I had been able to do would be content with an inferior position and no fun all his life."[44] In addition to having math virtuosos in astronomy posts in England, one also found theoretical astronomers holding chairs in applied mathematics. For example, James Jeans, who later became one of Britain's leading theoreticians in astronomy, began his career in the early 1900s as a lecturer in mathematics at Trinity College in Cambridge. Not being able to obtain a chair at the college, in 1905 he moved to Princeton University in the United States to take a chair in applied mathematics. Only when the incumbent holding the Stokes chair for mathematics at Christ's College retired could Jeans return to a Cambridge post.[45] Jeans played an important role in the 1920s in the theoretical side of research at Mount Wilson.

Hale was keenly aware of the European strength in theory. He was diligent in maintaining contact with European theorists, who in turn val-

Figure 1.7. Arthur Stanley Eddington, Plumian Professor of Astronomy, University of Cambridge. (Courtesy Niels Bohr Library, AIP.)

ued the observational strengths at Mount Wilson. During the fall of 1917, James Jeans was preparing his book on cosmogony, which became one of the classics in the field. He wrote to Hale for permission to include some Mount Wilson material. "On selecting photographs," he related to his friend, "I am not surprised to find that all the 16 which I should like to have permission to reproduce come without exception from Mount Wilson." Jeans admitted to feeling "a little embarrassed at asking for permission to illustrate my book entirely from Mount Wilson photogs., but venture to do so, as yours are so preeminently the best." About a year later, Hale visited Jeans at his estate at Bex Hill in England. He wrote enthusiastically to his wife: "I had a delightful visit with the Jeans at Bex Hill. You may remember that he was a member of the Princeton faculty some years ago. He is an extremely able mathematical physicist and astronomer—Schuster says Rayleigh compares him with Poincaré. He has recently worked out a theory of stellar evolution and we are cooperating with him in his studies of the nature of spiral nebulae." Four years later Hale made the informal collaboration with his theoretical colleague more official. He offered Jeans a Mount Wilson research associateship for the year 1923.[46]

The perception in America that theoretical expertise in physics must be sought in Europe persisted into the 1920s. It was reinforced in the middle of the decade when developments in quantum theory began to pour out of Germany.[47] The new ideas had important applications to astrophysics. The American theoretical astrophysicist Henry Norris Russell applied himself almost exclusively during this period to using quantum theory to calculate frequencies of spectral lines for chemical elements of astrophysical interest. American spectroscopists were very much dependent on theoreticians from abroad. William F. Meggers, at the Bureau of Standards in Washington, D.C., corresponded regularly with Russell about spectra of new elements. Yet he also kept in close touch with theoreticians like Arnold Sommerfeld in Munich, whom he invited in 1923 to speak at the Bureau on quantum theory and atomic spectra. Sommerfeld later sent Meggers his student, Otto Laporte, to spend some time there. After he left, Meggers thanked Sommerfeld profusely:

There has been such an avalanche of theoretical developments during the past year from [Wolfgang] Pauli, [Werner] Heisenberg, [Friedrich] Hunt, [Max] Born and [Pasqual] Jordan, [Erwin] Schrödinger, and others that Dr. Laporte has been more than busy keeping himself and others informed. We have depended upon him so much for this information and for its application, that we nicknamed him our Herr Geheimrat, and shall miss him very acutely in the future. Shortly after he came he organized a colloquim [sic] which has been maintained mainly by his energy and enthusiasm. All of us derived great benefit from these meetings, and we regret that the leading spirit is gone. Many of us recognize that lack of a permanent employee of Dr. Laporte's type is a serious defect in the organization of our Bureau and I have suggested to our Herr Geheimrat that we will try to get him a permanent appointment as soon as he becomes an American citizen.[48]

By this time (1926) the American astronomy community was so strong that European astronomers were coming to study and work with them more often than the reverse. In many cases, Europeans were coming to the United States solely to use the superior equipment there, particularly the large telescopes. Like Meggers and most other researchers at strong centers of experimental and applied physics, astronomers continually sought contact with European theoreticians.

This interplay between observational prowess and theoretical expertise runs like a thread through the story of how American astronomers judged Einstein's theory of relativity. Interaction between astronomers and physicists, Americans and Europeans, scientists and the public, all have a bearing on this basic theme. Relativity came from within the physics discipline, in particular from the German theoretical tradition. The Americans' lead in technological capabilities gave them a major role in relativity testing. The competition between them and the British often highlighted the perceived superiority of the Americans. Yet on the theoretical side, Americans depended on British theorists to explain what the theory was all about. When the public became obsessed with relativity after the war, astronomers had to cope with being cast in the role of expert in a domain that was largely unfamiliar to them.

CALIFORNIA ASTRONOMY: THE NATION'S LEADER

The successes of Hale and Campbell at Mount Wilson and Lick made an impression on the international astronomy community. The advantages of good seeing and large telescopes allowed these Western observatories to move quickly to the forefront of astrophysical research. During the war years, British observers noted this fact in their yearly reports of astronomical progress around the world. "The year 1915 is noteworthy for the increase of our knowledge of the motions of the nebulae, and this increase is in the main due to the activity of the Pacific observers." In 1917: "The year is remarkable for an outburst of activity among the astronomers of the Pacific Coast in the recognition of 'novae' in spiral nebulae."[49]

Even in the United States, the California community was an acknowledged leader. Hale's and Campbell's leadership and influence, both nationally and internationally, were important. Clear skies also had a lot to do with it, as Heber Curtis (fig. 1.8) realized after he had left Lick to lead the Allegheny Observatory in Pittsburgh: "Rotten weather, not very cold as yet, but cloudy most of the time. We had a record in October, 23 usable nights in succession, but November and December promise to be likewise

Figure 1.8. Heber Doust Curtis at the Crossley reflector. (Courtesy Mary Lea Shane Archives of the Lick Observatory, University Library, University of California–Santa Cruz.)

records, with the negative sign! I sure would like a good Mount Hamilton night and the use of the Crossley [reflector] once more!"[50]

By the 1920s, the California community was a significant force in educating astronomers. Lick's ties with the University of California at Berkeley ensured that the new generation of astronomers would be schooled in the astrophysics tradition carried on at the two great California observatories. In 1927, over 20 percent of the leading astronomers with doctorate degrees received them in California. Almost three-quarters of these had teaching or research fellowships while they were doing their studies—the highest rate of material support in the country. Students were trained in theoretical astronomy, practical astronomy of position, astrophysics, and modern physics. As part of their degree requirements they could carry out their own investigations at one of the top research observatories in the country, Lick Observatory. Their chances of getting a job after graduating were excellent. A large number of them found positions at Mount Wilson or at Lick. This excellence in astronomical research and education combined to give Pacific astronomy a leading place in the American community. In 1927, over half of the astronomers who were members of the prestigious National Academy of Sciences were located in California or were elected when they lived there.[51]

In the following story, California astronomers emerge as pacesetters in the research and discussion that surrounded Einstein's theory of relativity. Their great strength in observation and relative weakness in theory would have a major impact on how they judged Einstein.

Chapter Two

ASTRONOMERS AND SPECIAL RELATIVITY:

THE FIRST PUBLICATIONS

As news of einstein's relativity papers began to percolate from Germany to other countries, some astronomers tried to explain to their colleagues what was going on. The first articles on relativity that appeared in astronomical journals introduced the community to the new theory before astronomers began to conduct research on the subject. Since relativity eventually attracted enormous interest, expositions of the theory would later be in great demand. Astronomers mined journals for early articles on the topic. Hence these first publications exerted some influence on attitudes toward the theory and its comprehension.[1] In America, the first paper to deal explicitly with relativity in an astronomical journal was published in October 1911 by Lick astronomer Heber Curtis. In his bibliography, Curtis included four articles written on the subject in England, a few works by American physicists, and a large number of German publications. The British works reveal a lack of appreciation of Einstein's innovative approach and a desire to reinterpret his work into the more familiar ether.

Henry Crozier Plummer and the Problem of Aberration

The Oxford astronomer Henry Crozier Plummer was the first astronomer to publish an article dealing with the "principle of relativity" in a British astronomical journal. His father, William Edward, also an astronomer, was senior assistant at Oxford's University Observatory. Following in his father's footsteps, Plummer studied at Oxford, graduating with first-class honors in mathematics. He received his Oxford M.A. eleven years after his father. In 1901 he was hired as second assistant at the observatory. He was there when he learned of relativity. In 1912 he was appointed Astronomer Royal at the University of Dublin and director of Dunsink Observatory. Plummer's primary interest was dynamical astronomy. In 1918, he published a book on the subject, which is still used as a text and reference in theoretical and practical celestial mechanics.[2]

Plummer's paper was the second of two articles on the theory of aberration.[3] Astronomers noticed this effect in the eighteenth century. To ob-

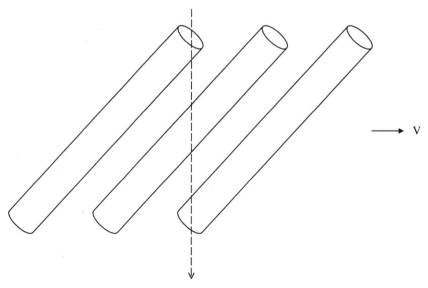

Figure 2.1. Aberration of starlight. Astronomers must tilt their telescope slightly forward in the direction of the Earth's orbital motion (V) so that starlight will pass through the length of the telescope from end to end as it moves with the Earth. (Tilt angle of telescope highly exaggerated.)

serve a star through a telescope, the instrument has to be inclined very slightly in the direction of the Earth's motion along its orbit. This minute tilt is required so that light entering the telescope will hit the other end dead center (fig. 2.1). In the split second it takes for starlight to go from one end to the other, the telescope has moved along the Earth's orbit. At the time, scientists believed that light is a wave disturbance in the ether, an invisible substance that permeates all space. As you move through the ether, scientists expected that your speed and the speed of light would add so that you would measure a slightly different light speed, depending on your direction of motion. As the Earth moves along its orbit, the position of nearby stars shifts slightly relative to the distant stars. Astronomers expected they could measure the speed of the Earth relative to the ether by comparing the aberration angle of nearby stars at different parts of Earth's orbit. All attempts to do the measurement failed. Theoreticians were able to prove that to a first-order approximation, "it is impossible for the observer to detect his own absolute motion in the ether." Plummer noted that the field of physics had recently escalated the problem. The Michelson-Morley experiment showed that even to second order, motion relative to the ether is undetectable. He related that the electronic theory of matter had explained this new finding. "The result is that the Principle of Relativity, with its far-reaching implications, has obtained a cardinal position in modern science."[4]

Plummer introduced the "principle of relativity" to his astronomical colleagues as "a product of the last decade" and "due chiefly to Professor Lorentz." In a footnote he explained that the "fundamental hypothesis concerning the contraction of matter in motion is due to [the Irish physicist George Francis] Fitzgerald." He credited Sir Oliver Lodge as being the first to mention Fitzgerald's suggestion in print in 1892. He thanked "Professor Whittaker" for this reference.[5] In the 1950s, Sir Edmund Taylor Whittaker perpetrated the belief that special relativity was largely the work of Fitzgerald and Lorentz, being merely amplified by Einstein.[6] Plummer's acknowledgment is the earliest indication that Whittaker might have developed his position as early as 1910. Indeed, we shall see in the next section that he had.

To Plummer, the important result was that everything could be expressed in terms of the effects of motion relative to the stationary ether. For him, the innovation had come directly from the electron theory of matter and ether. He referred to Einstein once in his article: "The laws of stellar aberration and of the Doppler effect are at once deducible, as Einstein has shown."[7] In a footnote, he cited Einstein's 1905 paper. Plummer was interested only in Einstein's mathematical derivations and had not worked through his basic ideas regarding the nature of space and time.

Of particular interest to Plummer were the implications for the theory of gravitation: "[I]n practice we do not actually observe the apparent motion of the Sun and use the result to correct our observed positions of the stars. The motion which we do use is derived by calculation from the theory of gravitation. Hence, if we are to be consistent, we must regard Keplerian motion as an appearance, not as a reality. And here we come in contact with the general problem of the dynamics of the electron, which in the historical sense is responsible for the introduction of the principle of relativity." Plummer saw the "principle of relativity" as signaling the ascendancy of electrodynamics, perhaps to encompass gravity. Referring to the relativistic transformations of length, time, and mass, he remarked: "The result of the work of Lorentz and others is to show that these transformations suffice to explain the complete compensation of effects arising from the motion of any system through space over the whole field of electrodynamics as well as of optics. The same will be true of gravitation if gravity can be expressed in terms of electrodynamic entities."[8] Later commentators on relativity would echo the hope that gravity might perhaps be incorporated into the electromagnetic view of nature.[9] Plummer concluded, "The developments of modern physical theory concern the astronomer no less than the physicist."[10]

In 1910, like most British physicists at the time, Plummer viewed the theory as elevating the status of the "principle of relativity." He interpreted the Lorentz transformations as "compensatory effects" due to matter-ether interactions, masking the optical effects thought to be taking place due to motion through the ether. Having emerged from electrodynamics to explain the indetectability of the Earth's motion through the ether, relativity might now lead to an electrodynamic theory of gravitation.

EDMUND TAYLOR WHITTAKER: RELATIVITY AND THE ETHER

The second paper on relativity published for the British astronomical community appeared the following month as a Council Note in the Royal Astronomical Society's *Monthly Notices*.[11] Its author, E. T. Whittaker, was Astronomer Royal of Ireland and Andrews Professor of Astronomy in the University of Dublin, though he was more a mathematician by training and profession. Like many mathematics specialists in England, Whittaker was active in the Royal Astronomical Society, acting as secretary from 1900 to 1906. During this period, he taught mathematics, theoretical physics, and astronomy at Cambridge, where he was regarded as "a mathematician with a strong interest in astronomy." Both James Jeans and Arthur Eddington studied with him at Cambridge. Whittaker became Astronomer Royal of Ireland in 1906.[12] The position brought with it a comfortable living arrangement at the Dunsink Observatory. His teaching requirements at the university were not demanding, giving him plenty of time to work on a comprehensive history of the ether. He remarked that this work "involved an immense amount of reading and historical research, which was made possible by the freedom and comparative leisure of Dunsink."[13] Whittaker's four years of intensive research and reading on the physics of the ether deeply influenced him. In the same year he penned the astronomical note, Whittaker published the first edition of his major historical work on theories of the ether from the time of Descartes to the end of the nineteenth century. He wrote a second, updated and more comprehensive version in 1953.[14] Science historians examining Whittaker's later monumental work on the history of the ether, published decades after Einstein's relativity theory first appeared, were perplexed by his minimizing of Einstein's contribution. This 1910 astronomical article indicates how early Whittaker's views on the matter had crystallized.

Whittaker began by noting that celestial mechanics, "based on the Newtonian laws of motion, is profoundly affected by discoveries which have been made in recent years regarding measurements of space, time and force."[15] He described how scientists had made three distinct attempts to detect "absolute motion" and "absolute velocity." To make his point, he distorted the facts somewhat, but it made a good story. The first stage, he claimed, was due to Newton, since the principle of relativity comes straight out of Newton's laws of motion. According to this principle, all frames of reference that have uniform velocities relative to each other are valid for expressing Newton's laws. It was therefore "hopeless to look to purely dynamical considerations for guidance in the recognition of absolute rest." In fact, Newton did not set out to prove the existence of absolute space, nor did he want to detect "absolute motion" relative to it. He had used the concept of absolute space to derive his equations of motion—"rest," "velocity," and motion in general were all defined relative to it. Yet one of the consequences of his laws was the principle of relativity, which stated that uniform motion (constant speed in a straight line) is in principle undetectable. These two ideas, one in a sense being a consequence of the other, contradicted each other in spirit.[16] Nonetheless, Newton retained the notion of absolute space as a standard against which acceleration could be measured. It was this very tension that Einstein addressed head on when he sought to generalize relativity and find a way to make acceleration relative. As early as 1907, he realized that this effort would require a theory of gravitation. Whittaker, however, was reconstructing history to suit his story.

The second "attempt" to detect absolute motion, according to Whittaker, was the astronomical study of proper motions in the nineteenth century.[17] He claimed that this attempt eventually failed, too. "Throughout the nineteenth century it was thought that absolute motion in space, or at any rate absolute motion relative to the general body of stars, could be determined by the astronomical study of proper motions; and the Sun was supposed to have 'an absolute velocity of about fifteen miles a second towards a point in the constellation Hercules.' . . . this result has been overthrown by the labours of [Jacobus C.] Kapteyn."[18] In fact, astronomers had been looking for ways to find out how stars are distributed in space. They were not looking for absolute motion relative to space. The "absolute velocity" of the Sun was merely a convenient expression astronomers used to indicate motion relative to the system of stars. The usual procedure was to assume that stellar motions are random on average. Then any systematic preferred motion of the stars would be due to the Sun moving in the opposite direction. According to R. L. Waterfield, "The best we can do is to take all the available proper motions of stars from all over the sky, find the average of them, and consider this as the speed and direction of the stellar system relative to the solar system, or, what comes to the same thing, of the solar system relative to the stellar system."[19]

William Herschel in England was the first astronomer to calculate the solar motion. As early as 1783, he had deduced that the Sun was moving toward the constellation Hercules. His analysis was based on the proper motions of fourteen stars, and the assumption that their motions were random.[20] At the turn of the century, the Dutch astronomer Jacobus C. Kapteyn challenged the assumption that the motions of all the stars in the stellar system are random. Since the 1890s, he had been conducting a statistical study of proper motions of stars. In 1904, he published results suggesting that there are two streams of stars drifting in different directions relative to the Sun. Eddington corroborated Kapteyn's theory of two star streams, or the two-drift hypothesis, and developed it. Karl Schwarzschild in Germany proposed a rival hypothesis of an ellipsoidal velocity distribution, based on radial velocities instead of proper motions. These researches ushered in a new field of investigations of stellar movements and the structure of the stellar system.[21] Whittaker referred to Kapteyn's research to show that astronomers could no longer use proper motion studies to determine the absolute velocity of the Sun in space. Again, he was reconstructing history to suit his story of the development of relativity.

For the third and last phase of scientists' search, Whittaker moved to electrodynamics: "But even when the failure of dynamics and astronomy to reveal absolute motion was admitted, it was still hoped that the solution might be found by aid of the theories of light and electricity."[22] Whittaker related how "numerous optical and electrical experiments" had been conducted to determine the "absolute velocity of the Earth." The expected effect "always failed to show itself." He told his astronomical audience that "at last physicists have been driven to the conclusion that a previously unrecognized compensatory influence must exist, which removes all effects of motion through the aether from the quantities which are measurable in the experiments."[23] Whittaker reported: "The nature of this compensatory influence was first discovered by Fitzgerald." In a footnote he added: "The hypothesis was adopted by Lorentz in a communication made to the Amsterdam Academy on November 26 of the same year [1892]."[24] According to Whittaker, Fitzgerald had "discovered" that "the dimensions of material bodies are slightly altered when they are in motion relative to the aether, the linear dimensions of a body in the direction of motion being contracted in the ratio $(1 - v^2/c^2)^{1/2} : 1$."[25]

Whittaker understood the contraction effect in terms of current theories of electrons and matter established by Lorentz and others, and the possibility of explaining natural phenomena in purely electromagnetic terms.[26] According to Whittaker, if one accepts that "the forces of cohesion which determine the size of material bodies are really electrical in their origin," then "the Fitzgerald contraction follows as a necessary consequence."[27] Einstein had derived the transformation equations (including the contraction effect) from considerations of space, time, and motion (kinematics). Then in an electrodynamical section, he used these purely kinematical transformations of space and time coordinates to derive transformations of the electric and magnetic field. He made no recourse to theories of matter.[28] The name of Albert Einstein did not appear anywhere in Whittaker's article. There is no indication that he had grasped Einstein's handling of basic kinematical concepts to arrive at his theory of relativity.

In the second half of his note, Whittaker introduced Hermann Minkowski's four-dimensional formulation of Einstein's theory. While the Göttingen mathematician's ideas attracted a wider scientific audience to Einstein's theory, Whittaker's 1910 exposition for astronomers contained nothing remotely close to Einstein's ideas. In fact, Minkowski did not really grasp the physical implications of Einstein's work, although he recognized that Einstein's theory penetrated to fundamental concepts of physics. Minkowski believed that he had discovered the absolute world that physicists had always hoped they would find in the ether. Only it was a four-dimensional space-time world.[29] Whittaker took the same approach, and he tied it explicitly to the ether.

Whittaker interpreted Einstein's second postulate—the constancy of light velocity—in terms of the ether that Einstein had completely rejected:

> We have already seen reason to suppose that the fundamental branch of physical science is the theory of the aether, and we are consequently led to measure space, time, and force in such a way as to give the simplest possible form to the laws of aethereal disturbance. Accordingly two philosophers, situated respectively on two stars which are in motion relative to each other, will not choose the same standards of length and time; each of them will in fact choose his standards so as to satisfy the condition that the velocity of propagation of aethereal disturbance, relative to a framework which moves with his own star, is to be reckoned equal in all directions.

For Whittaker, grounds of simplicity ruled that each observer chooses his coordinates so that he will measure the speed of light through the ether the same in all directions. "The projection of the four-dimensional world of space and time into the three-dimensional world of space and the one-dimensional world of time is therefore arbitrary." It was done in "an infinite number of ways," none having "absolute primacy over the others."[30]

Whittaker believed that Minkowski's formulation provided a way to transcend the arbitrary choices of various observers. "If we wish to describe natural phenomena in a way independent of the bias of the particular observer, we must have recourse to the language of four-dimensional analysis." His description reveals his belief that a method for dealing with absolutes had arrived: "We begin with a 'substantial point,' which represents the location of a definite particle, together with the instant at which the particle occupied this location. . . . We then proceed to define various four-dimensional vectors, the 'absolute velocity,' 'absolute acceleration,' and 'absolute force,' and formulate the law of motion in the form mass x absolute acceleration = absolute force."[31] Whittaker went on to illustrate how, using the four-dimensional "absolute" vectors, one can express the law of motion analytically. He concluded that the equations differ from those given by Newton's laws "owing to the presence of the factor $(1 - v^2/c^2)^{-1/2}$" and that "it thus appears that Newton's laws must henceforth be regarded as only approximately true."[32] Whittaker saw Minkowski's contribution as establishing an analytical tool of four-dimensional analysis and a conceptual framework for absolute space-time, of which the ether was the perfect embodiment.

Whittaker's 1910 paper reflects the prevalent British view that relativity grew out of ether physics, in particular a search for Earth's absolute motion through the ether. According to this view, the electronic theory of matter explained why it was not possible to measure this absolute motion. On this picture, the famous 1887 Michelson-Morley ether drift experiment plays a critical historical role. In fact, most books on Einstein's 1905 theory of relativity (renamed "special relativity" after 1916) state or imply that Einstein developed his theory in response to the Michelson-Morley experiment. Science historians have since shown that this view is historically inaccurate. A view of science that emphasizes observation and experiment as the precursor to any theory has reinforced this historical inaccuracy. The textbook account of relativity is nonetheless pedagogically useful. Both these factors were undoubtedly operating in the early years when physical scientists were trying to assimilate the new development. The practice has persisted in treatments of special relativity to this day, whether they were written by critics or proponents, as popular or technical treatises, in the 1920s or the 1960s or later.[33]

Thirty years later Whittaker published a revised and enlarged edition of his history of the ether, covering the twentieth-century developments around relativity up to 1926.[34] In this version, he proposed that Einstein had merely amplified work done by Lorentz and French mathematician Henri Poincaré. Max Born who knew Einstein's and Lorentz's work intimately, was in Edinburgh when Whittaker was preparing his manuscript.[35] Despite evidence to the contrary that Born showed him, Whittaker insisted on publishing his version. He also retained his assessment in a biographical memoir on Einstein that he wrote several years later for the Royal Society of London.[36]

There is evidence that Whittaker was motivated by a distaste for relativity, in particular the cosmological interpretation of Einstein's general relativity theory postulating that the universe is expanding.[37] In November

1953, he wrote to cosmologist George McVittie: "If Vol. II of my 'History of the Theories of Aether and Electricity' is ever published, it will blast some reputations. The 20th century is quite as bad as any preceding age in attributing discoveries to the wrong people." A month later he wrote again, asking McVittie for clarification on the early history of theoretical work on the expanding universe. "You will see that I am becoming rather skeptical about the expanding universe, and indeed about General Relativity altogether."[38] Whittaker was not pleased with the way science was heading, and he felt it his duty to set the record straight.

> Theoretical physics is at present in such a chaos that it will be difficult for me to write Vol. III of the "History of the Theories of Aether & Electricity." Vol. II, which covers the years 1900–1928, has already gone to the printers and should be published in the late spring. It deals with a lot of interesting things—the discoveries of special & general relativity, the quantum theory, modern spectroscopy, matrix mechanics, & wave mechanic, & I think it will be a better book than Vol. I, which was a history based on documents, whereas Vol. II is an account of events I have myself seen, and of men I have known personally. Most of them are now dead, so I can tell the candid truth about them.[39]

Whittaker clearly believed that he was making an important contribution. As we shall see, it was not uncommon that antipathy toward Einstein's later work on the general theory influenced commentators on relativity.

RELATIVITY AND SUBJECTIVISM

In November 1910, Gavin Burns presented a paper on "The Principle of Relativity" to the British Association. The published version of his talk contained reference neither to Einstein, Lorentz, Plummer, or Whittaker, nor to any Europeans at all. Burns cited only American papers for those "who wish to pursue this interesting question."[40] Burns observed:

> The advocates of the principle of relativity do not appear to assert that the alterations in the units of mass, space, and time are real physical changes. The assumed changes are of the nature of a device for the purpose of bringing the observed phenomena under ordinary mechanical laws.... Physical science assumes that the physical world is an absolute reality, and all explanations of phenomena are based on that assumption. The principle of relativity brings home to us very clearly the philosophic truth that the physical world is not an ultimate reality, but is that which the human mind conceives it to be.

This purely subjectivist view was typical of many later treatments of relativity. Burns may have found support for his interpretation in the writings of American chemists Richard Chase Tolman and Gilbert Lewis, whom he cited. Lewis and Tolman had contrasted Lorentz's early idea that the length contractions were real with what they believed was the correct situation, namely, that the distortions in a moving body were not real physical changes in the body itself. Yet they were careful not to fall into the subjectivism that Burns favored. Imagining an electron and a number of observers moving in different directions, Lewis and Tolman described how each observer would consider himself at rest. They would see the electron distorted in different directions and by different amounts. They emphasized that "the physical condition of the electron obviously does not depend upon the state of mind of the observers." They went on to state: "Although these changes in the units of space and time appear in a certain sense psychological, we adopt conceptions of space, time and velocity, upon which the science of physics now rests. At present, there appears no other alternative."[41] Burns chose to emphasize the psychological aspect. He insisted that laws of science, including the principle of gravitation, are not objective, but are creations of the human mind. Plummer and Whittaker believed that relativistic effects such as length contraction were real. Unlike them, Burns did not use ether-based interpretations to present the relativity development. He believed the effects were not real.

Einstein had a view different from all of these. He characterized theory as "free inventions of the human intellect." Science correlates such inventions with empirical facts.[42] He did not conclude that this process reveals no objective reality. On the contrary, he believed that only by building theory from a few principles of far-reaching consequence could one glean the secrets of Nature. He viewed the theory of relativity as a step toward a clearer understanding of the physical world. It revealed what is invariant in this world (independent of an observer), rather than descending into pure subjectivism. To his dismay, this philosophical debate regularly intruded into discussions of relativity during Einstein's life.

In further discussing examples of "the application of the principle," Burns displayed an empirical bent similar to that which characterized American physicists' discussions of relativity before 1910. He cited the Michelson-Morley experiment of 1887 as "direct experimental evidence of the shortening of a body in the direction in which it is moving." He referred to experiments by Alfred Bucherer in 1908 as "direct experimental evidence that the mass of a body is variable." He also pointed out a method C. V. Burton had proposed of "determining the absolute value of the Sun's motion through space from observations of the eclipses of Jupiter's satellites." He noted that according to the principle of relativity, "such a determination is impossible."[43]

Burns was not a significant player in the scientific community. Nonetheless, his reading of relativity as demonstrating the primacy of pure subjectivism foreshadowed a dominant theme that would be adopted into the general culture. To this day, the notion that "everything is relative" is commonly—and erroneously—attributed to Einstein.

Using Relativity to Calculate Planetary Orbits

In 1911, the Dutch astronomer Willem de Sitter published the first article in an English astronomical journal that went beyond merely describing relativity to astronomers. He noted that the principle of relativity, "first developed in connection with the electromagnetic theory of light, has in recent years been more and more considered as of universal application, and the claim has been made that the whole of our physical sciences should be framed in conformity with it."[44] De Sitter presented a detailed treatment of planetary orbits "from the point of view of the practical astronomer, investigating only such effects as may be expected to yield the possibility of an empirical verification of the principle." He referred to the previous papers of "Messrs. Plummer and Whittaker" and added a footnote chastising them for their reliance on the ether concept:

> Both authors make free use of the word "aether." As there are many physicists nowadays who are inclined to abandon the aether altogether, it may be well to point out that the principle of relativity is essentially independent of the concept of an aether, and, indeed, is considered by some to lead to a negation of its existence. Astronomers have nothing to do with the aether, and it need not concern them whether it exists or not. All Mr. Plummer's results remain true, and retain their full value, if the "aether" is eliminated from his terminology. And also in Mr. Whittaker's note the word "aether" is not essential, except, of course, from an historical point of view.[45]

De Sitter did not mention Einstein. "The starting point of my investigation," he stated, "has been the papers by [Henri] Poincaré and Minkowski." He attributed his derivations of the equations of motion to Poincaré's paper and added: "I also owe much to conversations with and advice from my colleague Professor Lorentz."[46]

De Sitter proceeded with a detailed treatment of planetary motion viewed from a general or arbitrary system of reference ("what Newton would call absolute space, and . . . absolute time") and a special one selected for convenience. He emphasized, however, that the general system of coordinates was not "absolute" in the sense that Newton—and the ether advocates—had used it.

> The "general" system only differs from any other possible system in that no convention is made as to its origin or the direction of its axes. It is thus not possible, for the point of view of the principle of relativity, to speak of "absolute" velocity or position otherwise than as velocity or position relative to any, not specified, "general" system of reference. If the word "absolute" is used in this sense, it is unobjectionable, but unnecessary. The laws of nature must, of course, be primarily framed with respect to the general system of reference. This does not mean that they assert anything about "absolute" motion or "absolute" time, but only that they must be so built up as to be true in whatever system of reference we choose to use.[47]

De Sitter asked, "What is the law of force that must replace Newton's law and what is the motion of a planet under this law?" If the law differs from ordinary Keplerian motion, "we shall have to consider the question whether the differences are large enough to be verified by observation."

De Sitter was able to show that the principle of relativity predicted slightly different motions of the planets than Kepler's laws. The axis of elliptical orbits will rotate very slowly so that the point of closest approach to the Sun (the perihelion) will move. All the differences between Newtonian theory and the principle of relativity turned out to be "too small to be detected from observation" except for the planet Mercury.[48] However, the German astronomer Hugo von Seeliger had already postulated that dust distributed through the solar system could account for all the residual motions of planetary perihelia unaccounted for by Newtonian theory. Such dust was not directly observable. Von Seeliger suggested the zodiacal light as indirect evidence for its existence, due to scattering of sunlight by his proposed dust rings. De Sitter noted: "Until we have some independent means of accurately determining this mass [of dust]—which seems a very remote possibility indeed—any motion of the perihelion of Mercury within reasonable limits can be so explained."[49] De Sitter showed that a combination of the principle of relativity and two out of three of von Seeliger's suggestions would satisfy many of the specific residuals for the different planets. By varying the density of one or another of von Seeliger's proposed dust rings, de Sitter could match most (though not all) of the perihelion motions of the various planets. The situation was inconclusive.

De Sitter examined C. V. Burton's proposal "to determine the velocity of the solar system 'with respect to the aether' from observations of eclipses of Jupiter's satellites."[50] He showed that, in principle, the experiment could be used to prove or disprove the principle of relativity, but found that the amounts to be measured were too small to provide a practical test. De Sitter also considered whether or not the velocity of the solar system relative to the fixed stars "is the same in any system of reference."[51] He concluded that "the problem is not changed in aspect by the introduction of the principle of relativity, and we need not further refer to it here."[52] He came to a similar conclusion regarding the definition of astronomical time. He calculated the Earth's motion two ways: time measured from the Earth's reference frame, and from the Sun's reference frame. The difference "has been found to be insensible."[53] Again, from the viewpoint of practical astronomy, it was impossible to determine any observable differences between the old and new theories. "So far as the interpretation of observations is concerned, we can identify astronomical time with the variable t of any system of reference we wish to use."[54]

De Sitter's thorough treatment showed that only one observable effect might be worthwhile for deciding on the validity of the principle of relativity. Mercury's perihelion motion showed enough of a difference between Newtonian theory and observation that it required explanation. Seeliger's dust hypotheses were ad hoc and difficult to verify by observation. De Sitter left the question open. As Einstein developed his theory of relativity further, Mercury's perihelion motion would play a critical role.

AMERICAN ASTRONOMERS' INTRODUCTION TO RELATIVITY

Heber Doust Curtis of Lick Observatory wrote the first article on relativity to appear in an American astronomical journal. Unlike Plummer, Whittaker, and de Sitter, Curtis was not theoretically inclined. He was an impeccable observer and instrument maker. He started his career as a professor of Greek and Latin at Napa College in California. He was drawn to astronomy when he discovered a small Clark refracting telescope on campus. In 1896, when the college merged at San José with what became the College of the Pacific in 1911, Curtis acquired use of a small observatory.[55] He changed his title to professor of mathematics and astronomy, and in 1900 he obtained a Vanderbilt fellowship to study for a doctorate at the University of Virginia. He spent a summer vacation at Lick Observatory and volunteered as an assistant for two eclipse expeditions—one with Lick in 1900 and the other with the U.S. Naval Observatory in 1901. Campbell invited Curtis to join the Lick staff after he received his degree, and after his graduation in 1902, he accepted Campbell's offer. For eight years he continued Campbell's studies of radial velocities of brighter stars, half the time at Lick and half at Lick's southern station near Santiago, Chile. In 1910 Curtis returned from Chile to take up the program of nebular photography with the Crossley reflector that had been started by the former director, James E. Keeler, a couple of years before his death in 1900.[56]

Curtis returned to Lick about the time that interest in relativity was mounting. He came across the papers in British astronomical journals and also read Einstein's papers and those of others. In 1911 he published a review article for astronomers containing an extensive bibliography, including the three British articles discussed previously, all the American sources cited by Burns, and papers by Einstein, Lorentz, Poincaré, Max von Laue, Minkowski, Planck, Arnold Sommerfeld, Emil Wiechert, and Paul Ehrenfest.[57] He began his treatment with a literary flourish:

> [Sir Walter] Scott describes the noble Saracen as receiving with politely concealed disbelief the Crusader's statement that water, when cold enough, would become so solid that an army could march over it; doubtless the keenest minds of the Middle Ages would have met with incredulity the assertion that the dimensions of bodies change with their temperature. The newer theories of matter and mass and the results of radioactivity have only recently torn us from the moorings of beliefs which had come to be regarded as no less immutable than eternal truth, and have left us in a receptive mood for any changes in physical theory, no matter how startling. Nevertheless, it is with considerable of a shock [sic] to our conservatism that we accept some of the conclusions of the new theory of relativity, that system of physical theory which has been developed within the past few years, and is today accepted by many of the keenest minds among the physicists of the world, some going so far as to call it the greatest advance in physical theory since the days of Newton.[58]

Curtis told his readers that the theory's conclusions "seem no less than revolutionary"—the rate of a body's motion through space affects its dimensions; mass and time are also affected; and no velocity can exceed that of light. Length contractions will distort the shape of a rotating disk. Energy and mass are equivalent. Curtis pointed out "if the theory of relativity is true, Newtonian dynamics must be abandoned." He quoted "no less an authority than Poincaré" as raising the possibility that gravitation propagates at light velocity, "bringing the principle of gravitation from its present mysterious isolation into kinship with light and electricity."[59] "It is evident that a physical theory with such possibilities as this may conceivably have many points of interest for the astronomer. In any event, the attention which it has been and is attracting is sufficient reason for its consideration here."[60] Curtis cautioned readers: "Past histories of physical theories thought at one time to be supported by the full testimony of experiment and rigorous mathematics is sufficient to cause some conservatism." While the "names of those who have accepted the theory form some guarantee of its value," he quoted Sir J. J. Thompson that "a physical theory is to be regarded as a policy, rather than as a creed."[61]

Curtis gave a brief summary of Michelson's unsuccessful attempts to measure second-order effects due "the relative movement of aether and matter" that theory had predicted. He noted that the adopted explanation for the negative result "was independently suggested by Fitzgerald and by Lorentz without regard to the then undeveloped theory of relativity." He attributed the latter to Einstein: "Suggested in part by the need for a physical system of laws which should conform to modern electrical theories of matter and which should explain these negative results, the theory of relativity has been developed, mainly by Professor Einstein of Zürich. Its list of adherents is a long one, with such names as those of Lorentz and Poincaré at the head; objections have been raised to it by Michelson, Larmor, and others, but it may fairly be said to be a widely accepted theory to-day."[62] Curtis clearly appreciated Einstein's methods. On length contraction and time dilation, he wrote: "While some of the above ratios can be derived geometrically, any one who appreciates a beautifully worked out and clearcut mathematical treatment will refer to the original papers by Einstein, or to [Emil] Wiechert's excellent résumé of the theory,[63] where the system is built up from the simple fundamental assumption and carried up through to the electromagnetic theory of light and the electronic theory of matter."[64]

Curtis derived the transformation equations between two systems in uniform motion relative to each other, following Einstein's method. He explained: "This transformation has received the name of Lorentz-transformation and is of great importance. It has been shown to be the only transformation which leaves the laws of the electromagnetic theory unchanged."[65] After deriving the Lorentz transformations, Curtis did the same for the contraction and time dilation relations, the law of addition of velocities, "to which some physicists have taken objection," and the fact that the speed of light is a limiting velocity. He pointed out that only ß-rays had physical velocities capable of testing the truth of the latter assertion, and that these "range from one third to nine tenths of the velocity of light."[66]

Curtis presented the theory thoroughly, exhibiting a working acquaintance with Einstein's basic assumptions and fundamental results. He refrained from expressing an opinion, relying on assertions from the community of physicists interpreting and developing the theory. His appreciation of the differences between Lorentz and Einstein is particularly remarkable in light of the general response to relativity in America and Britain. Curtis's almost unique position was largely due to his having systematically read the original works of Einstein and most of the German-speaking elaborators of the theory.

Curtis concluded his review with a summary of points against and for relativity. On the con side: (1) "Some physicists find the addition theorem difficult of acceptance," and (2) except for Bucherer's and Kaufmann's experimental verification of the increase of mass of ß-rays with velocity, "it does not appear possible at present to prove the theory one way or the other." On the pro side: (1) it "is at variance with no known facts," (2) it "gives a satisfactory explanation for the negative results secured by Michelson and others," (3) it "is directly supported by Kaufmann and Bucherer's experimental proof that the transverse mass of a moving electron is a function of its velocity," and (4) "It may possibly afford a theory of gravitation which shall bring this force into the realm of the other physical forces moving with the velocity of light."[67]

As for astronomical consequences of relativity, Curtis remarked there was "but little to be said." He referred to de Sitter's recent article showing that relativity predicted a periodic deviation from ordinary Keplerian motion that was too small to detect. Curtis also reproduced de Sitter's table of perihelion motions for Mercury, Venus, Earth, and two comets. The only appreciable amount, at 7.15 seconds of arc per century for Mercury, "just represents the well-known excess of observation over theory, explained by Hugo von Seeliger as due to the attraction of the mass forming the zodiacal light." Curtis concluded: "There is no other effect due to relativity which can be determined astronomically at present."[68] This situation was soon to change.

PART TWO

1911–1919
ASTRONOMERS ENCOUNTER EINSTEIN

Chapter Three

THE EARLY INVOLVEMENT, 1911–1914

Einstein's Two Predictions

In June 1911 Albert Einstein submitted a new paper to the prestigious *Annalen der Physik* entitled "On the Influence of Gravitation on the Propagation of Light."[1] For four years he had been thinking about the implications of his theory of relativity for gravitation. The fundamental tenet of his 1905 theory of relativity was that any observer in uniform motion (constant speed, moving in a straight line) could assume he is at rest, and the laws of physics would be the same for all such observers. No experiment could be performed that would tell whether or not the experimenter is at rest or in uniform motion. Could this principle be extended to acceleration? We feel the effects of acceleration, so an accelerating observer knows he is moving. He can't assume he is at rest. For this reason, Newton had felt the need to retain the idea of absolute space to account for acceleration. Einstein wanted to see if he could eliminate absolute space entirely. He soon realized that to extend relativity to acceleration would involve gravitation. His first published discussion had been in 1907 within a review article he wrote on relativity. There he first introduced a novel idea, which he called the "equivalence principle."[2]

Einstein imagined himself falling from a roof. During the fall, all bodily sensations due to gravity would disappear. Remove all contact with the surroundings by encasing him in a closed box, and he will have no visual or other sensory clues to the fact he is falling: he feels as if he is stationary and gravity has disappeared. Today, we call this state of falling "free fall" or "zero gravity." During the fall, the person feels as if he is simply floating in one spot. Einstein considered the converse situation. He imagined a place where gravity is not present, way off in outer space. If he were in a closed vehicle that was accelerating at the same rate that a person falls in Earth's gravity, he would be pressed against one side, which he would call "the floor." He would feel exactly the same as if he were motionless on Earth's surface, under the influence of Earth's gravity. Einstein took a bold step and postulated that acceleration and gravitation are equivalent. His "equivalence principle" states that a uniform and stationary gravitational field is physically indistinguishable from a system moving with a constant acceleration without any gravitation. Here was his opening to generalize relativity to acceleration. An experimenter has no way of telling whether he is at rest in a gravitational field or accelerating at the rate of a "free fall" with no gravity present.

Einstein used his equivalence principle to explore the effect of gravitation on light. He showed that a clock on the Sun's surface runs slower than a clock 93 million miles away on Earth's surface. Consequently, light emitted by an atom at the Sun's surface would have a lower frequency than light emitted by a similar atom on Earth. Passing sunlight through a spectroscope and comparing the lines of the spectrum with the same lines from a terrestrial source, the solar lines would be shifted toward the red end of the spectrum—a gravitational redshift. Einstein next imagined his accelerating room at the rate of "free fall" with no gravity present. A light on one wall sends a beam across to the opposite wall. In the instant it takes the light to cross the room, the room has accelerated "upward" a tiny amount. The light hits a spot on the opposite wall a smidgen "lower" than the light source. The light curves "downward" ever so slightly. Using his equivalence principle, Einstein predicted that the same would happen in a stationary room in Earth's gravity. Light bends in a gravitational field.

In his 1911 paper, Einstein returned to this subject.

> In a memoir published four years ago, I tried to answer the question whether the propagation of light is influenced by gravitation. I return to this theme, because my previous presentation of the subject does not satisfy me, and for a stronger reason, because I now see that one of the most important consequences of my former treatment is capable of being tested experimentally. For it follows from the theory here to be brought forward, that rays of light, passing close to the sun, are deflected by its gravitational field, so that the angular distance between the sun and a fixed star appearing near to it is apparently increased by nearly a second of arc.[3]

Einstein derived the gravitational redshift of spectral lines in a new way. "The spectral lines of sunlight, as compared with the corresponding spectral lines of terrestrial sources of light," he concluded, "must be somewhat displaced toward the red." This redshift could be measured if the conditions under which the solar bands arise were exactly known; but as other influences such as pressure and temperature affect spectral line position, "it is difficult," he concluded, "to discover whether the inferred influence of the gravitational potential really exists."

Einstein then described the gravitational light bending, and suggested how astronomers could find out if light was bent by the Sun's gravity. First, astronomers would photograph stars in the vicinity of the Sun during a solar eclipse and measure their position in the sky. Then they would wait a few months for the Earth to move along its orbit and photograph

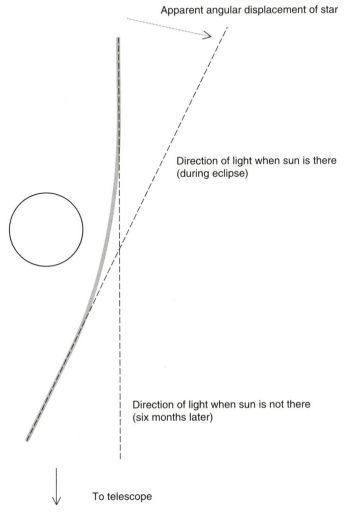

Apparent angular displacement of star

Direction of light when sun is there
(during eclipse)

Direction of light when sun is not there
(six months later)

To telescope

Figure 3.1. Bending of starlight in the sun's gravitational field. The observer sees the star displaced outward away from the Sun's position in the sky compared to the star's position when the Sun is not there.

the same stars at night when the Sun is out of the way. The stars should appear slightly farther from the Sun during the eclipse than they were when they were not nearer the Sun. This effect is due to the light from the star bending slightly toward the Sun on its way to Earth. (Fig. 3.1.)

Einstein derived a formula for the angular deviation a (in seconds of arc):

$$a = \frac{2kM}{c^2D}, \qquad\qquad (1)$$

where k is the gravitation constant, M the mass of the Sun (or any gravitating object), c the velocity of light, and D the distance from the path of the ray to the center of attraction. The farther away the light ray passes from the solar center (larger D), the smaller the angular deviation. For a ray just passing the limb* of the Sun (D = solar radius), Einstein calculated that the angular deflection would amount to about 0.83 seconds of arc toward the Sun's center. Stars in the solar vicinity during an eclipse should appear to be displaced outward from the Sun by that amount, compared to their position when the Sun is out of the way.

At the end of the paper Einstein appealed to astronomers to take up the observational challenge of looking for these effects. "It is greatly to be desired that astronomers should undertake this investigation, although the foregoing reasoning may prove to be insufficiently founded or even entirely illusory. For, aside from any theory, the question must be considered, whether with our present resources an influence of the gravitation field on the propagation of light can be established."[4]

Within a few years, two observatories in America—Lick and Mount Wilson—would absorb Einstein's predictions into their research programs. They did not, however, initiate these programs to verify relativity. The work had been going on for some time, having started years earlier. When some researchers realized they could apply ongoing techniques to test Einstein's predictions, they incorporated the test into their work.

SOLAR ECLIPSES, "VULCAN" AND THE PRINCIPLE OF RELATIVITY

The development of astronomical photography in the nineteenth century revolutionized solar research. Until then, studies of the Sun's surface had consisted primarily of observing sunspots. Photographic plates revealed fine structure in the solar surface, and observatories soon established new programs to monitor daily changes. Around this time, amateur astronomer Francis Bailey made a discovery at the "annular eclipse" of 1836 that stimulated interest in observing luminous phenomena around the Sun at eclipses. Bailey observed that just before the Moon became centrally placed over the Sun, the Sun's limb covered by the Moon appeared to break up into dazzling drops or beads of light, a phenomenon called "Bailey's beads" today. At the eclipse of 1842, astronomers systematically studied the corona and the prominences for the first time. During this

---

* The outer edge of the solar disk.

eclipse, observers tried photography without success, and they had only limited results at an eclipse in 1851. Photography was first used on a large scale and with great success at the total solar eclipse of 1860. Photographic plates confirmed the solar origin of the prominences, previously thought to be emanations from the lunar atmosphere. From then on, astronomers traveled far and wide to the zone of totality of every solar eclipse, to maximize the few minutes available to make as many observations as possible.[5]

Mounting an eclipse expedition requires long planning and great expense. Centers specializing in this line of research quickly emerged around the world. In 1894, British astronomers centralized their administrative apparatus for planning eclipses. They established the Joint Permanent Eclipse Committee (JPEC), run by the Royal Society and the Royal Astronomical Society. The JPEC organized all the main British eclipse expeditions and coordinated the securing of government funds to finance them.[6]

In the United States, it was individual observatories that financed and sent out parties to observe eclipses. The Lick Observatory rose to prominence in eclipse hunting soon after its founding. Less than one year after the observatory had gone into operation, its director, Edward S. Holden, sent an expedition to observe the total solar eclipse of January 1, 1889, which was visible in California. The well-publicized expedition boosted public interest in astronomy and helped launch the Astronomical Society of the Pacific.[7] After Campbell became director in 1901, he headed a series of expeditions that placed the Lick among the leading institutions specializing in eclipse work. As in Campbell's large radial-velocity program, his advantageous use of the new technologies of spectroscopy and photography characterized his contribution to the eclipse field.

One purely photographic problem that Campbell included in the overall Lick eclipse program was the search for the hypothetical planet "Vulcan." In the first half of the nineteenth century, the French astronomer Urbain Jean Leverrier had predicted the existence of a large planet on the outskirts of the known solar system based upon an analysis of the observed motion of the planet Uranus. In 1846, astronomers discovered Neptune using Leverrier's calculations. Yet even taking this new member of the solar family into account, Leverrrier could not fully account for the observed planetary motions using the Newtonian laws of gravitation. The largest discrepancy was for Mercury: the perihelion of its elliptical orbit advances more rapidly than could be accounted for by gravitational forces from the other planets. In 1859, Leverrier hypothesized a planet orbiting close to the Sun to account for the remaining discrepancy. In the same year, a French amateur astronomer reported a black point in transit across the face of the Sun. Leverrier assumed that his hypothetical planet had been discovered, and he christened it "Vulcan." Leverrier calculated its orbit and predicted other transits. None were ever observed, although two hundred spurious sightings of "Vulcan" were reported between 1859 and 1878.[8]

At an eclipse in 1878 two American astronomers, Lewis Swift and James Craig Watson, reported two bright starlike objects near the Sun. Neither could be identified with any of the fixed stars. Watson and Swift were careful observers, and their colleagues assumed that they had found intramercurial planets. Their "discovery" rekindled interest in the search for Vulcan, yet it was never found again. Years later, the astronomer Samuel Alfred Mitchell complained: "The reputations of these two astronomers for careful observing were so great that it cost the science of astronomy a quarter of a century of eclipse observations before it was finally decided that no intra-Mercurial planets exist which are as large or as bright as the objects supposed to have been seen."[9]

Campbell established a systematic program to resolve the issue. He ordered special lenses for a photographic search of the solar vicinity during an eclipse. He put Charles Dillon Perrine (fig. 3.2) in charge of what he called the "Vulcan Problem." Perrine worked three eclipses, in 1901, 1905, and 1908, and found only well-known stars, at least three hundred or four hundred of them on the plates of 1908. In August of that year, Campbell reported his final conclusion on the Vulcan matter: "It is felt that the Lick Observatory observations of 1901, 1905 and 1908 bring definitely to a conclusion the observational side of this problem, famous for half a century."[10] Campbell left open the possibility that instead of planets, matter too small to be detected directly might be spread uniformly around the Sun, an idea systematically developed by Hugo von Seeliger of Berlin. Von Seeliger postulated a ring of tiny particulate matter, or dust, equivalent in effect to the hypothesized Vulcan but spread out and therefore not immediately observable. For observational support, he appealed to the existence of zodiacal light, which he attributed to reflected sunlight by his postulated dust ring. He developed several forms of his idea to explain different residuals in the orbits of the inner planets, Mercury's being only the largest.[11] Campbell wrote von Seeliger after Perrine finished working on the 1908 eclipse plates: "During the past year I have been very much interested in your paper demonstrating that the outstanding residuals in the motions of Mercury and the other smaller planets are due to attractions by the material responsible for the zodiacal light. I have read your paper carefully and can make no adverse criticism. Neither have I seen criticism by others. . . . It seems to me that our observations and your theoretical deductions lend renewed interest to the zodiacal light as a subject for investigation." Von Seeliger answered that as far as he knew, his work had not been criticized, but largely ignored, especially in England. He was delighted that it had aroused interest in America.[12]

Figure 3.2. Charles Dillon Perrine when he was at Lick Observatory, ca. 1897. (Courtesy Mary Lea Shane Archives of the Lick Observatory, University Library, University of California–Santa Cruz.)

Campbell and von Seeliger did not correspond further on the subject, though Campbell remained interested in it. In 1910, Yale theoretician Ernest William Brown published a criticism of von Seeliger's work in England. Brown pointed out that Leverrier and the American astronomer Simon Newcomb had suggested similar solutions before von Seeliger, and that none was satisfactory. He maintained that von Seeliger's were no different and that his different hypotheses were ad hoc, being designed just to explain each anomaly. In April 1911, Campbell attended a meeting in which Brown presented his criticism of von Seeliger's dust hypothesis. Not having seen the published paper, he asked Brown for the reference. In correspondence, Brown told Campbell that his main criticism of von Seeliger's ideas was the "attempt to explain *several* of the anomalies by means of *several* [dust] hypotheses. . . . The existence of so much matter round the sun is [not] by any means certain on account of the large amount of light which it would reflect." In deference to his observational colleague, he added, "However, I am not an expert in that matter."[13]

Hearing that Campbell had "been interested in the elusive 'Vulcan,' " Brown asked him to help check if the Vulcan hypothesis might "account for the great inequality in the moon's motion," a subject on which Brown was an acknowledged expert. Campbell replied that earlier searches for Vulcan, done by visual methods, had led to "purely negative results, and to much wrangling," and that the "photographic method of search" in Perrine's hands had come out negative. Campbell nonetheless offered to search any regions where Brown might direct on the "magnificent large scale photographs" in the Lick collection. Brown answered that the Lick investigations "put the non-existence of such intra mercurial planets as I have postulated beyond much doubt. I hardly think that it would be worth while to re-examine the plates with such a faint hope."[14]

About this time, astronomers' discussions of the "principle of relativity" began to impinge on anomalous perihelion motion. The Dutch astronomer Willem de Sitter's 1911 article on relativity (see Chapter Two) referred to von Seeliger's work in the context of calculating the astronomical consequences of the principle of relativity. The motions of planetary perihelia so determined were of the order predicted by von Seeliger's hypothesis. De Sitter showed that by combining the principle of relativity and two out of three of Seeliger's suggestions he could account for many of the discordances observed.[15] Two years later de Sitter expanded his work on planetary perihelia in an article in a series on "Some Problems in Astronomy." Campbell monitored these developments and published a short note, summarizing attempts to account for discrepancies between the observed positions of bodies in the solar system and their predicted positions on Newtonian gravitation theory. He mentioned that "[Fritz] Wacker, [Hendrick Antoon] Lorentz, and de Sitter have discussed the principle of relativity as a possible important factor in explaining existing residuals in the motion of planetary perihelia."[16]

By the time Campbell published these remarks, he had taken an active interest in the observational side of an entirely new kind of research on the "principle of relativity." The "magnificent" collection of eclipse plates

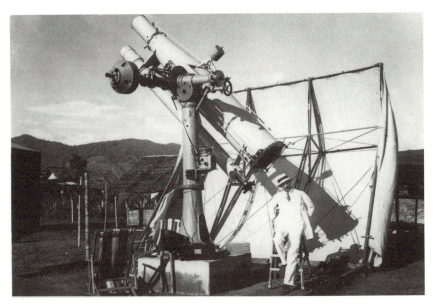

Figure 3.3. Erwin Finlay-Freundlich at a solar eclipse expedition in 1929. (Courtesy Astrophysikalisches Institut Potsdam.)

that had convinced Brown to drop the Vulcan idea in relation to the theory of the Moon's motion, and the lenses that had taken these plates, were the new research tools; but the project had nothing to do with anomalies in planetary perihelia motion. The eclipse plates and lenses helped in the first search for light bending in a gravitational field. This work evolved into a new problem, the "Einstein Problem," on the research agenda of Lick eclipse expeditions after 1912.

EINSTEIN FINDS AN ASTRONOMER

Erwin Finlay-Freundlich, a junior observer at the Berlin Observatory at Babelsburg, was probably the first astronomer to hear about Einstein's new work on the astronomical consequences of relativity (fig. 3.3). In August 1911, Leo Wentzel Pollak, a *Demonstrator* at the Institut für Kosmische Physik at the German University in Prague, paid a visit to the observatory. It was part of Freundlich's job to show visitors around. Pollak was friendly with Einstein, who had recently come to Prague as professor of theoretical physics. He told Freundlich about Einstein's new paper on the influence of gravitation on the propagation of light and his appeal to astronomers to test his astronomical predictions. Freundlich was captivated by the news. His job at the observatory consisted largely of routine observation. He had trained in mathematics and astronomy from 1905 to 1910 at Göttingen under brilliant teachers, including astrophysicist Karl Schwarzschild and mathematician Felix Klein. On Freundlich's graduation, Klein secured him a position at the observatory as assistant to the director, Herman Struve. When Freundlich protested that he knew little practical astronomy, his teacher told him, "You did not come to university in order to learn everything but in order to learn how to learn everything. You are to go to Berlin." Freundlich's work at the observatory was not intellectually challenging. It included compiling a zone catalog of pole stars, photometric observations, and working with Leo Courvoisier on the meridian circle. Pollak's news about Einstein's work promised a refreshing change.[17]

Freundlich wrote Einstein that same night, offering to help develop ways to look for light bending near the Sun or the planet Jupiter. Back in Prague, Pollak told Einstein about the young Berlin astronomer, and Einstein gave him permission to send Freundlich proofs of his article. "Prof. Einstein has given me strict orders," wrote Pollak, "to inform you that he himself very much doubts that the experiments could be done successfully with anything except the Sun." He urged Freundlich "to send further reports to me, or perhaps to Prof. Einstein, about your views on an astronomical verification."[18]

Freundlich and Einstein began corresponding on the subject, thrashing out the various possibilities for measuring the bending of light. Freundlich was concerned that the Sun's atmosphere might make observations difficult and hoped that the effect might be detectable near Jupiter. Einstein felt certain that the planet would be too small to measure. "I know very well that to obtain an answer through experience is no easy matter," he wrote, "since refraction in the solar atmosphere might interfere. Nevertheless, one can say with certainty: If no such deflection exists, then the assumptions of the theory are not correct. One must keep in view that these assumptions, even if they seem obvious, are nonetheless rather daring. If only we had a much bigger planet than Jupiter. But nature has not made it her business to make the discovery of her laws convenient for me."[19] Einstein suggested that Freundlich measure plates from a past solar eclipse taken by astronomers at the observatory in Hamburg, and Freundlich took up this line of investigation with old eclipse plates later that year.

Hoping to avoid having to wait for an eclipse, Freundlich suggested that it might be possible to photograph stars in the solar vicinity at any time. Einstein was skeptical. To his close friend, Heinrich Zangger, he mentioned Freundlich's plan "to measure the apparent positions of fixed stars in the sun's proximity in bright daylight using a shrewd method. But I do not yet believe it." He politely asked Freundlich if it "is really possible," noting that, if so, "then you shall surely be successful in determining whether the theory is valid."[20]

As luck would have it, a visitor to the Berlin Observatory in November of that year opened for Freundlich another avenue of research on the problem. Charles Dillon Perrine, who had successfully resolved the "Vulcan problem" while at Lick Observatory, had left Lick in 1909 to become director of the Southern Hemisphere observatory in Cordoba. When Freundlich told him about Einstein's light-bending prediction, Perrine suggested that he write to various astronomers who might have old eclipse plates on which star images might be measured for deflection. Naturally, he mentioned the Lick Vulcan plates. Freundlich immediately drafted a circular letter, which he sent to several observatories, including Lick, asking for "support of astronomers, who possess eclipse-plates" to test Einstein's predicted light deflection by the Sun.[21]

The first response came from Edward Charles Pickering, director of Harvard College Observatory. He credited his brother William with devising the "best method of photographing the eclipse of the Sun and stars on the same plate." William had tried it out during the eclipse of 28 May 1900, but it had obliterated images of faint stars. Pickering told Freundlich that Samuel Pierpont Langley of the Smithsonian Institution had improved on his brother's method, and that Campbell at Lick had also done so "with still better results." He did not elaborate how Langley and Campbell had improved the technique, but suggested to Freundlich that their plates might have recorded the information he wanted. Pickering did not think that the Harvard eclipse plate collection would be useful, but he mentioned that his observatory had many photographs of stars "passing very near to Jupiter," noting that these might also show the effect.[22]

Freundlich told Pickering that the effect for Jupiter would be too small to measure, but he pursued the matter of the eclipse plates.[23] He sent another copy of his circular letter to the Smithsonian Institution's secretary, who rewarded him with copies of Langley's old plates. The U.S. Naval Observatory supplied two photographs taken at the eclipse of August 30, 1905.[24] Freundlich was thrilled with the response from the American observatories, since "in Europe there hardly exist any [such plates] at all."[25]

Einstein was delighted at Freundlich's success in enlisting support from astronomers. "I am extremely pleased that you have taken on the question of light deflection with such enthusiasm and am very curious what the examination of the available plates will yield. This is a very fundamentally important question. From the theoretical standpoint, it is quite probable that the effect really exists."[26]

Freundlich's circular letter never reached the appropriate desk at Lick. Receiving no reply, he wrote a second time directly to Campbell, mentioning Perrine's advice "to apply especially to you." He asked for glass copies of all eclipse plates that showed the Sun and stars in the same field. "I need not assure you that the result of my whole investigation depends to the greatest part from your kind support and you can imagine how much I shall be obliged to you." He told Campbell that he had already heard from Pickering, Abbot, "the Naval Observatory, and the English Observatory," all of whom had promised "to support my investigation."[27]

Campbell told Freundlich that he had seen Einstein's paper and that "we shall be glad to assist you as far as possible in testing the question." He referred Freundlich to three Lick publications describing Perrine's photographs taken "in his search for intramercurial planets." Campbell promised to send positives on glass of all Perrine's plates that seemed applicable and of the corresponding chart plates of the same stellar regions taken several months before the eclipses with the same instruments; but he had reservations about the suitability of the plates. "Unfortunately none of these plates has the sun's image in the center of the field; in every case, I believe, the sun's image is on or near the edge of the plate." Campbell feared the resulting aberrations would be "troublesome," but he went on to give Freundlich the dimensions of the large plates so that he could devise a suitable measuring apparatus.[28]

In view of the likely unsuitability of the Vulcan plates for the task at hand, Campbell offered to lend the Vulcan cameras to Perrine to try Freundlich's problem at the eclipse in Brazil on October 9–10, 1912. The photographs would be taken with the Sun's image in the center of the field. Campbell advised Freundlich to write Perrine, and he, too, wrote Perrine the same day, urging him to go and suggesting the procedure to use.[29]

Freundlich was delighted with Campbell's response and eagerly awaited the arrival of the Lick plates. "Fortunately," he wrote Campbell, "the effect pronounced by Mr. Einstein decreases so rapidly with increasing distance from the Sun, that by measuring the distance of two stars more or less distant from the Sun, I am able to measure at least the simple effect nearly quite undiminished. Perhaps there will be found also a few stars both north and south from the Sun."[30] As for the coming eclipse, Freundlich accepted Campbell's offer with thanks from himself and from his director, Hermann Struve, who "also would be pleased, if one could get good plates for my purposes at the future eclipses." Struve, however, would later turn sour on Freundlich's project.

Perrine agreed to enlarge his eclipse program to include Freundlich's investigation and to take the photographs himself. Campbell sent the lenses down via the astronomer William Joseph Hussey. Perrine left Buenos Aires on September 13, 1912, and the eclipse took place on October 10. A few days later Campbell received a telegram from Edward C. Pickering of Harvard, the communication center for American astronomy: "Perrine cables from Brazil, rain."[31]

Meanwhile, measuring of the old Vulcan plates got off to a slow start. Plates sensitive enough for copying had to be shipped from New York. They arrived at Lick on April 30, 1912, and Campbell assigned Heber Doust Curtis to produce the copies.[32] Curtis had followed Einstein's work and knew the formula for light bending by the Sun. He scribbled it down on a scrap of paper (fig. 3.4a) and calculated values for the deviation for three different angular distances from the Sun:

at the limb	0.83 seconds
1 degree from the limb	0.28 seconds
6 degrees from the limb	0.06 seconds

"It should be possible," he noted, "to get this with some certainty from several hundred stars." The paper containing these calculations is pinned to another with Curtis' assessment of the plates (fig. 3.4b). Curtis concluded: "*For the purpose of Freundlichs investigation* doubt if more than 6 or 8 [plates] would be really useable."[33]

Curtis selected some of the plates from the Spain and Flint Island eclipses. He rejected those taken at Aswan in Egypt since "only a few stars were shown on these plates and the images had trailed to such an extent as to make them useless" for Freundlich's measures. On June 6, 1912, he sent the plates to Berlin via the Smithsonian Institution's Bureau of International Exchange. Curtis warned Freundlich that the measures would be difficult to make: "Even on the original negatives many of the star images are excessively faint, and can be made out only with the greatest difficulty; I fear that you will not be able to make them out at all on the copy in many cases, though I have purposely made the positives rather thin, so as not to blot out these exceedingly faint images by over-exposure." To make matters worse, the images were double on the 1908 Flint Island plates, and Curtis cautioned Freundlich "to avoid setting on small spots or defects instead" of the star images. Curtis noted that for an adequate treatment of the problem, "plates should be taken with the Sun central, and the cameras should be rated to a stellar rate, instead of to the solar rate, as was the case in all these eclipse plates."[34]

The day after Perrine's eclipse expedition had been rained out, an optimistic Freundlich, unaware of the failure in Brazil, wrote Campbell that the Vulcan plates had arrived. The plates from the Smithsonian Institution had come several months earlier, and Freundlich had been measuring the coordinates of all the stars on one of the plates, using apparatus borrowed from Karl Schwarzschild of Potsdam. The plates from the

Figure 3.4a. Notes by Heber Curtis in spring 1912 calculating light bending at different angular distances from the sun. "☉" means "Sun." (Courtesy Mary Lea Shane Archives of the Lick Observatory, University Library, University of California–Santa Cruz.)

Naval Observatory had been splintered in transit. Freundlich, sanguine about the possibilities with the Smithsonian and Lick plates, had not requested replacements.[35]

The old eclipse plates proved as disappointing as the news from Brazil. On all the plates Freundlich received, including "the very valuable ones of the Lick Observatory," the insufficiently sharp star images made a "suc-

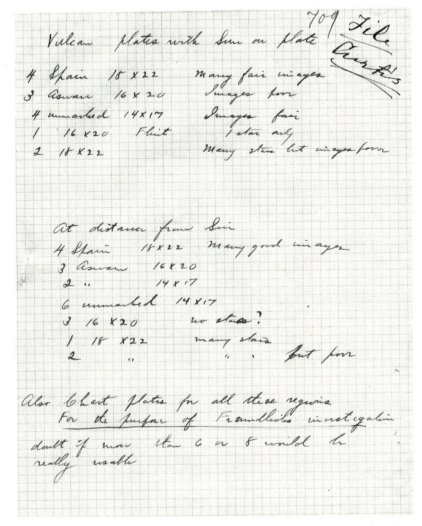

Figure 3.4b. Curtis's assessment of the Vulcan plates for measuring the "Einstein effect." (Courtesy Mary Lea Shane Archives of the Lick Observatory, University Library, University of California–Santa Cruz.)

cessful measuring of the plates illusory." Karl Friedrich Küstner of Bonn examined one of the better Lick plates and concurred that because they were taken to find intramercurial planets they were useless for the light-bending problem.[36]

Freundlich's approach to measuring the eclipse plates differed from what later became the accepted procedure. He measured the rectilinear coordinates of each star on an eclipse plate, measured the coordinates of the comparison stars on the chart plate, and then calculated deflection from the difference between the two measures for each star. This absolute method of measurement was typical of the Toepfer machines of this period, which Freundlich had borrowed from Schwarzschild.[37] Although the trailing of the images and the reliance on copies aborted Freundlich's enterprise, his absolute method of measurement would in any case have produced intolerably large errors. Curtis also favored the absolute method, but Campbell preferred a differential approach, where the positive of an eclipse plate is superimposed upon the negative of a comparison plate, and only differences between star images are measured. This issue was to become critically important in subsequent attacks on the problem.

Einstein thanked Freundlich "for your detailed reports and for the unusually lively interest you are taking in our problem. It is a terrible pity that the heretofore existing photographs are not sharp enough for such a measurement." He had been thinking about Freundlich's earlier suggestion to photograph stars near the Sun during daytime. He now thought it might be possible, even though his astronomer colleague at the university "rejected the idea on the grounds that the brightness of the atmosphere increases very rapidly with proximity to the sun." Einstein thought that perhaps in very dry latitudes at higher altitudes "it should be possible to observe, during the day, stars that are close to the sun, and to perform measurements on them."[38] Einstein told his young astronomer collaborator that his theoretical studies "are progressing briskly after indescribably painstaking research, so that the chances are good that the equations for the general dynamics of gravitation will be set up soon."

Einstein's reputation in German-speaking Europe had grown tremendously. Several institutions had courted him over the past year and he had finally accepted a position in Zurich. His old friend and former classmate Marcel Grossmann, who had been appointed dean of mathematics in 1911, was the one who lured him back. As soon as he arrived in Zurich in August 1912, Einstein began collaborating with Grossmann on trying to develop a generalized theory of gravitation (figs. 3.5, 3.6). While at Prague, he had an inkling that gravitation had something to do with geometry. He knew that Gauss's theory of surfaces had mathematics that would be useful. He needed a mathematician to help him find the right geometry, and Einstein turned to his friend. "Grossmann, you must help me or else I'll go crazy!"[39] And help he did. Grossmann introduced Einstein to Riemannian geometry, which, although difficult, promised a way to express the laws of physics in a way that was entirely independent of the observer's coordinates. In the theory that the two men worked out together, gravitation is expressed as a four-dimensional entity called a tensor. A tensor has sixteen components. An equation with tensor quanti-

Figure 3.5. Einstein in 1912 when he returned to Zurich as professor at the Zurich Polytechnik. (Courtesy Image Archive ETH-Bibliothek, Zurich.)

Figure 3.6. Mathematician Marcel Grossmann, Einstein's school friend who brought him back to Zurich and collaborated with him on general relativity. (Courtesy Image Archive ETH-Bibliothek, Zurich.)

ties in it is complicated. If you multiply two tensors, you have to keep track of the sixteen components in each one. In Einstein's equations, one tensor represents mass and energy (stress-energy tensor). The other represents the geometry of space-time. Einstein wrote to physicist Arnold Sommerfeld around this time that he was exclusively occupied with gravitation and was confident that he would master the difficulties "with the help of a mathematician friend of mine here." The experience had imbued him with "an enormous respect for mathematics, whose more subtle parts I considered until now, in my ignorance, as pure luxury!" "Compared with this problem," he told Sommerfeld, "the original theory of relativity is child's play."[40] Sommerfeld, who wanted Einstein to come to Göttingen to talk about quantum theory, told mathematician David Hilbert that "Einstein is evidently so deeply immersed in gravitation that he is deaf to everything else."[41]

It would be three more years before Einstein was able to come up with a satisfactory formulation for a generalized theory of gravitation. Meanwhile, Freundlich turned his attention to eclipse observations to find Einstein's light bending. He also turned to solar spectroscopy to find ways to test Einstein's gravitational redshift prediction.

PUZZLES IN THE SUN'S SPECTRUM

Ever since Henry Rowland began mapping the Fraunhofer lines of the Sun's spectrum in the late nineteenth century, investigators had noticed that there was a systematic, though complicated, shift of lines to the red. The American spectroscopist Lewis E. Jewell demonstrated in the late 1890s that the shifts were systematic, due to some physical cause other than Doppler effects. Jewell compared the wavelengths of solar absorption lines with corresponding emission lines in the electric arc. He was able to show that with few exceptions the solar lines were displaced toward the red end of the spectrum relative to the arc lines, hence the term "redshifts." Fairly soon after Jewell announced his result, laboratory work by others revealed that pressure, more than temperature, appeared to be the cause for the redshifts. More than a decade later, French spectroscopists Charles Fabry and Henri Buisson used an interferometer to examine the spectral lines of neutral iron in the Sun, the electric arc in air, and the vacuum arc. They were able to establish that pressure tended to broaden and shift the arc lines in a complex manner comparable to what they observed in the Sun.[42]

Arthur Scott King at Mount Wilson studied how pressure affects lines of several different elements, especially iron, produced in the electric furnace. Over a period of two decades (1908–1929) he established that the pressure effect was independent of the furnace temperature, the density of the iron vapor, or impurities in the vapor sample.[43] When King began this work, the Mount Wilson Solar Observatory still engaged primarily in solar work. The 60-inch reflector was not operating yet and most of the staff did daytime work on the Sun. The separate parts for the telescope mounting, including the 5-ton fork, were on the mountain by

June 1908, and the 60-inch mirror was finally installed safely in the telescope by 7 December. On the 20th, the first photographs were taken with the instrument.[44] Walter Sydney Adams had worked in stellar spectroscopy at Yerkes Observatory when Hale was still director there, then he decided to follow Hale to Mount Wilson in 1904. When the 60-inch reflector went into operation, Adams returned to stellar spectroscopy, a field in which he distinguished himself by discovering the spectroscopic method for measuring stellar distance. During the four years before his return to stellar spectroscopy, Adams worked with Hale and others on solar spectroscopy.[45]

One of the problems that Adams investigated was the comparison of wavelengths of solar lines from the limb and center of the Sun. Relative comparisons of lines from different parts of the Sun, without reference to laboratory wavelengths, avoided ambiguities in the lab spectra.* Jacob Halm, at the Cape of Good Hope Observatory, first reported in 1907 that lines of iron at the limb of the Sun tend to be redshifted relative to those at the center. Two years later, Fabry and Buisson confirmed this effect for elements other than iron; but astronomers drew no definite conclusions pending improved equipment with much better resolution.[46] Measuring limb-center shifts in the Sun became an important research problem.

At Mount Wilson, the principal instrument for solar spectroscopy in the early years was the Snow telescope (fig. 3.7a), a horizontal instrument Hale had designed at Yerkes and brought to California in the move to found the Mount Wilson Solar Observatory.[47] Hale decided that a vertical solar telescope would be superior. The optical parts would be at the top, projecting the solar image straight down into a subterranean temperature-controlled observing room containing the spectrograph. He designed a 60-foot tower telescope, which went into operation in 1908 (fig. 3.7b).[48]

Among the research projects carried out with the tower telescope in its first year was Adams's investigation of the limb-center shifts of solar spectral lines. The new instrument, with its high dispersion** and steady solar image due to the height of the optical parts above thermal effects on the ground, allowed for decisive measurements. In 1908, Hale reported: "The comparative study of the spectra of the limb and center of the sun favors the conclusion that the relative displacements of the lines near the limb (after eliminating the Doppler effect) are due to pressure."[49] In 1910 Adams published the final results of his study.[50] The situation was somewhat more complex than Hale had described two years earlier. For most

---

* King's and others' laboratory research with the iron arc was revealing complexities in the lab spectra that introduced possible sources of error when comparing them directly with solar spectra.

** Spreading out of the spectrum, making it easier to measure individual lines.

Figure 3.7a. Horizontal Snow telescope on Mount Wilson. (Courtesy Hale Observatories, Carnegie Institution of Washington.)

Figure 3.7b. Sixty-foot tower telescope on Mount Wilson. (Courtesy Hale Observatories, Carnegie Institution of Washington.)

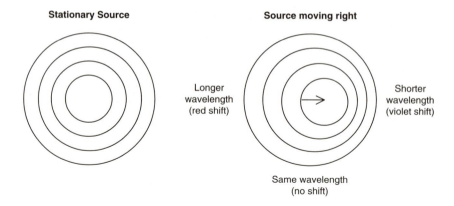

Stationary Source   **Source moving right**

Longer wavelength (red shift)

Shorter wavelength (violet shift)

Same wavelength (no shift)

The Doppler Shift. A stationary source (left-hand diagram) sends out successive waves that arrive at regular intervals in all directions. Waves from a moving source (right-hand diagram) bunch together in the direction of motion and spread apart in the opposite direction. Viewed at right angles to the motion, the waves arrive at the same intervals as if the source is stationary.

Figure 3.8a. Astronomers use the Doppler shift to determine the velocity of a star along the line of sight (radial velocity). They photograph the spectrum of a star and compare the stellar spectral lines with those of an iron arc in the spectrograph. Light from a star that is moving toward Earth will be shifted toward the violet compared to the arc lines, whereas a star moving away will have lines shifted toward the red. The amount of the shift allows astronomers to calculate the star's velocity relative to Earth.

spectral lines, the variation of redshift with wavelength favored a pressure interpretation, but for spectral lines due to cyanogen the pressure interpretion did not work. Laboratory experiments had established that cyanogen lines are insensitive to pressure, yet they had non-zero limb-center shifts in the Sun. To explain this positive result, Adams appealed to velocity (Doppler) effects (fig. 3.8a). He assumed that cyanogen gases are rising radially outward from the Sun. From Earth, gases at the center of the solar disk would be approaching the telescope, whereas gases at the limb would be moving across the line of sight. The lines from the center would be shifted toward the violet, while the lines from the limb would not (fig. 3.8b). This pattern would result in a relative redshift of lines at the limb compared to lines at the center.

While Adams was doing his pioneering work with the powerful instrument at Mount Wilson, another investigator was considering the same

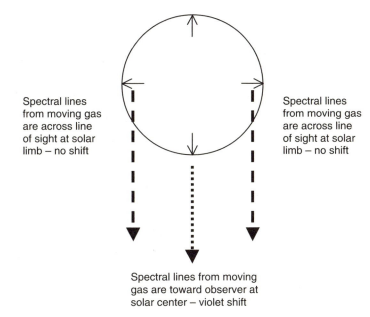

Spectral lines from moving gas are across line of sight at solar limb – no shift

Spectral lines from moving gas are across line of sight at solar limb – no shift

Spectral lines from moving gas are toward observer at solar center – violet shift

Figure 3.8b. Velocity interpretation of non-zero limb-center shifts for cyanogen gas. If cyanogen gas is ascending, then the ascending gas at the solar center will be moving toward the observer and the absorption lines will be shifted toward the violet. The cyanogen gas at the limbs will move across the line of sight and the absorption lines will not be shifted. Relative to lines from the center, lines from the limb would appear to be displaced toward the red, hence a non-zero redshift.

problem in a solar observatory on the other side of the world—the British spectroscopist John Evershed at the Kodaikanal Observatory in southern India. Evershed began his scientific career as one of many amateurs in the United Kingdom. With an industrial job in London, Evershed was able to pursue spectroscopic research at a private observatory that he had established in Kenley, Surrey. In 1906 Evershed accepted an offer of the assistant directorship at the Kodaikanal Observatory. It was his first professional post as an astronomer, obtained at the age of forty-two. On his way to India he visited astronomers in the United States, staying at Mount Wilson for a month. He arrived at Kodaikanal on 21 January 1907.[51]

Within a year, Evershed had taken up several lines of spectroscopic investigation. He included the "determination of relative shifts of certain lines in [sun]spot and in limb spectra; the lines chosen being those subject to large pressure shifts." He also was looking at "the determination of the amount and probable cause of the general shift towards the red of the

Figure 3.9. John Evershed at Kodaikanal Observatory in India, ca. 1909. He was director from 1911 to 1923. (Courtesy Indian Institute of Astrophysics Archives.)

lines at the sun's limb discovered by Halm."[52] His preliminary observations on limb-center shifts indicated complexity. On the whole "the general shift of all the lines at the limb towards the red is clearly brought out by the measures." However, certain lines most affected by pressure were shifted in the opposite direction—toward the violet—compared to the same lines at the center. Evershed was not yet able to determine the precise amount of the redshift.[53]

While conducting his studies of sunspot spectra, Evershed made an important discovery that all previous investigators had missed. Others had usually studied sunspots when they were near the central meridian of the Sun. Evershed observed spots at various positions, up to 50 degrees on either side of the central meridian. He found that the closer a spot was to the limb, the more pronounced the displacement of spectral lines in the outer regions of the spot. The pattern suggested an outward flow of material radially from the center of the spot, parallel to the surface of the Sun.[54] The discovery of the so-called "Evershed effect" established Evershed's reputation as a careful observer (fig. 3.10).

In the same year, 1909, Evershed came out with results that challenged the accepted interpretation that pressure was the dominant force behind the redshifts of spectral lines in the Sun.[55] He used the most recent data on lab and solar spectra. For the iron arc, he had Heinrich Kayser's tables

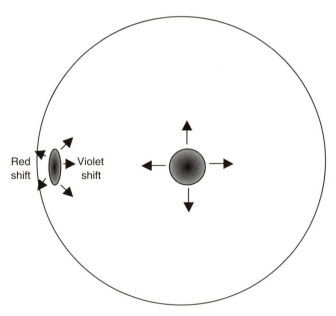

Figure 3.10. The Evershed effect. Gases moving radially outward from sunspots at the solar center are moving across the line of sight, hence no shift of spectral lines. Near the limb, the spot is at an angle to the observer. Gases on the far side of the spot are moving away from the observer, and on the near side of the spot are approaching the observer; hence the shifts in spectral lines.

of iron lines in the arc at normal pressure. For the Sun, he used Henry Rowland's preliminary tables of solar spectrum lines. Evershed divided the spectral lines into two groups of lines, one being least affected by pressure and the other being most affected.[56] He then determined the Sun-arc differences for two groups. Any difference between the two groups could only be due to pressure, since any relative displacements due to causes other than pressure would apply equally to all lines. They would be eliminated in the comparison of displacements for both groups. Evershed found the differences between the two groups to be small. A pressure in the solar atmosphere of less than one atmosphere (0.13) would account for them. All other investigators had obtained pressures in the Sun ranging from two to seven atmospheres.

By 1912 Evershed, now director of the observatory, could write in his annual report: "The general result of the whole investigation, although far from being completed, appears to throw great doubt on the usual interpretation of the line displacements, which ascribes the general shift of the solar lines, as well as the relative shift of the lines at the limb, to the effect of pressure."[57]

Einstein was introduced to the complexities of the Sun's spectrum by the Dutch physicist Willem Julius, who wrote to him in August 1911 to find out if he might be interested in taking a position on the faculty at Utrecht. Einstein ultimately declined, opting for Zurich, but the two men struck up a lively correspondence about the Sun. Julius had recently proposed dispersion as a mechanism to explain the solar redshifts, and Einstein sounded him out about his new gravitational redshift. "It would be of great importance to know exactly whether this shift *in the magnitude observed must* occur as a consequence of dispersion," he wrote. "If yes, then my darling theory must go in the wastebasket."[58] Julius assured Einstein that dispersion and gravitation need not be mutually exclusive. He admitted that the "astrophysicists do not 'believe' in a great influence of anomalous dispersion on the phenomena, or are perhaps afraid of such an influence."[59]

Julius was interested to know if gravitation could account for the increase in displacements from the center of the solar disk to the limb. Einstein assured him that his theory "leads to a constant displacement of about 0.01 angstrom, independently of the position on the sun's surface." Julius told him that Hale and Adams at Mount Wilson found that emission lines from the Sun's chromosphere displayed a very small displacement relative to absorption lines from the limb. Einstein misread Julius's letter and thought the chromosphere lines were compared to terrestrial lines rather than lines from the limb. A very small displacement would mean essentially a zero shift between Sun and Earth. "If these lines are very fine," he asserted, "then I believe that my theory is refuted by these observations." Einstein wanted Julius's frank opinion.

> After all, I know very well that my theory rests on a shaky foundation. What attracts me to it is the fact that it leads to consequences that seem to be accessible to experiment (mainly the refraction of light rays by the gravitational field), and that it provides a starting point for the theoretical understanding of gravitation. It would be important to show that my result is incompatible with experience—precisely because of its theoretical significance; the road I took might be the wrong one, but it had to be tried out.

Julius quickly corrected Einstein's misunderstanding. He showed that in fact "the mean displacements of the chromosphere lines would have exactly the magnitude demanded by your theory."[60]

Einstein and Julius continued their correspondence and even had a chance to meet each other to continue their discussions. Both recognized that explaining the solar redshifts was complex. They also acknowledged that aspects of their respective theories were incompatible. Einstein urged Julius to write to Adams at Mount Wilson about their discussions and to help get to the bottom of the issue. "It would be so interesting if we could learn something reliable about the reason for the solar shift."[61] Julius was impressed with the quality of the photographs that Adams was producing with Hale's tower telescope, but he was frustrated at his refusal to accept his dispersion theories. "It would be of enormous interest if, at Mount Wilson, the rich observational material were, for once, also investigated with respect to the possible consequences of gravitation and anomalous dispersion instead of being investigated only from the points of view of the pressure, and the Doppler and Zeeman effects." Einstein agreed. "By all means we must point out everything to Adams," he wrote, "because the questions are extremely important, and the pressure hypothesis is certainly inadequate, if not totally worthless."[62]

By 1912, the year that Evershed denounced Adams's pressure interpretation of the redshifts, Einstein's discussions with Julius had convinced him that it would be very difficult to demonstrate his predicted gravitational redshift in the Sun. He noted to Freundlich that "the broadening of the spectral lines toward both sides depends on a variety of causes (pressure—dispersion of light (Julius)—motion (Doppler), so that a compelling interpretation can hardly be achieved." He asked his colleague whether there were any extremely sharp lines in the Sun that might be utilized to test his theory.[63]

Interest in Einstein's gravitational explanation of the redshifts began to mount about this time, starting in Germany. In the fall of 1912, the mathematical physicist Max Abraham suggested in a letter to astronomer Karl Schwarzschild that relativity might be called upon to explain some of Adams's earlier results on limb-center shifts in the Sun. Abraham noted that while Adams favored a pressure interpretation, he felt that "it could quite well be the gravitational-theoretical shift, which in the middle would be in part compensated by motion." Abraham admitted that it would be very difficult to distinguish between the three effects: motion, pressure, and gravitational redshift.[64] About a year later, Schwarzschild measured Sun-arc displacements at five different positions from center to limb, and found that the shifts were smaller than the relativistic prediction. Einstein himself presented Schwarzschild's work on 5 November 1914 to the Academy of Sciences in Berlin.[65]

Meanwhile, Evershed had been continuing his work disproving the pressure interpretation of the center-limb shifts. In 1913 he published a paper suggesting that the relative shifts could not be due to pressure. Charles E. St. John at Mount Wilson had published results correlating line intensities to levels in the solar atmosphere. Spectral lines are created when gas in the solar atmosphere absorbs certain frequencies of light streaming outward from the Sun's center. Viewed in the spectrograph, the Sun's spectrum has a dark line at that particular frequency, where light has been absorbed. Lines due to absorption by gases high in the solar atmosphere (near the solar surface) are under less pressure than lines produced at lower levels. St. John assumed that the greater the pressure, the stronger in intensity the spectral line. So weak lines would be due to absorption by gases higher in the solar atmosphere where pressure would be less, while strong lines would come from lower levels, presumably under greater pressure. Evershed found that the mean Sun-arc shifts for weak lines (higher) were greater than the shifts for strong (lower) lines—exactly the opposite of what one would expect if the shifts were due to pressure.[66]

Evershed also criticized recent work by Fabry and Buisson presenting evidence for a solar atmospheric pressure of about 5.5. He attributed differences between his and their data to differences in the arc. He claimed that the French investigators had distorted their final result by neglecting lines most affected by pressure. He concluded that the pressure in the reversing layer must be less than one atmosphere. At the time, Evershed was not aware of Einstein's prediction, and having ruled out pressure, he was left with "the only possible explanation of the shifts of the solar lines to be motion in the line of sight." He did not elaborate his own theory, which "involves an apparent influence of the earth on solar phenomena analogous to that which affects the distribution of sunspots and prominences." For the moment, his main objective was to refute the pressure interpretation of the redshifts.

About this time, Freundlich published an article drawing attention to Einstein's gravitational prediction. He pointed out that Fabry and Buisson and Evershed had obtained values for the solar redshifts roughly the same as Einstein's predicted values for the relativistic effect.[67] Evershed took notice in a subsequent paper describing his remarkable motion theory of limb-center shifts, published with Thomas Royds in 1914.[68]

Having ruled out pressure as the cause of the limb-center shifts, Evershed was left with motion. The only motion explanation Evershed could come up with was "a motion parallel to the solar surface, and directed away from the Earth at all points of the solar circumference. This suggests that the Earth itself controls the movement, exerting a repelling action on the solar gases."[69] Evershed admitted that "the pressure theory presents a much more rational explanation," but he presented detailed observations and arguments that were decidedly against it. In particular, he analyzed carefully the main results of Adams's 1910 paper.[70]

Adams had found that the limb-center shifts for neutral iron lines increased with wavelength more rapidly than a simple linear increase. A linear wavelength dependence would imply a velocity shift. So Adams interpreted the higher-power law as evidence of pressure effects. Evershed and Royds checked Adams's conclusion by measuring arc-center shifts. They found that the arc-center shifts decreased with increasing wavelength. They then added the arc-center and the limb-center shifts together to determine limb-arc shifts. Adding the two effects together, the total limb-arc shift remained constant for all wavelengths. If pressure were an important factor, there should be a marked increase with wavelength.

Furthermore, Adams had found a small positive limb-center shift for the cyanogen bands. He attributed it to ascending movements of the cyanogen gases at the center of the solar disk, implying a violet shift at the center or a redshift at the limb relative to the center. The Kodaikanal observers directly measured the center-arc shifts and found that the cyanogen at the center of the disk was descending, not ascending. This contradicted Adams's suggestion that violet shifts at the center could explain the limb-center shifts. Evershed and Royds added the limb-center shifts and center-arc shifts and found that the cyanogen at the limb was receding parallel to the solar surface. This supported their hypothesis that solar gases were receding from the Earth.

Evershed and Royds were reluctant to accept their own hypothesis, but felt compelled to do so. "While fully appreciating the absurdity of this idea," they wrote, "we feel that there may be some justification for it in the apparent influence of the earth on the distribution of sunspots on the visible disk and of the prominences on the east and west limbs." The authors felt there was "no alternative to the earth-effect hypothesis, unless we assume some cause for line shifts other than motion or pressure." And here they appealed to relativity. "According to the 'Theory of Relativity' of Einstein, the sun's gravitational field should diminish the frequency of the light emitted, and the mean shift to the red found by us at the centre of the disk agrees very closely with the theoretical gravitational shift calculated by E. F. Freundlich." However, they noted that the shifts varied from line to line at the center and that limb shifts did not vary with wavelength, "facts which would apparently offer serious difficulties to this explanation."[71]

At Mount Wilson, Evershed's research stimulated intensive investigations to check his observations and conclusions. When Evershed began to come out with results contradicting Adams's and others' pressure interpretation of limb-center shifts, Adams was already moving into stellar spectroscopy. The solar work was taken over by a relative newcomer, Charles Edward St. John.[72] St. John joined the staff at Mount Wilson just about the time Adams was beginning his work with the solar tower telescope and King was starting his laboratory investigations of pressure shifts. After receiving a Ph.D. from Harvard in 1896 and then teaching for a year at Michigan, St. John taught physics and astronomy at Oberlin College for eleven years. St. John spent many a vacation doing astrophysics research at Yerkes Observatory, where he became friendly with Hale, Adams, and others who would later follow Hale to Mount Wilson. St. John was over fifty years old when he decided to leave teaching and a dean's position at Oberlin to devote the rest of his life to research.[73] At Mount Wilson, St. John went on to become one of the world's leading solar spectroscopists.

Two years after he came to the solar observatory, St. John began a series of investigations on the recently discovered Evershed effect. In confirming and amplifying the British spectroscopist's discovery, he must have come to appreciate the other man's skill and ingenuity.[74] In response to Evershed's criticisms of Adams's work, St. John began an extensive study of the cyanogen bands to test Evershed's observations of the non-zero shifts. The results of this investigation, not published until 1917, became intimately connected with testing Einstein's gravitational redshift.

From 1914 on, studies of line shifts in the solar spectrum incorporated attempts to include the gravitational redshift as a possible contributing mechanism besides pressure, velocities, and anomalous dispersion. The instrumentation and techniques existed before the researchers involved became aware of the relativity prediction. They adopted the Einstein test into a research program developed for other problems. The same was true at Lick Observatory, where they adapted the eclipse test for intramercurial planets to look for Einstein's light-bending prediction at a solar eclipse.

THE RUSSIAN ECLIPSE OF 1914

After his failure to obtain any useful measurements from old eclipse plates, Freundlich turned his attention to the coming solar eclipse, which would be visible from nearby Russia on 21 August 1914. In an article describing his negative results, he announced his willingness to collaborate with anyone wishing to take photographs at the eclipse and detailed the requirements for taking adequate observations. Freundlich asked Campbell whether he would join in. "You have been so kind to send your cameras to Prof. Perrine at the last eclipse and I would be most thankful for your kind support to get plates for my problem." Campbell promised that if, as he hoped, an expedition were sent from Lick, then "we shall plan to obtain photographs meeting the requirements of your problem; that is, with the solar image in the center of the field and with the driving clock adjusted to follow the stars." He would "be glad to send them at once to the Berlin Observatory for your use."[75] The Lick director attached as a condition to the collaboration that "any results obtained by you might be announced by you in a preliminary way in the Nachrichten, or otherwise, and your full paper on the subject be published as a Lick Observatory Bulletin."

As he would not know for two months whether funding for an expedition would materialize, Campbell made light of making detailed plans.

"As we say in this country," he remarked to Freundlich, "it is well not to count the chickens before they are hatched; that is, we may not go to the eclipse, or, if we go, clouds or other factors may prevent the success of our plans."

Freundlich was hoping that several groups, including Lick and his own observatory, would obtain results and that he would be able to compile all of them together. He asked Campbell if he would consider publishing a full paper containing the results from all plates obtained. Freundlich hoped that Campbell would agree that a companion paper should appear at the same time in German. "The paper would be of greater scientific value," he suggested, "if based upon the whole material and you will understand that I would like to publish a full paper also in my mother language." Campbell was amenable, and in the end, both men left the matter open.[76]

Freundlich tried to enlist the Greenwich Observatory in his project, but its director, Frank Dyson, declined: "It would be an extremely delicate research to undertake at an eclipse, if not quite beyond present possibilities, and so I cannot venture to promise to take photographs specially for this purpose." Dyson did refer Freundlich's appeal to the JPEC, of which he, as Astronomer Royal, was chairman. The committee decided "that a special equipment would be required to carry out this work satisfactorily and . . . that no suitable instruments were at its disposal."[77] In fact the JPEC financed more than one expedition to Russia, and instruments could have been adapted to Freundlich's problem if someone had wanted to do so.

Campbell monitored the scientific literature from Europe on the astronomical consequences of relativity. In May 1913 he wrote a brief report noting that the principle of relativity might be an important factor in explaining existing residuals in the motions of planetary perihelia.[78] In the same month, he received word that funding from one of Lick's longtime benefactors, William H. Crocker, a trustee of the University of California, had come through. Campbell was planning to visit Germany the following August to attend the International Solar Union meeting in Bonn and the meeting of the Astronomische Gesellschaft in Hamburg. He suggested to Freundlich that they meet to discuss the eclipse observations "so that they will best meet your requirements." As the junior member of the observatory, Freundlich was not going to the meetings, so Campbell made the extra trip to meet him.[79] Campbell was favorably inclined toward Freundlich, and toward Germany in general. He was perfectly content to contribute observations, having his German colleague measure and write up the results.

Campbell's visit to Germany coincided with a high point in international relations among astronomers. The Berlin Observatory, which had originally declined to join the International Union for Solar Research, accepted election on the proposal of British and American members. At the meeting, it was Campbell, as a member of the U.S. National Academy of Sciences, who seconded the motion by Herbert Hall Turner of Britain.[80] The German astronomers were extraordinarily hospitable at the Bonn and Hamburg meetings. "Not only were the scientific sessions successful," Campbell remarked in his report to Hale, "but the social arrangements were on a very extensive scale and extremely happy. The hosts at later meetings are going to have hard work to live up to the high standard set at Bonn." On his return to the United States, Campbell published an extensive, glowing account of the meetings. Turner wrote a similar report for the British.[81]

Einstein was delighted with the response Freundlich was getting from astronomers abroad. "It is thanks to your zeal," he wrote in August, "that the astronomers have now also started to show interest in the important question about the bending of light rays." Einstein was enjoying spirited exchanges in German scientific journals with physicists about relativity, and he outlined his "view about the other current theories of gravitation" for his astronomer colleague. While discounting theories by Max Abraham and Gustav Mie, he noted that a competing theory by Gunnar Nordström "is very reasonable and points to a contradiction-free way in which to succeed without the equivalence hypothesis." Luckily, Nordström's theory, while also predicting a gravitational redshift, did not demand a bending of light rays in the gravitational field. "The investigations during the next solar eclipse will show which of the two conceptions corresponds to the facts."[82]

Einstein's life had taken a dramatic turn that year. As early as January, the German chemist Fritz Haber had begun to investigate ways to bring Einstein to Berlin. He recruited two of the most powerful scientific men in Berlin, the physicist Max Planck and the chemist Walther Nernst, to visit Einstein in Zurich in July. Their mission was to find out if Einstein would be willing to move to Berlin. They offered the thirty-four-year-old physicist an astounding package—membership in the Prussian Academy of Science, which came with an honorarium of 900 marks and the highest prestige possible on the continent; the maximum salary for professors of 12,000 marks, half from the Prussian government and half from a benefactor, the industrialist Leopold Koppel, which the physics-mathematics section of the academy would administer; a professorship at the University of Berlin with the right but no obligation to teach; and the directorship of a new physics institute to be created under the auspices of the Kaiser-Wilhelm-Gesellschaft. Einstein's only responsibilities would be to live in Berlin and attend meetings of the Prussian Academy. The deal had taken months to put together, and Einstein took a day to think about it. In the meantime, Planck and Nernst went for a hike in the Alps. The three men had decided that when Einstein would meet them at the station on their return, he would be carrying white flowers if he declined their offer, and red flowers if he accepted. When the two Berliners pulled into the station, they were pleased to see red ones.[83]

During their visit, Einstein told his Berlin colleagues about the new theory of gravitation he was developing with Marcel Grossmann. Planck was skeptical and "as an older friend" advised the younger Einstein against continuing, "for in the first place you will not succeed; and even if you succeed, no one will believe you."[84]

By August, Nernst could write to the British physicist Frederick Linde-mann that he and Planck had succeeded in getting Einstein to come to Berlin. "The Academy has already elected him; we have great hopes of this." Einstein remarked wryly to a friend that the Berlin scientists seemed "like people anxious to acquire a rare postage stamp."[85]

Toward the end of August, Einstein told Freundlich that from a theoret-ical standpoint "the matter is now settled more or less. Privately, I am quite certain that light rays do undergo bending." He was nonetheless very interested in all possible tests.[86] Freundlich was getting married the following month and would be spending his honeymoon in the Alps. Ein-stein insisted that they arrange to meet, for the first time. "This is wonder-ful," Freundlich wrote excitedly to his fiancée, "because it fits in with our plans." When Freundlich and his bride pulled into the train station in Zurich, Einstein was waiting for them sporting "a very conspicuous straw hat." Einstein and Fritz Haber immediately whisked the young couple off to Frauenfeld, a few miles from Zurich. Einstein was delivering his ad-dress with Marcel Grossmann on their new generalized theory of relativ-ity at a meeting of the Swiss Society of Natural Sciences. From the po-dium, Einstein singled out Freundlich in the audience as "the man who will be testing the theory next year." Einstein and Freundlich discussed gravitation all the way back to Zurich while the new Frau Freundlich admired the scenery.[87]

Einstein and Freundlich had been toying with the possibility of measur-ing the displacement of stars near the uneclipsed sun. When Einstein had taken up his new professorship in Zurich in 1912, he had told his col-league Julius Maurer about the idea. Maurer was discouraging, but he suggested that Einstein consult Hale at Mount Wilson. Not long after his meeting with Freundlich, Einstein finally did so, in a letter now well known; and Maurer graciously added an appeal that the famous Ameri-can astronomer consider Einstein's question about the possibility of test-ing light bending with daylight observations of stars in the vicinity of the Sun.[88] Hale saw no hope for the endeavor and referred the problem to Campbell, who the physicist Paul Epstein told him was interested in the eclipse method. Campbell wrote both Hale and Einstein of his plans to obtain photographs for Freundlich using the intermercurial lenses. Hale explained in detail to Einstein why the daylight method would not work.[89] "The eclipse method, on the contrary, appears to be very promising," he told Einstein, "as it eliminates all of these difficulties, and the use of photography would allow a large number of stars to be measured. I therefore strongly recommend that plan."

Meanwhile, Freundlich was encountering difficulty in raising funds for his own expedition. His boss, Hermann Struve, refused to support the project financially, so Freundlich applied to the Berlin academy for funds. In December there was still no word, and Freundlich needed to order photographic plates and begin making arrangements for the trip. Einstein was furious with Struve's attitude and tried to influence the academy's decision. "After the receipt of your last letter," he told Freundlich, "I immediately wrote to [Max] Planck, who applied himself seriously to the matter and took it upon himself to talk about the matter with Schwarzschild. *I shall not write to Struve.* If the Academy shies away from it, then we will get that little bit of mammon from private individuals." Einstein assured Freundlich that the money would be raised somehow. "Should everything fail," he offered, "then I will pay for the thing myself. . . . So, after careful consideration, just go ahead and order the plates, and do not let the time be squandered because of the money problem." In the end Einstein did not have to dig into his own pocket. Due in large part to support from Planck and Nernst, the academy came through with 2,000 marks for scientific equipment. Freundlich also contacted the chemist Emil Fischer, who gave him 3000 marks for the journey and supplied him with another 3000 marks from the firm of Krupp in Essen. Freundlich borrowed some equipment from Perrine, who was also sending an expedition from Cordoba, but would not be attempting the Einstein test himself.[90]

Toward the end of 1913 Curtis published an article summarizing the status of the Einstein problem. With physicists "divided into two warring camps on the subject of the theory of relativity," Curtis told astronomers that "whether it will ever become generally accepted is an open question, particularly as Einstein has recently been forced to make certain rather radical alterations in the theory as originally stated by him." The light-bending test gave astronomers a chance to contribute to a controversial theory. Curtis related Freundlich's abortive attempt to measure the Lick Vulcan plates ("Probably the only observational material at present available to test the truth or falsity of this hypothesis") and explained why the plates proved to be unsuitable. He provided details on how to make observations at the coming eclipse in Russia. Using the intramercurial plates taken at Flint Island in 1908, he estimated the limiting magnitudes of stars at different distances from the Sun's limb. He discussed the star field that would be visible around the eclipsed Sun and suggested that observers could use the bright star Regulus in the constellation Leo to guide their telescopes in right ascension.[91] Curtis's publication gave all parties going to Russia the necessary information to make the Einstein test themselves. Except for Freundlich and Campbell, none did.

Freundlich left Berlin on 19 July 1914 with Walther Zurheilen and a Zeiss technician. A week later they joined up with the Cordoba party at Feodosiya in the Crimea. Together with the items from Perrine, Freundlich had assembled a battery of four astrographic cameras to photograph the eclipse.[92]

Campbell made elaborate travel plans around the Lick expedition to give his three sons an unforgettable experience. Curtis would accompany the instruments from New York to Libau and from there to the eclipse station, about 15 miles from Kiev. Campbell and his family would make the trip via Gibraltar, Naples, and Vienna, reaching Kiev about 20 July. They would make their own sleeping and dining arrangements, and if all went well with the eclipse, they would leave Kiev about 27 August. Campbell then wanted to attend the Astronomische Gesellschaft meeting, set to take place in St. Petersburg and at the Russian National Observatory at Pulkova; then to return home via Berlin, Paris, and London, to give his sons a chance "to see the important observatories in those cities."[93]

The outbreak of war blighted the fruition of these plans. As the international situation worsened, at first it was not certain how the eclipse party would be affected. Richard Hawley Tucker, acting director of Lick during Campbell's absence, wrote: "We are imagining that the threatened war will make things exciting for your eclipse party, but we hope on every account that Russia will keep out of the complications."[94] A week after their arrival in Kiev, Campbell's party was oblivious to the growing tensions. Curtis had arrived several days earlier and had found "a splendid location for the observing station, had rented the place, including a fine house, had installed a cook, etc." Campbell was delighted, especially as the work was going ahead of schedule; but though the clouds of war were not yet apparent, those of nature were not encouraging. "The weather has been pretty unpromising," Campbell wrote, "raining nearly every day, and cloudy the most of the time. We hope for clear sky on the afternoon of the 21st. But nothing will surprise us."[95] (Figs. 3.11a, b.)

Curtis and Mrs. Campbell visited Kiev for supplies and found out that on 30 July Austria had declared war against Serbia and that Russia had begun mobilizing. The next day the reserve at the town of Brovary, where the eclipse camp was set up, mobilized. On 2 August the astronomer Robert Filopowitch Foghel of the University of Kiev called on Campbell with the news that Germany had declared war against Russia. Campbell wrote immediately to the British Consul in Kiev to ask for protection for the

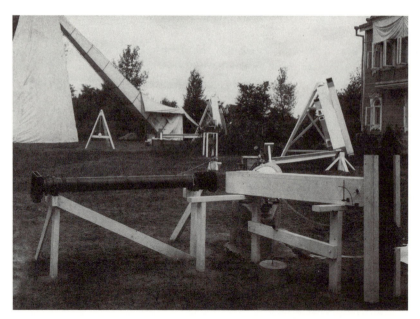

Figure 3.11a. Lick Observatory Vulcan cameras (top right) set up at eclipse camp in Russia, 1914. The long, slanted camera in the left background is the Schaeberle 40-foot camera. (Courtesy Mary Lea Shane Archives of the Lick Observatory, University Library, University of California–Santa Cruz.)

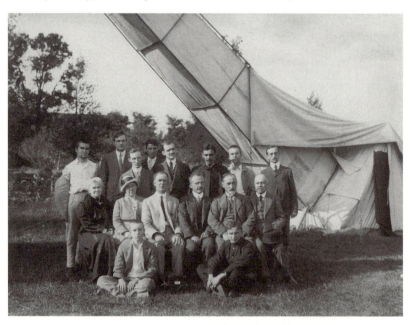

Figure 3.11b. Campbell and his family with H. D. Curtis and astronomers at the Russian eclipse camp. *Back row*: Douglas Campbell (left), Wallace Campbell (second from right); *middle row, from left*: Mrs. Thompson (Mrs. Campbell's mother), Mrs. Campbell, Campbell, Curtis; *front row*: Kenneth Campbell (left). (Courtesy Mary Lea Shane Archives of the Lick Observatory, University Library, University of California–Santa Cruz.)

expedition; he requested "a detail of two policemen in uniform here at our station, from August 20th to August 25th, to guard the expedition against the acts of ignorant or excited people who may be inclined to connect the eclipse with events occurring around them." The American Embassy in St. Petersburg advised him that, "as England, France, Germany, Austria, Russia and Belgium are all at war and nearly all other countries mobilizing, the neutral ships leaving neutral ports are naturally few." The chargé d'affaires assured Campbell that the U.S. government would probably charter ships for Americans to return back home.[96]

Einstein was concerned the war would adversely affect the German eclipse expedition. "My good old astronomer Freundlich will experience captivity instead of a solar eclipse in Russia. I am concerned about him." Indeed, the Russian authorities allowed the American party to observe the eclipse but arrested Freundlich's. They deported the older members immediately and held Freundlich and his younger colleagues as prisoners of war in Odessa until they could exchange them for Russians caught in Germany when hostilities broke out. They confiscated Freundlich's equipment. Perrine had not arrived early enough to take over Freundlich's light-bending test, but it would have been useless in any case. The day before the eclipse, the weather improved somewhat, but at the time that the eclipse would occur the next day, the Sun was not visible through clouds. Mrs. Campbell's diary notes for the day of the eclipse speak eloquently: "Total failure. Thick gray cloud at eclipse time and lovely clear sunshine afterward."[97]

Campbell's party avoided Odessa because of expectations that Turkey would declare war. They chose not to subject themselves to the long railway trek to Siberia; so they returned via the Baltic to London, threatened by German mines and submarines. "The war is a terrible thing," Campbell remarked to Hale. "I fear that we, in this country, do not appreciate how very real the war is to people in the countries involved." He worried about the next International Solar Union meeting, planned for 1916 in Rome, and that the war would inject "intense international hatreds" into science. Hale agreed that the situation was grave. "The war grows worse and worse," he answered, "and I greatly fear from what I hear from Germany that all international relations may be seriously interrupted, perhaps for years. I agree with you in thinking that we should make unusual efforts to assist the next Solar Union meeting, and only hope it may not prove necessary to postpone it."[98] Hale's fears were soon realized. All activities of the Union were suspended for the duration of the war.

Considering the gravity of the war situation, Campbell felt that his party had gotten off rather lightly; but he was crushed by the scientific failure. He wrote Hale:

If the critical two minutes had been clear so that we could have brought valuable results home with us we would have thought nothing of the inconveniences which later greeted us at various points. I never knew before how keenly an eclipse astronomer feels his disappointment through clouds. Eclipse preparations mean hard work and intense application, and I must confess that I never before seriously faced the situation of having everything spoiled by clouds. One wishes that he could come home by the back door and see nobody.[99]

From then on, Campbell applied himself single-mindedly to the Einstein problem independent of his former colleague from Germany.

Chapter Four

THE WAR PERIOD, 1914–1918

TROUBLES WITH FREUNDLICH

The international hostilities quickly destroyed the congeniality that had existed among scientists in the warring countries. Lines of communication were severely disrupted as borders shut down and the Atlantic turned into a battleground. Einstein had moved his family to Berlin the previous spring. Within months his wife, Mileva, heartbroken by lost affection from Einstein, had returned with their two sons to Zurich. Now separated from his family and living on his own in wartime Berlin, Einstein watched the unfolding tragedy with dismay. "Europe in its madness has now embarked on something incredibly preposterous," he wrote to Paul Ehrenfest in Leiden. "At such times one sees to what deplorable breed of brutes we belong. I am musing serenely along in my peaceful meditations and feel only a mixture of pity and disgust."[1] Yet Einstein was relatively content with his newfound bachelorhood. "I am residing all by myself in my large apartment," he told Ehrenfest, "in undiminished tranquility."

While Europe raged, Einstein threw himself into his work. For almost two more years, he continued his struggle with relativity and gravitation. During this time he continued to work with Freundlich on trying to find experimental tests for his developing theory. Freundlich focused on the search for light bending and the gravitational redshift, but his routine job as assistant at the observatory made it difficult to find the time. In March 1915 he wrote to Struve asking for a position that would allow him "to dedicate my labors to problems that aim at supporting modern physical theories," meaning relativity. Struve categorically refused. Einstein tried to use his influence to help, asking Planck to talk to Struve. Einstein reported back to Freundlich that his boss had "railed about you thoroughly. He said you do not do what he asks of you, etc."[2] In October 1915, Einstein tried to intervene a second time, writing to Otto Naumann, the ministerial director for educational matters in the Prussian Ministry of Education. He asked the senior bureaucrat to consider releasing Freundlich from his duties "for some years, without his salary as assistant being cut off." Einstein met with Naumann in November to try to convince him to promote Freundlich to observer, so he could focus on observational tests of relativity—to no avail.[3]

Freundlich did not make things easy for his champion. In the same year he drew the ire of Hugo von Seeliger, an influential member of the German astronomy establishment. His first transgression was publishing a criticism of von Seeliger's dust hypothesis which von Seeliger had proposed as an explanation for the excess advance of Mercury's perihelion. Von Seeliger, a veteran of scientific debates, published an aggressive response. Writing to Arnold Sommerfeld, Einstein penned a postscript: "Tell your colleague Seeliger that he has a ghastly temperament. I relished it recently in a response he directed against the astronomer Freundlich."[4]

A more serious situation arose from Freundlich's attempt to verify Einstein's gravitational redshift in stellar spectra. Campbell's systematic study of the radial velocities of stars had revealed an unexplained excess radial velocity for B-type stars compared to the other spectral types. Freundlich did a statistical study of the radial velocities of fixed stars, using Campbell's data and those of others. Consistent with the prevalent belief at the time that the universe is static and uniform in all directions (isotropic), the average spectral line displacement for all stars should have been close to zero. Freundlich found a non-zero average redshift, which was most pronounced for the B stars. Interpreted as a Doppler shift, it would represent an average recession of B stars away from Earth at almost 5 kilometers per second. He attempted to explain this excess as a gravitational redshift. Campbell and others had estimated the mass of B stars in binary systems, using Kepler's laws, to be about fourteen times the mass of the Sun. Freundlich used Einstein's gravitational redshift formula to calculate the average mass of the B stars, assuming that the average redshift was entirely due to gravitation. Assuming a density for B stars of about one-tenth the density of the Sun, he obtained a mass twenty times the mass of the Sun. This value was the right order of magnitude, and he interpreted it as demonstrating the reality of the gravitational redshift.[5]

Freundlich sent a preliminary paper for publication on 1 March 1915. He asked Einstein if he would present his work to the Prussian Academy later that month. Einstein agreed to do so only if Freundlich would address some points he considered weak or incomplete. He was also worried about Freundlich's quantitative treatment: "*It is a shame that your descriptions are not detailed enough to be able to estimate the uncertainty attached to your estimates.* Thus a nonspecialist cannot get a notion of the reliability of your calculations. A much more in-depth presentation would be desirable. The worst in this regard is the specification of the mean densities." Freundlich made a few changes, but he did not address all of Einstein's points. Einstein did not present the paper, but he mentioned it when he presented a paper on general relativity and its application to astronomy later that month. He indicated that Freundlich had just published a new paper showing that redshifts of the order predicted by relativity are observed for B and K stars.[6]

Einstein saw Freundlich's work as "surely fundamental,"[7] but von Seeliger thought otherwise. As Einstein had feared, it had to do with Freundlich's calculations. Von Seeliger found an error in Freundlich's determination of the mass of the B stars, which compromised his conclusion favorable to relativity. If corrected, the calculated mass of B stars using the gravitational redshift formula would be much too large. Rather than communicating the error directly, von Seeliger sent the correct formula to Struve, who passed it on to his assistant. Freundlich was mortified that "such a gross and unnecessary mistake could still happen" and set out to put things right. Von Seeliger explained to Struve why he deigned to respond to the younger Freundlich, who was after all only an assistant. His reason: he didn't like Einstein's relativity theory. "I happened to hear that Einstein sets great store in Dr. F's line of reasoning. You know that I am extremely skeptical of many hypotheses of the latest physics, and that is why the question under discussion seemed to me to be of some interest."[8]

Meanwhile, Einstein made a stupendous breakthrough on the theoretical front that would escalate von Seeliger's opposition but launch a scientific revolution.

EINSTEIN'S BREAKTHROUGH

For several years Einstein had been trying to find a theory of gravitation that was entirely independent of an observer's coordinate system, building on the tensor formulation that he had developed with assistance from Marcel Grossmann in Zurich. As he struggled with the complexities, he published various versions of the theory, alternating between elation at solving the problem and frustration that he didn't have it quite right. By the beginning of 1915, he had come up with a solution that he thought worked, although it didn't have all the features he would have liked. For one, it predicted too small a perihelion shift for Mercury. It also was not completely independent of the observer's coordinates. Nonetheless, he thought he had succeeded in generalizing relativity to incorporate gravitation. "To have now really reached that objective," he told a former student, "is the greatest satisfaction of my life, even though none of my colleagues in the field has as yet been able to realize the depth and the necessity of this road."[9] Einstein went to Göttingen to explain his theory to the mathematicians Felix Klein and David Hilbert. By the end of August he had them "completely convinced."[10]

Not long afterwards, Einstein himself became unconvinced. In September 1915, he wrote Freundlich that his equations yielded a "blatant contradiction" and he wondered if he was making an error. "Either the equations are already numerically incorrect . . . or I am applying the equations in a principally incorrect way. I do not believe that I myself am in the position to find the error, because my mind follows the same old rut too much in this matter." He appealed to Freundlich: "If you have time, do not fail to study the topic."[11]

Einstein soon discovered that he hadn't made a mistake—his equations were just not the right ones. He came back to a solution using a specific formulation involving the so-called Ricci tensor. He had abandoned it years earlier, albeit "with a heavy heart," because the equations had the desirable feature of being completely independent of the observer's coordinates. Einstein had rejected them on other grounds. Now, his earlier objections resolved, he reinstated them and started anew.[12] His revised equations immediately bore wonderful fruit. When he sat down in November 1915 to calculate the astronomical predictions, he found twice the light bending. Instead of a deflection of 0.85 seconds of arc, light grazing the Sun would be deflected 1.7 seconds of arc. The calculation of Mercury's perihelion motion was much more complex. When Einstein finished, he was astounded. He later reported that his heart had started to palpitate. The shift came out precisely at 43 seconds per century—almost the exact amount measured by astronomers and unexplained until Einstein's latest effort. "The result of the perihelion motion of Mercury gives me great satisfaction," he wrote to Arnold Sommerfeld. "How helpful to us here is astronomy's pedantic accuracy, which I often used to ridicule secretly!"[13] By 25 November 1915, Einstein had the final equations. Matter and energy were intimately linked with the geometry of space-time. Planets orbit Suns because the geometry of space-time around these massive bodies is curved, just as two drivers starting at the Earth's equator going north gradually approach each other because the Earth's surface is curved. There is no "force" pulling them toward each other. Einstein showed that a similar, albeit more complex kind of geometrical curvature can explain gravity. "For a few days," Einstein recalled to Ehrenfest several months later, "I was beside myself with joyous excitement."[14]

Einstein's desire to find a theory that was independent of the coordinate system had guided his arduous search. In physical terms, he explained to Ehrenfest, "the reference system has no real meaning." This notion runs counter to common sense. The Newtonian idea of objects moving and interacting in absolute space is closer to our intuitive understanding of the world. We imagine space as a container that persists even if we are not there. This idea is so strong that Einstein took years to shake it. In discussing a point with Ehrenfest he remarked: "I cannot hold it against you that you have not yet understood the admissibility of generally covariant equations, because I myself needed so long to arrive at total clarity on this point. The root of your difficulty lies in that you instinctively treat the reference system as something 'real.' "[15]

Karl Schwarzschild took an immediate interest in Einstein's new theory. After Einstein's 1911 predictions, he had tried to find the gravitational redshift in the solar spectrum without success. At the outbreak of war, he took a leave from his position as director of the Astrophysical Observatory in Potsdam to volunteer for the army. He received Einstein's reports to the Prussian Academy at the Russian front, where he was engaged in calculating artillery trajectories. In his spare time, he worked on Einstein's theory. As a preliminary exercise "to become versed in your gravitation theory," he derived a complete solution to the Mercury perihelion problem, where Einstein had only solved it to first-order approximation. "As you see," he wrote Einstein, "the war is kindly disposed toward me, allowing me, despite fierce gunfire at a decidedly terrestrial distance, to take this walk into this, your land of ideas." Einstein was surprised with Schwarzschild's exact solution. "I would not have expected that the exact solution to the problem could be formulated so simply. The mathematical treatment of the subject appeals to me exceedingly." He presented Schwarzschild's paper to the academy a few weeks later. "I am very satisfied with the theory," he wrote the astronomer. "It is not self-evident that it already yields Newton's approximation; it is all the more gratifying that it also provides the perihelion motion and line shift, although it is not yet sufficiently secure. Now the question of light deflection is of most importance."[16]

Schwarzschild tackled another problem "in order to acquaint myself with your energy tensor." He calculated the behavior of a mass point in the gravitational field of an incompressible fluid sphere. "I would not have done so," he admitted to Einstein, "had I known that it would cause me so much trouble." Schwarzschild discovered that if you squeeze the fluid sphere down in size, there is a specific radius that depends on the mass, for which the density inside the sphere goes to infinity and where the equations become meaningless. For the mass of the Sun, the so-called Schwarzschild radius is only 3 kilometers. For the Earth it is less than a centimeter. Einstein presented Schwarzschild's remarkable paper to the Prussian Academy in February 1916.[17] It was not until the 1960s that astronomers realized that massive stars at the end of their lives could collapse inside their Schwarzschild radius, creating a black hole.

Schwarzschild's theoretical forays into Einstein's general relativistic world were cut short by an illness he contracted at the Russian front. He returned home in March 1916 with a rare, incurable skin disease, and died on 11 May. Einstein was asked to deliver the eulogy to the academy in June.[18]

Besides Freundlich and Schwarzschild, no German astronomer showed interest in pursuing the theoretical or observational side of Einstein's work. With Schwarzschild gone, Einstein had only Freundlich, who increasingly became a liability.

The "Freundlich Affair"

Einstein had been looking to Freundlich to spearhead the observational testing of relativity. His latest theory predicted the same gravitational redshift as his earlier version, so he viewed Freundlich's work on the redshift of B stars as another verification of his theory. He told his friend Heinrich Zangger: "The general relativity problem is now finally dealt with. The perihelion motion of Mercury is explained wonderfully by the theory. . . . Astronomers have found from observations $45'' \pm 5''$. I have found with the general theory of rel. $43''$. Add to this the line shift for fixed stars which, as you know, has also been established securely, thus this is already considerable confirmation of the theory. For the deflection of light by stars, the theory now provides an amount twice as large as before."[19]

Freundlich suggested a new method of observing stars near Jupiter to test for light bending. Although Einstein had initially rejected the notion, he was now keen on it. Freundlich wanted to use a method the Dutch astronomer Jacobus C. Kapteyn had developed to measure stellar parallaxes. The method would involve taking photographs of stars as they disappear behind Jupiter (occultation) and comparing their positions when Jupiter is in another part of the sky. He consulted with the Zeiss Company about constructing an instrument to measure the plates.[20] But Freundlich's boss, Struve, would have nothing of the project. Einstein complained to Sommerfeld: "Freundlich has a method of measuring light deflection by Jupiter. Only the intrigues of pitiful persons prevent this last important test of the theory from being carried out. But this is not so painful for me anyway, because the theory seems to me to be adequately secure, especially also with regard to the qualitative verification of the spectral-line shift."[21]

Einstein stepped up his efforts to liberate Freundlich from his mundane observing chores. He again asked Planck to talk to Otto Naumann about "the Freundlich affair." At Planck's suggestion, Einstein followed up with a long letter to Naumann describing the important work that Freundlich would be able to undertake. Einstein's letter was a crash course on relativity and its testing for the most senior bureaucrat in the Education Ministry: "The issue involves testing the so-called theory of relativity. This theory bases itself on the assumption that no physical reality can be lent to time and space; it leads to a very specific theory of gravitation, according to which Newton's classical theory is valid only in admittedly superb first-order approximation. Verification of the results of this theory can be done only with methods used in astronomy."

Einstein outlined the three astronomical tests, starting with the gravitational redshift. "This result," he noted, "was confirmed qualitatively by Mr. Freundlich using the already available observational data gathered mostly by American observatories. Such a redshift was shown to be present on average, especially for fixed-star classes in which significant stellar masses had been determined through astronomical means." Einstein pointed out that "measurement of numerous stars was necessary" in order to obtain an average value of the stellar redshift. "The unknown individual motions of the separate stars also produce line shifts," he explained, and the averaging of many stars eliminated the individual velocity effects. Freundlich wanted to try a "more stringent test of this prediction of the theory ... on a binary star composed of two stars of substantially different sizes. We would then no longer be dependent on establishing a mean value, but could test the theory directly from the observational data gained from a single binary star of that type."

Einstein referred to the light bending prediction as "the most interesting and astonishing of all, and probably also undoubtedly the theory's most characteristic one; and precisely this consequence has not yet been subjected to any test." He referred to Freundlich's aborted attempt to study the effect at the Russian eclipse, but then moved on to his newest idea: "Through close examination of the available observational material, Mr. Freundlich now arrives at the result that the light-deflection effect could be demonstrated also with the planet Jupiter, although only with the subtlest of photographic measurements and through an increased number of observations. The most important specialists in the pertinent field of precision technology confirm that the observational method conceived of by Mr. Freundlich ought to lead to the objective." This would be Freundlich's "most import project" and it would "just be necessary that Mr. Freundlich be released for a few years from the routine duties of measuring star positions so that he could devote himself in peace to the projects indicated here."[22]

When Naumann approached Struve about the idea, the astronomer informed the administrator that the gravitational redshift in B stars was "not proved by the totally superficial investigations made so far." He claimed that the observatory did not have the means to make observations on stellar redshifts. He also belittled Einstein's accounting of Mercury's perihelion motion, insisting that Newton's theory of gravitation could explain the effect. Struve was adamantly opposed to Freundlich working on the Jupiter project: "Even a 'multitude of the most sophisticated measurements' by expert observers, let alone by those who do not come under this heading, will not yield any useful result and merely cause a needless expenditure of time and effort."[23]

Struve's caustic jab at Freundlich's observational abilities reflected the escalating battle that von Seeliger was waging over Freundlich's stellar redshifts. Freundlich published a more detailed treatment of his work in the *Astronomische Nachrichten*, mentioning his previous calculation error but omitting to acknowledge von Seeliger. He compounded this slight by taking some liberties with the data and revising his estimate of the density of B stars. With this much smaller density estimate, he obtained a mass for B stars of "approximately 25–30 solar masses," which was again of the right order of magnitude to be consistent with the relativity prediction.[24] Von Seeliger was incensed and wrote indignantly to Struve that Freundlich's paper was "scientifically *disingenuous*—to put it mildly—the likes of which has never happened before in my 40 years of experience in science."

> Scientific decency, which is otherwise practiced without exception, would have required that the source of the correction be indicated, and scientific honesty should have prevented such distortions of the real state of the facts. The other salvaging attempts with the obvious tendency to cover up have completely miscarried as well. . . . But in the interest of our science I cannot silently accept Mr. F's entire outrageous proceedings, and I will bring it out before the public in some form or other, even though I keenly regret charging one of your official employees with such a serious reproach.

As promised, von Seeliger blasted Freundlich publicly in the *Nachrichten* and tore his argument to shreds. He showed that Freundlich's original density estimate yielded a mass for B stars of sixty-five solar masses based on a gravitational redshift, completely contrary to the observed masses of B stars. He also elaborated how Freundlich had essentially picked a density that would get the result he was seeking using the corrected equation. He concluded: "The result of the whole study is thus, that . . . not only can an indication of the presence of a gravitation effect not be proved, but on the contrary, the latter can only be completely contradicted. Criticizing the data used is as little the intent of these lines as it is criticizing Einstein's theory. The object is only to present Mr. Freundlich's method of observation and the manner in which he evaluates clarifications which are brought to his attention."[25]

Von Seeliger was actually correct about Freundlich's shoddy approach to the problem, although his public attack went beyond the normal decorum of the scientific community. However, Sommerfeld advised Einstein against continuing to help Freundlich. Einstein acknowledged Freundlich's weaknesses, but emphasized the important role he was playing in raising awareness of relativity among astronomers. He wrote to Sommerfeld:

Fr. is of the "greyhound" breed. . . . His way of bolting is also not particularly distinguished. I have known this person's weaknesses for a long time—and have been more or less irritated by him. . . . I would *not* choose Fr. as an intimate friend of mine but would always maintain a healthy distance from him either way. . . . On the other hand, Freundl. offers something worth its weight in gold—an enthusiastic dedication to the problem; that is a rare trait he does not share with very many. . . . Freundl. is not really creatively talented, but he is intelligent and resourceful. The greyhound nature . . . comes to a large part from his pounding heart when he investigates a scientifically important issue. We must not forget that Fr. had devised the statistical method that makes possible using fixed stars in addressing the line-shift question. Although the nasty calculation error did slip by him and some other things there are greyhound-like as well (density definition), the overall value of the matter ought not to be forgotten because of it.

Einstein reminded Sommerfeld that Freundlich "was the only colleague in that profession until now to support my efforts effectively in the area of general relativity. He has devoted years of thought, and of work as well, to this problem, as far as was possible beside the exhausting and tedious duties at the observatory. What a pitiful rascal I would be if, now that the idea has become accepted, I dropped the man on the thought that I am now no longer dependent on him."

In actual fact, Einstein still depended on Freundlich for access to the world of astronomy. The ongoing war cut Germany off from the international community and the German astronomy establishment was hostile to relativity. Freundlich was all he had.

Fr. has shown that modern astronomical tools are good enough to demonstrate light deflection by Jupiter, which *I* would not have thought possible despite having considered the case years ago. I simply lack contacts in astronomy. Now, I gladly admit that Fr.'s weaknesses do not make it seem at all desirable that the execution of this important matter be placed solely in his hands. But up to now no one has made any effort toward participating in the undertaking, so that I am *de facto nolens volens* dependent on Freundlich alone in this endeavor of promoting the resolution of this eminently important question.[26]

Einstein was very keen that astronomers take up the observational challenge of testing his predictions. He was frustrated that Freundlich, the only German astronomer committed to the problem, was being stymied.

Einstein's exchange with Sommerfeld occurred when Karl Schwarzschild was still alive at the Russian front. Einstein mentioned the troubles he was having with Freundich to Schwarzschild while corresponding with him about his new theory. Schwarzschild pointed out difficulties in Freundlich's idea of using Jupiter to test for light bending. He also suggested that there might be a clique against Freundlich and advised Einstein to distance himself from him. Einstein addressed Schwarzschild's concern about Jupiter. "It *has* to work! Jupiter's moons could serve in studying closely the systematic errors of which you speak; for the apparent displacement of Jupiter's moons through light deflection is entirely negligible owing to the smallness of the moon-Jupiter distance. The angle to be confirmed amounts to 2 · 0.02" and is thus within the order of currently attainable precision." As for Freundlich, Einstein emphasized that he "was the first astronomer to understand the significance of the general theory of relativity and to address enthusiastically the astronomical issues attached to it. *That is why I would regret it deeply if he were deprived the possibility of working in this field.*"

Schwarzschild checked out the Jupiter problem with one of his astronomers back home and told Einstein that Jupiter was too much to the south, suggesting that southern observatories should take up the problem. He advised Einstein that "we shall not be able to agree easily on Freundlich. . . . Debating this way and that about him is pointless. I just think that he has already fallen out with Struve to such a degree that it would be best if you exerted your influence toward obtaining another occupation for him." Einstein tried to get Freundlich a position in Göttingen and other places, as well as continuing to try to convince Struve to promote him. He did not succeed.[27]

NEWS OF EINSTEIN'S BREAKTHROUGH SPREADS

Wartime Berlin was virtually cut off from England and America, where scientists were oblivious to Einstein's revolutionary work of 1915. It was via a small network of theorists in neutral Holland that news of general relativity spread first to England and then across the Atlantic to American astronomers. Theoretical physicists H. A. Lorentz and Paul Ehrenfest were intimate friends of Einstein. He sent them proofs of his articles, and immediately after his breakthrough he corresponded with them about the details of his new theory.[28] Einstein also wanted to spread the news widely to scientists to stimulate more interest in developing and testing the theory. The best way would be to write a book on relativity, "although, as with all things that are not supported by a fervent wish, I am having difficulty getting started. But if I do not do so, the theory will not be understood, as simple though it basically is."[29] Einstein admitted to Lorentz that his "series of gravitation papers are a chain of wrong tracks, which nevertheless did gradually lead closer to the objective." He suggested that his Dutch colleague write the book he wanted, since his letters on the theory were so clearly written. "It would certainly be of tremendous benefit to the issue if you made your considerations available to other physicists as well, by writing an article on the foundations of the theory. . . . I could do this myself, of course, to the extent that it is all clear to me. However, nature has unfortunately deprived me of the talent of being able to communicate in writing, so that what I write may well be correct but is quite indigestible."[30]

Lorentz did publish a series of four articles on general relativity for the Amsterdam Academy,[31] but Einstein had to write the comprehensive paper himself for the German scientific community. The obvious place to publish was the *Annalen der Physik*. Einstein asked the editor, Wilhelm Wien, whether he could also approach a publisher to bring his article out as a book. Planck intervened, arranging to have Barth, the publisher of the *Annalen*, produce the book version. As part of the deal, Einstein requested only twenty offprints instead of one hundred, knowing that the book version would ensure a large audience.[32]

Willem de Sitter, director of Leiden Observatory, had been following the development of general relativity closely. His colleagues Lorentz and Ehrenfest shared the papers that Einstein sent them, and he received an early copy of Einstein's *Annalen* paper. De Sitter also had the advantage of conversation with Einstein in the fall of 1916 during a visit to Leiden.[33] Knowing that German periodicals were not reaching Britain, de Sitter sent a copy of Einstein's paper to Arthur Stanley Eddington, then a secretary of the Royal Astronomical Society (RAS). Realizing that Eddington could not very well reprint the work of a German in an English journal during wartime, de Sitter offered to write an article on the theory himself for publication in the Society's *Memoirs*. Eddington, who had been following the relativity developments with interest before the war, immediately appreciated the revolutionary implications of Einstein's latest work. He told de Sitter that he was "immensely interested in what you tell me about Einstein's theory. . . . Hitherto I had only heard vague rumors of Einstein's new work. I do not think anyone in England knows the details of his paper."[34] (Fig. 4.1.)

Eddington approved de Sitter's plan but advised publication in the RAS's frequently published and widely distributed *Monthly Notices*, rather than its infrequent *Memoirs*. He reiterated that "no one in England has yet been able to see Einstein's paper and many are very curious to know the new theory." De Sitter told Einstein that he was writing "a small paper . . . on the new theory of gravitation and its astronomical consequences" for the British astronomers. "Your theory still seems to be almost entirely unknown in England." In the end, de Sitter prepared a series of three detailed articles, including his own original contributions. He submitted the first to Eddington in late August 1916. Eddington de-

Figure 4.1. This network spread news of Einstein's theory in wartime. *Back row, left to right*: Einstein, Paul Ehrenfest, Willem de Sitter; *front row*: Arthur Eddington and Hendrik Antoon Lorentz. (Courtesy of the Niels Bohr Library, AIP.)

cided to fast-track de Sitter's paper, which he considered to be "of exceptional importance" (fig. 4.2).[35] He sent it immediately to the printer to appear in a planned supplement to the *Notices* in October rather than waiting for the November council meeting. The next two papers came slower, owing to a fire at the printer's. Eddington undertook to master de Sitter's and Einstein's papers, and soon he too was interpreting and elaborating general relativity.[36]

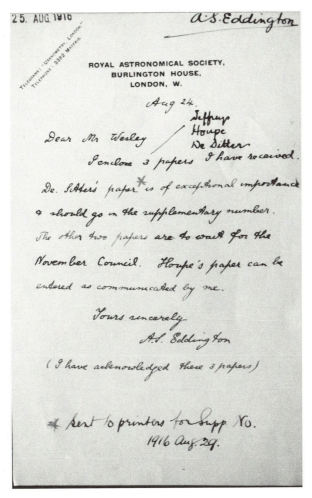

Figure 4.2. Eddington's letter to RAS secretary Wesley, instructing him to fast-track de Sitter's paper due to its "exceptional importance." (Courtesy Royal Astronomical Society.)

Nor did de Sitter and Eddington neglect more general audiences. Writing in the semipopular *Observatory*,[37] de Sitter emphasized the requirement that the laws of nature should be independent of the choice of coordinates, introduced the concept of an invariant tensor, and related Riemann's metrical tensor to Minkowski's four-dimensional time-space. Using the latter notion, he introduced the picture of the worldline of a material particle. Einstein's theory of gravitation states that worldlines in this "time-space" (de Sitter's term) are geodesics, de Sitter explained, lines traced by particles subject to no forces. Gravitation is a property of space:

the coefficients determining the metric properties of time-space also determine the gravitational field. Einstein's equations for the metric and gravitational field give in first approximation Newton's law of gravitation. "If
the approximation is pushed one order further [de Sitter wrote] the well-
known anomaly in the motion of the perihelion of Mercury is exactly
explained, and no other effects at present within the reach of observation
are produced in the motions of the planets or the Moon."[38]

American astronomers had no one within their ranks who did the same
kind of elaboration of general relativity. They learned about Einstein,
whose papers did not reach them owing to the British blockade, from de
Sitter and Eddington.[39]

Mixed Reactions to a Complicated Theory

At least one reader of *Observatory* did not think that Einstein had done
anything new and true. In a letter to the editor, Thomas Jefferson Jackson
See, a pompous plagiarizer employed as an astronomer at the U.S. Naval
Station on Mare Island near San Francisco, objected to the "metaphysical" foundations of general relativity. He complained that while de Sitter
"carefully discusses the *analysis* of Einstein's treatment, he so completely
passes over all *physical considerations* as actually to convey the impression that gravitation is not a *physical problem*, but only an *analytical
one*." See expressed amazement that de Sitter "actually states that under
this theory gravitation is not a 'force,' but 'a property of space." See cited
Sir Oliver Lodge's calculations of the equivalent attractive forces of the
Earth on the Moon, and the Sun on the Earth, in terms of the breaking
strength of enormous number of steel pillars. Evidently, he concluded,
gravitation is "an influence exerted by *matter*," and Einstein's unique
merit was to demonstrate "the extent to which purely mathematical
reasoning may be misapplied by those who ignore appropriate physical
considerations."[40]

T.J.J. See had a reputation among astronomers on both sides of the
Atlantic. While on the staff of the astronomy department at the University
of Chicago in the 1890s, he secretly tried to prejudice President William
Rainey Harper against paying salaries at the Yerkes Observatory. Harper
eventually fired him when he realized that his behavior was that of an
immoral egomaniac.[41] See went on to work on the staff of the Lowell
Observatory but caused similar disruptions, alienating most of the staff
there. He was dismissed in July 1898.[42] He ended up with the U.S. Navy
job at the isolated naval station in northern California. By about 1910,
See had succeeded in alienating himself from all the American theorists
working on problems of cosmogony and the origin of the solar system by

plagiarizing their work and claiming it for himself. He had the habit of sending out elaborate press releases of his discoveries and cabling news of his achievements simultaneously to the *Astronomische Nachrichten* and the president of the United States. In 1899 Herbert Hall Turner published a remarkable satirical note in his "Oxford notebook," announcing the recent discovery of the "law of increase of gaseous reputation":[43]

$$T = J/J_1 \ C.$$

Despite See's reputation, his remarks against the "metaphysical" foundations of general relativity struck a responsive chord. His comments drew a reaction from James Jeans, who worriedly pointed out that See's letter "gives rise to a fear that Einstein's Theory may meet with an unfavourable reception on account of the somewhat metaphysical—one might almost say mystical—form in which his results have been expressed."[44] He offered assurances that "the more concrete part of Einstein's work is quite independent of the metaphysical garment in which it has been clad." Say astronomers on Jupiter had determined an empirical law of refraction in the Jovian atmosphere by observing Zenith* transits of the Jovian satellites. What would the Jovian astronomers have to do to their law of refraction if a revolutionary mathematician in their midst demanded a new law incorporating observations outside the vertical? The revolutionary mathematician, if a relativist, would assert that "the true laws of refraction can have no reference to chosen directions in space; technically speaking, they must be invariant for all changes of axes." Jeans insisted that if the new laws of refraction were verified, the Jovian mathematician might suggest that the result "proved that refraction was a 'property of space,' or that 'horizontal and vertical had separately vanished to shadows.' These suggestions, however interesting they might be to metaphysicians, might conceivably prejudice the scientific acceptance of the new laws of refraction, but we can see that such prejudice would be unwarranted." Jeans claimed that Einstein's "crumpling up of his four-dimensional space" may likewise be considered fictitious. Though the theory brought the true laws of gravitation, it did not explain the nature of gravitation. "Gravitational theory has, so far, been trying to fit round pegs into elliptical holes: the new theory finds a new round hole into which the pegs will fit perfectly, but this does not explain the physical nature of the pegs."

Eddington took issue with Jeans. "[It] is very well put," he wrote de Sitter of Jeans's reply, "though he rather understates (probably not in his own mind but in the wording of the note) the so called metaphysical aspects of the theory." In *Observatory* Eddington warned that Jeans' "ex-

---

* The point in the sky directly overhead.

cellent analogy" might lead to the "new conceptions of space and time being underrated," and he set out to defend "the 'metaphysical' garb in which the theory is usually clothed." Building on Jeans's analogy, Eddington pointed out that Jovians who had developed a law of refraction having special reference to the vertical might believe in a flat Jupiter. They would then hold the vertical to be a fundamental concept of geometry, with "*up* and *down* as different as right and wrong." In contrast with Jeans, Eddington took the notion of a curved space as a physical insight. "It is not a mystical theory," he asserted, "but a mere matter of obtaining sufficiently delicate measurements, to show that the space we have hitherto supposed we were measuring gets crumpled in a gravitational field." Eddington pointed out that the new law of gravitation could be compared only with difficulty to the old theory. To decide whether the new law conforms with the old idea of a force proportional to the inverse square of the distance, he noted, it is necessary to know what is meant by distance. "Neither the new conceptions nor the old suggest any one value from the various possibilities as being the true distance," Eddington emphasized. "Thus we are led to the metaphysical question immediately—What is distance? What is space?"[45]

The exchange between See, Jeans, and Eddington anticipated a major line in subsequent discussions of general relativity. See's questions were not silly. The geometricization of physics that general relativity called for, while one of its transcendental beauties for de Sitter and Eddington, was metaphysical nonsense for others. Jeans urged an instrumentalist interpretation, perhaps more as a tactic than as a conviction. As he wrote Oliver Lodge, "I fear de Sitter has rather prejudiced the reception of Einstein's theory by a too abstruse presentation."[46] Discussion of these points of view, as well as attempts at classical explanations of relativistic effects, occupied British astronomers and physicists for several years.[47]

The discussion in Britain was taken up in the United States by the ever-vigilant Curtis, who noted that "the theory as originally put forward by Einstein has been subjected to so many alterations as now to be referred to as the 'old' theory of relativity." The change, he explained, was that "all laws of nature must be invariant for *any* transformation of coordinates, instead of for one particular class of transformations as in the earlier Einstein theory." Curtis quoted directly from de Sitter: "Gravitation is thus put on quite a different footing from all other forces of nature—it is no longer a 'force,' properly speaking—it is more of the nature of a property of the four-dimensional time-space, imparted to it by matter or energy located in it." Curtis referred to the replacement of "ordinary three-dimensional Euclidean space by a four-dimensional time-space, where time is the fourth parameter." He noted that the "mathematics of such a physical universe is somewhat complicated, but it seems to fit well with all observed phenomena."[48]

It was much to swallow:

> Many will feel that the idea of a four-dimensional time-space is fully as difficult of comprehension as was the mystery of gravitation, all-pervading, inexplicable, in our classical physical theories. While the mathematician is willing to admit that many other forms of space or geometries of space would satisfy physical science as well as the Euclidean, we must confess that we are still of the point of view of the mathematician who stated that, while it would be possible, in a four-dimensional universe, to turn an egg inside out without breaking its shell, still he realized that there were many practical difficulties in the way of the accomplishment of this feat.[49]

Although in this context Curtis felt inclined to "sympathize somewhat with Professor See's point of view," he cautioned "the impressive quantities which he quotes as examples of the mighty force of gravitation, tho often used in this connection, are apt to be misleading." Curtis reminded readers that compared to electrostatic forces between two masses, the gravitational force is negligible. Curtis was mathematician or philosopher enough to find general relativity attractive because of "its unification of all matter and all forces in a single simple and homogeneous system." While he noted that "its future development and possible applications will be followed with great interest,"[50] we shall see that in the end, he came out with See, rejecting the geometricization of gravitation and seeking mechanical explanations for relativistic effects in the astronomical domain. Two aspects of Curtis's later position characterize the climate of opinion regarding relativity among American astronomers around 1920: mistrust of the geometrical notions of the theory and avoidance of its details.

In Germany, scientists had the benefit of reading Einstein's papers and the comments of others in great detail. As a consequence, reactions were more informed—and more diverse. Positive reactions from Einstein's small circle of colleagues in Germany and the neutral Netherlands and Switzerland were encouraging. Yet Einstein wanted others to understand the theory and work with it. When Max Born wrote a survey article "devoid of all mathematical form" on the theory for physicists, Einstein was thrilled to have been understood thoroughly and acknowledged "by one of my most highly qualified colleagues." "I was also filled with the happy sensation of cheerful goodwill that emanates from the paper and that otherwise so rarely lingers undiluted in the pale light of the study lamp."[51] Freundlich also published a semipopular review article with reduced mathematics, for which Einstein penned a brief foreword. He praised Freundlich's work for "making the basic ideas of the theory accessible to anyone who has some kind of acquaintance with the reasoning methods of the exact sciences." Freundlich noted "the extraordinary unity and logic of its [Einstein's gravitation theory] basic tenets. With one stroke it truly solves all the puzzles that have arisen since Newton's time."[52]

As word of Einstein's accomplishment spread in Germany, more sensational articles appeared in newspapers. They kindled darker reactions to Einstein's theory, fueled by political unrest and the duress of war. One popularizer, Max Weinstein, claimed that general relativity had removed gravity from its earlier isolated position and made it into a "world power" controlling all laws of nature. He warned that physics and mathematics would have to be revised. Popular reactions to Weinstein's writings prompted Wilhelm Foerster, emeritus professor of astronomy at the University of Berlin, to urge Einstein "to find a way of addressing the German public" to allay anxiety and skepticism stirred up by "doubts about previously held basic tenets of our knowledge of the world." He attributed this agitation to an almost psychopathic state among the populace. To illustrate, he noted: "Some are happy that you had now put an end to the global confusion caused by the Englishman Newton, etc. Surely you will find words free of scholarly jargon to introduce the German public to a sound and sober-minded interpretation of your so extremely important ideas and problems; but there really is a need for this now."[53] Einstein decided to write a popular book, although more for the scientifically and mathematically literate layperson. *On the Special and the General Theory of Relativity, Generally Comprehensible* was published early in 1917. Einstein joked to his friends that it really should have the subtitle "generally incomprehensible."[54]

As German physicists began to weigh in for and against the theory, Einstein welcomed each new supporter and parried thrusts from the critics. When Hermann Weyl "welcomed the general theory of relativity with such warmth and enthusiasm," Einstein was very pleased. "Although the theory still has many opponents at the moment, I console myself with the following situation: The otherwise established average brain power of the advocates immensely surpasses that of the opponents! This is objective evidence, of a sort, for the naturalness and rationality of the theory."[55] He was well aware that nonscientific issues motivated some of his critics. Ernst Gehrcke, for example, dredged up a forgotten publication by the German physicist Paul Gerber, who in 1898 had derived a similar formula for the perihelion advance using Newtonian theory. He published an attack on Einstein's equivalence hypothesis, implying that Einstein had been guided by Gerber's work in developing his theory. Einstein chose "not to respond to Gehrcke's tasteless and superficial attacks, because any informed reader can do this himself."[56] The physicist Max von Laue, however, did publish an effective rejoinder in 1917 when Gerber's paper was reprinted with an introduction by Gehrcke.[57] The philosopher Eduard Hartmann addressed Gehrcke's objections in a paper on general relativity for a philosophical readership, using arguments that Einstein shared with him in correspondence.[58] Gehrcke would later align himself with a nationalist, anti-Semitic campaign against Einstein in the 1920s.

Einstein was accommodating to philosophers and other nonspecialists who began to expound on his theory and its implications for philosophical and other readers. Many wrote to him for feedback and he graciously responded, pointing out misconceptions and clarifying details of his theory.[59] Nonetheless, he maintained a distance from philosophical speculation. "From reading philosophical books I had to learn that I stand there like a blind man before a painting. I grasp only the inductive method at work . . . but the works of speculative philosophy are inaccessible to me."[60] Ironically, as more nonspecialists waded into general relativity, many authors drew on Einstein's success with building a theory on one principle—relativity of motion for all observers—to claim the primacy of speculation over empiricism. Einstein objected strenuously to this claim linked to his theory. In August 1918 he griped to his friend Michel Besso: "On rereading your last letter, I discovered something that downright annoys me: speculation allegedly had revealed itself to be superior to empiricism. In this regard you are thinking of the development of the theory of relativity. But I find that this development teaches something different that is almost the opposite, namely, that in order to be reliable, a theory must be built upon generalizable *facts*. . . . No genuinely useful and profound theory has ever really been found purely speculatively."[61] To this day, the glib notion that "everything is relative" permeates popular accounts of Einstein's contribution. This misconception stalked Einstein all his life.

CONSTRUCTING THE UNIVERSE

While others grappled to understand his theory and debated its merits, Einstein continued to extend it further. In this work, he found a willing and able astronomer to wrestle with him intellectually. Willem de Sitter mastered the details of general relativity after hearing about it from Lorentz and Ehrenfest. In elaborating Einstein's theory for Dutch and British astronomers, de Sitter came to appreciate its power and usefulness. He began to make original contributions. Einstein "was delighted that you like the general theory of relativity so much."[62] The astronomer proved to be a valuable foil for the bold-thinking physicist. In a 1916 paper de Sitter developed a new formulation of Einstein's field equations, which Einstein used to predict the existence of gravitational waves.[63] He wrote de Sitter telling him that the equations produced three different kinds of gravitational waves, only one of which carries energy. The other two, Einstein surmised, "do not exist but are simulated by the coordinate system's wavelike motions against Galilean space." De Sitter was suspicious of any talk of motion relative to absolute space. He penned a marginal note on Einstein's letter: "What is this 'Galilean space'? Could one not just as well say the 'Aether'?"[64] Einstein and de Sitter engaged in an animated debate about this theme as they ventured into using general relativity to describe the universe as a whole.

Einstein's geometrical view of gravitation connected space, time, and matter in an intimate way. In his view, space and time did not exist without matter. Today, popularizers of general relativity often illustrate Einstein's gravitation theory with a ball bearing resting on a rubber sheet. The ball bearing creates a curved depression in the sheet. A smaller, lighter ball passing close to the ball bearing will curve around it in an orbit. If you remove the ball bearing from the sheet, it becomes flat. By analogy, if you remove all matter from the universe, you might think that space would become flat again. It would revert to the "Galilean space" of Newton. Einstein rejected the notion of a flat space without matter in it. "The essence of my theory is precisely that no independent properties are attributed to space on its own. It can be put jokingly this way. If I allow all things to vanish from the world, then following Newton, the Galiliean inertial space remains; following my interpretation, however, *nothing* remains."[65] According to this view, "inertia is simply an interaction between masses, not an effect in which 'space' of itself were involved, separate from the observed mass."[66]

Einstein first presented this idea in his 1916 review article on general relativity by considering a rotating sphere in empty space. A rotating sphere has a bulge at the equator because of the apparent outward (centrifugal) force due to inertia. When a car turns, we experience a similar centrifugal force pushing us outward along the radius of the turn. This "force" is due to inertia—our tendency to continue moving in the direction we were moving before the turn. According to Einstein's theory, the inhabitants of the sphere can consider themselves to be stationary. They could attribute the bulge to gravitational effects from distant masses.

Einstein applied this reasoning to the entire universe. The current view from astronomy was that the Sun is one of millions of stars in an immense disk over 200,000 light-years across called the Milky Way. Scattered among the stars are thousands of nebulae of different shapes and sizes. Many are spiral shaped and some astronomers believed that they are distant stellar systems like our own Milky Way. Others felt they are inside our stellar system, which is confined to the disk. Einstein postulated that unobservable "distant masses" at the boundary of the universe determine the curvature of space and produce the observed structure of the stellar system, maintaining its shape.[67] When he visited Leiden in the fall of 1916, he discussed his ideas with de Sitter. The astronomer didn't like what he heard. "I have been thinking much about the relativity of inertia and about distant masses, and the longer I think about it, the more troubling your hypothesis becomes for me." De Sitter was "bothered" by Einstein's "conviction that the boundary, the 'envelope,' will always remain hypothetical and will never be observed."

> Now we can say: the sources of inertia lie beyond the Milky Way, but when our grandchildren make an invention that enlarges the known world in the same proportion that it was enlarged 300 years ago through the invention of the telescope, then the envelope will simply have to shift farther outwards again. From this I conclude that the envelope is *not* a physical reality.
>
> If the hypothesis is accepted, one would first want to get a [crude] idea of *where* these distant masses are and of what they are composed; second, *how* the inertia comes over here from there. An artificial mechanism will be invented. The coordinate system with reference to which the envelope and this mechanism are at rest will also be defined. Although the principle of relativity will still hold *formally*, effectively, we shall have the old absolute space with the ether back.[68]

De Sitter's criticisms were friendly. He was an ardent fan of general relativity, yet he believed that with this added hypothesis, "your theory will have lost much of its classical beauty for me. With it an 'explanation' of the origin of inertia is gained, which is actually not an explanation, for it is not an explanation from known or verifiable facts but from masses invented *ad hoc*. I am convinced . . . that these masses will go the way of the 'ether wind.' "[69] Eddington, who read de Sitter's arguments in his second paper for the British *Monthly Notices*, also objected to Einstein's "ad hoc" masses. He had begun to work with general relativity, and de Sitter supplied him with Einstein's papers as he received them. Einstein was delighted that de Sitter was educating the British about his theory despite the war with Germany. "It is a fine thing that you are throwing this bridge over the abyss of delusion." He was happy to send papers for Eddington, and promised, "When peace has returned, I shall write to him."[70] After some digging in of his heels, Einstein finally accepted de Sitter's criticism. He thrived on the thrust and parry of scientific debate and was delighted with the interest his theory was receiving in the Netherlands. After his visit, he wrote Besso: "The general theory of relativity has already come very much alive there. Not only are Lorentz and the astronomer de Sitter working independently on the theory but a number of other young colleagues as well. The theory has also taken root in England."[71]

By February 1917 Einstein was working on another tack. "Presently I am writing a paper on the boundary conditions in gravitation theory," he wrote de Sitter. "I have completely abandoned my idea . . . which you rightly disputed. I am curious to see what you will say about the rather outlandish conception I have now set my sights on."[72] To Ehrenfest he wrote: "I have perpetrated something again . . . in gravitation theory, which exposes me a bit to the danger of being committed to a madhouse. I hope there are none over there in Leyden, so that I can visit you again safely."[73] To avoid the troublesome boundary conditions, Einstein came up with a spherical universe that is finite and unbounded. Its volume is finite, as are all other dimensions, including length, breadth, depth, and time, yet it has no boundary. Light can travel indefinitely through this universe. Like an ant crawling on a huge sphere, it covers a finite territory but never hits an edge. Light from a star could, in principle, be seen from two directions. Einstein calculated that his universe would have a radius of about 10 million light-years, or one thousand times that of the Milky Way. According to Einstein's original field equations, his spherical, closed universe would either contract or expand. Einstein introduced a "cosmological term," lambda (λ), into his equations to keep the universe static. He did so because astronomical observations indicated that stellar velocities are small and random, averaging to zero. In his paper he noted that introducing lambda did not change any of the astronomical predictions of general relativity. Ironically, the uniformly distributed matter of the spherical universe now became equivalent to Newton's "absolute space." Einstein noted this irony to Ehrenfest: "I am sending you my new paper. My solution may appear adventurous to you, but for the moment it seems to me to be the most natural one. From the measured stellar densities, a universe radius of the order of magnitude 10^7 light years results, thus unfortunately being very large against the distances of observable stars. The odd thing is that now a quasi-absolute time and a preferred coordinate system do reappear in the end, while fully complying with all the requirements of relativity. Please show the paper also to Lorentz and de Sitter."[74]

Einstein's cosmological theory provided a relation between the cosmological constant, λ, the average mass density of the universe, ρ, and the radius of the universe, R. In principle both R and ρ could be measured. A study of the statistical distribution of stars could determine the average mass density. "The star statistics question has become a burning issue to be addressed now," Einstein told Freundlich. He suggested they write a joint paper, and he returned to his hope that he might find a way to free up some of Freundlich's time for the work. "We must act cautiously in doing this, though, so that your position is not jeopardized." Einstein pointed out to his colleague that for the first time his theory provided a way for astronomers to study the universe as a whole. "The matter of great interest here is that not only R but also ρ must be individually determinable astronomically, the latter quantity at least to a very rough approximation, and that then my relation between them ought to hold. Maybe the chasm between the 10^4 and 10^7 light yearscan be bridged after all! That would mean the beginning of an epoch in astronomy."[75]

Einstein admitted to de Sitter that "I have erected but a lofty castle in the air." With the observable universe so small compared to his calculated size, his model was untestable by observation.

> I compare the space to a cloth floating (at rest) in the air, a certain part of which we can observe. This part is slightly curved similarly to a small section of a sphere's surface. We philosophize on how we must construe the continuation of the cloth so that an equilibrium is reached in its tangential tension, whether it is fastened in position at the edges, extends infinitely, or has a finite size and is a closed unit. Heine has provided the answer in a poem: "And a fool waits for an answer." So let us be satisfied and not expect an answer, and rather see each other again as soon as possible in acceptable health in Leyden!

To which de Sitter replied: "Well, if you do not want to impose your conception on reality, then we are in agreement. I have nothing against it as a contradiction-free chain of reasoning, and I even admire it. I cannot concur completely before having calculated with it, which is not possible for me to do right now."[76]

Five days later, de Sitter had done his homework. In working through Einstein's cosmological model, he discovered another model that worked for an empty universe "*without matter.*" A four-dimensional space-time hypersphere now replaced Einstein's three-dimensional spatial hypersphere. "I personally much prefer the *four*-dimensional system," he told Einstein, "but even more so the original theory, without the undeterminable λ, which is just philosophically and not physically desirable."[77] Einstein replied: "In my opinion, it would be unsatisfactory if a world without matter were possible. Rather, the . . . field should *be fully determined by matter and not be able to exist without the matter*. This is the core of what I mean by the requirement of the relativity of inertia. One could just as well speak of the 'matter conditioning geometry.' To me, as long as this requirement had not been fulfilled, the goal of general relativity was not yet completely achieved. This only came about with the λ term."[78] De Sitter objected strongly to Einstein's assumption that the universe was static, which had been his reason for introducing λ.

I must emphatically contest your assumption that the world is mechanically quasi-stationary. We only have a snapshot of the world, and we cannot and must *not* conclude from the fact that we do not see any large changes on this photograph that everything will always remain as at that instant when the picture was taken. I believe that it is probably certain that even the Milky Way is *not* a stable system. Is the entire universe then likely to be stable? The distribution of matter in the universe is extremely *inhomogeneous* (I mean of the stars, not your "world matter"), and it cannot be substituted, even in rough approximation, by a distribution of constant density. The assumption you tacitly make that the mean stellar density is the same throughout the universe . . . has no justification whatsoever, and all our observations speak against it.[79]

De Sitter presented Einstein's and his cosmological models in the third paper in his series on relativity for the British *Monthly Notices*. His model predicted that test particles would move away from one another at speeds that increase with the distance between them. This phenomenon would mean that the spectra of distant objects would appear redshifted, and that the redshift would increase with distance. De Sitter referred to observations of spiral nebulae by Vesto Melvin Slipher at Lowell Observatory in Arizona. Slipher had found that thirteen spirals had large redshifts. Could this be evidence of de Sitter's cosmological effect?[80]

Einstein and de Sitter continued to debate their cosmological models for the next few years. Hermann Weyl and Felix Klein entered the fray and ended up supporting de Sitter.[81] Their work marked the beginning of an entirely new field of astronomy—relativistic cosmology. English-speaking astronomers became aware of Einstein's and de Sitter's early cosmological considerations through de Sitter's discussions in the *Monthly Notices*.

While Einstein continued to enjoy theoretical triumphs, astronomers involved in testing his predictions soon brought difficulties from the observational side.

CHALLENGES FROM SOLAR OBSERVATIONS

Astronomers outside Germany who had taken up observational tests of relativity since 1911 continued their work regardless of whether they had heard of the newest breakthrough. Einstein's full general theory of relativity called for a gravitational redshift of the same magnitude as his theory of 1911, and light bending double the amount predicted by the equivalence principle. The gravitational redshift is calculated in general relativity by inspecting one of the sixteen components of the metric tensor—the g_{44}

(time) component. In the first approximation, this component is just the Newtonian potential, and it turns out that the slowing of clocks in a gravitational field will emerge from any derivation that combines special relativity and Newtonian theory. The doubling of the light bending arises from a combination of time curvature and space curvature. The first, which dominates at small velocities, is equivalent in first approximation to Newton's law of gravitation; it emerges from any joining of special relativity and Newtonian theory, and so appeared in Einstein's theory of 1911. The second part of the light bending was entirely new in the general theory, and cannot be derived on Newtonian theories.

While Campbell and Curtis pursued light bending at Lick, their colleagues to the south at Mount Wilson looked for the gravitational redshift in the Sun. By 1915, Julius and Einstein had their wish. The Mount Wilson research program now included "observations bearing upon possible evidence of anomalous dispersion and the Einstein effect in the solar atmosphere."[82] Comparisons of the solar and terrestrial sources had to be pushed to the third decimal place. Both the 60-foot and the new 150-foot tower telescopes were equipped especially for the work.[83] An interferometer was installed in the 60-foot instrument to provide an additional means for checking the previous observations and making new measures.

Charles St. John first tackled the anomalous dispersion problem. By 1915, he could definitely rule it out as an explanation for the redshifts. Evershed agreed.[84] St. John then began observations of redshifts of forty-three cyanogen bands in the solar spectrum. From laboratory experiments, he knew these lines are insensitive to pressure effects. St. John selected them to check Evershed's contention that pressure played a minor role, and also to check the relativity hypothesis. If Evershed was right that pressure was not important, and if Einstein was correct, then these lines should show a redshift. At a meeting of the National Academy of Sciences on 5 June 1917, Hale read a preliminary announcement of results from St. John's investigations that would have a profound effect on opinions and attitudes toward Einstein's theory.[85]

St. John declared on the authority of Einstein's work of 1911 that "the equivalence principle of generalized relativity" led to predictions of redshift and light bending, the amount of the latter for a star near the limb of the Sun amounting to 1.75 seconds of arc.[86] This mixing of the 1911 mechanism with the 1915 deflection indicates the tendency of the Mount Wilson astronomers to gloss over theoretical details and concentrate on the results to be measured.[87] St. John noted that "confirmation of either of these consequences" would have an important bearing on the interpretation of astrophysical data. In particular, it would complicate matters a great deal for astronomers, especially in spectroscopy. "The problem of determining stellar motions in the line of sight, a matter of fundamental importance, would be confronted with difficulties of a high order, depending as it does upon line displacement in stellar, relative to terrestrial spectra. Our knowledge of the motions, pressure, and many other phenomena in the solar atmosphere must be obtained from line displacements in the spectrum, but here it would be possible to apply definite corrections, this would in many cases, however, modify our interpretations."[88] Presumably with some relief, Hale read out: "The general conclusion to the investigation is that within the limits of error the measurements show no evidence of an effect of the order deduced from the equivalence relativity principle."[89]

St. John measured solar line displacements (presented to four decimal places) at the limb and at the center, compared to the carbon arc. He made the limb measures at the pole to eliminate Doppler shifts due to the Sun's rotation. St. John found the center lines slightly shifted toward the blue (negative redshift), indicating a slight upward movement of the vapor over the center of the solar disk. If a relativistic redshift were in operation, he could explain this result with solar vapors rising quickly enough to cause a Doppler effect that was masking the gravitational redshift. Then at the limb, where the rising vapors would be moving across the line of sight, he should observe the full gravitational redshift. However, St. John's measures indicated essentially zero shift at the limb, ruling out the relativity prediction.[90]

St. John discussed one complication. Unlike most spectral lines, the cyanogen bands are not sharp, individual lines. They occur in series of lines closely packed together in bands. Relying on the superior equipment at his disposal, St. John examined high-dispersion spectrograms of excellent definition in order to pick out lines that he felt were sufficiently free from neighboring ones. He further divided the final forty-three lines into two groups, one with high weight (containing the narrowest lines) and one with lower weight (broader lines). In this way, St. John felt that he had successfully dealt with the problem of isolating clear lines.

Evershed in India was not as confident in St. John's use of the cyanogen lines. In particular, he disagreed with St. John's weighting system. St. John gave the highest weight to narrow lines of low intensity because he could measure their positions precisely. These lines, claimed Evershed, were more easily affected by blends with other lines, whereas the broader lines were not so easily shifted by the presence of intruding lines.[91] Evershed felt that the highest weights should be given to the stronger lines. Using a revised technique of measurement, he reported results on thirty cyanogen lines and bands.[92] Where St. John found zero shifts at the limb, Evershed found a positive result. "But it appears that the shift at the limb is perhaps on the average not much more than half of the predicted gravitational effect," he noted, "whilst for the iron lines it is in many cases twice as great at the limb as is required on the relativity hypothesis." He noted that "if we exclude relativity, we are 'up against' the Earth effect, and even if we assign the shifts to a combination of relativity and motion, we cannot by that means escape the Earth effect, for the motion component would still be in the direction of the Earth."[93] Evershed concluded that "whatever way we may eventually interpret the limb-arc shifts, we seem able even now to state, with St. John, that our results are distinctly unfavourable to relativity."[94]

The news about cyanogen distressed Eddington, who was busy preparing his extensive "*Report on the Relativity Theory of Gravitation*" for the Physical Society of London. "St. John's latest paper has been giving me sleepless nights," he complained to Adams, "chasing mare's nests to reconcile the relativity theory with the results, or vice versa. I cannot make any headway."[95] Nor did he. He wrote in his report: "The difficulties of the test [the measurement of the predicted redshift in the solar spectrum] are so great that we may perhaps suspend judgment; but it would be idle to deny the seriousness of this apparent break-down of Einstein's theory."[96] Eddington emphasized that the redshift was "a necessary and fundamental condition for the acceptance of Einstein's theory; and that if it is really non-existent . . . we should have to reject the whole theory constructed on the principle of equivalence."

Sommerfeld asked Hermann Weyl if his new theory of gravitation might be better than Einstein's: "Up to now no indication of this [gravitational redshift] has appeared. Schwarzschild did not find any; new careful American measurements at Mount Wilson also did not. I would be interested to find out from you when there is a chance, whether redshift is unavoidable in your theory as well." Unlike Einstein, Sommerfeld discounted Freundlich's work on the B stars. "What Freundlich has published on this, as a purported verification of the same, is more or less a fraud."[97]

Weyl told Einstein about Sommerfeld's reference to the Mount Wilson results, asking "what is the story there?" Einstein replied that he had "heard about the American measurements and discussed them with Freundlich." He was unconcerned about the results from America. "They do not seem to prove anything yet. No flawless measurements have been carried out yet on terrestrially generated lines; the electrical arc used up to now is unsuitable. We are now in the process of procuring or begging for the funds for an electrical furnace that generates flawless lines *thermally*. Only in this way will secure results be attainable. In a few years the verdict will be in."[98]

For astronomers, the prestige of the Mount Wilson group lent a great deal of weight to St. John's conclusions. But the observations were difficult. Evershed's proposal of a strange Earth effect on solar spectral lines mitigated his criticisms of the Mount Wilson data. As early as 1914 he had entertained this idea, but he did not like it. "This is very puzzling," he wrote to Campbell at Lick, "as it seems to indicate a repulsion of the gases by the earth." Evershed proposed photographing the spectrum of Venus alongside the daylight spectrum, to "have a comparison of Sunlight from different sides of the Sun." As Venus moved around its orbit, the reflected Sunlight would alternately be from the side facing the Earth or the side facing away. If the Earth effect was real, then a systematic error should show up in comparing the Venus spectrum with the solar spectrum.[99] Evershed asked the Lick director to check old Venus spectra, taken for other purposes, to see what he could find.

"Your guarded hypothesis as to the increase in wave lengths at the Sun's edge does not appeal very strongly to me," Campbell replied. He pointed out that if Evershed were right, "the other planets, and especially Venus which is almost a duplicate of the Earth as to diameter and mass and is closer to the Sun, should have a repulsive action as well as the Earth." Nonetheless, Campbell analyzed Lick plates of Venus that had been taken as part of a solar parallax program, and he provided Evershed with the data. He told his British colleague that he was contemplating a program to test the rotation period of Venus, and if he proceeded, he could use the plates for Evershed's problem.[100]

Evershed was grateful for Campbell's analysis of the old Lick plates. To him, the data appeared to favor an Earth effect, "but perhaps I am misinterpreting the meaning of the sign of your residual, and of course everything depends on your methods of reduction." He hoped that Campbell would be able to carry out the Venus rotation program, and in particular, to obtain spectra when Venus was receiving light from the part of the Sun turned away from the Earth. "My hypothesis of a repulsion of the solar gases by the earth appeals to no one, and least of all perhaps to myself," he admitted, "but for this very reason, and because my results are so extremely difficult to explain otherwise, I am anxious to put it to the test." Evershed noted that his equipment "is unfortunately not adapted for photographing Venus, but I would gladly do my part, with your approval, in measuring any plates you might get."[101]

No collaboration ensued, and Evershed finally decided to try the Venus observations himself. In December 1916 he modified the spectrograph that he had been using for his studies of limb and center solar line displacements for photographing the spectrum of Venus.[102] By February 1918 Evershed could report results from two series of photographs obtained in February and October of the previous year. Both sets of data seemed to favor "the motion interpretation of the shifts, involving an Earth effect," though the February plates might have been affected adversely by "pole effect" in the iron arc. This effect was discovered during the early war years. Light emanating from the poles of the iron arc is shifted in a complicated way that may lead to systematic errors when solar and arc wavelengths are compared. The usual procedure for eliminating this source of error was to try to use light near the center of the arc, thus avoiding the light from the poles. Evershed announced plans to obtain a third series of spectra during June and July of 1918, and he hoped that "a decisive result" might then be forthcoming.[103]

The negative results of the various searches for a gravitational redshift in the Sun put a damper on the reception of Einstein's new theory of gravitation. Early in 1917 the Nobel Prize committee received three nominations for Einstein to receive the coveted award for that year. Although the committee's report referred to Einstein as "the famous theoretical physicist" and praised his work, it concluded that St. John's negative results precluded the prize going to Einstein.[104] Astronomers familiar with the technicalities acknowledged that the observations were difficult. The problems with the gravitational redshift test threw into relief the importance of the search for gravitational light bending.

Campbell had been continuing to pursue the problem well before news of general relativity reached the United States. A solar eclipse was visible from Colombia and Venezuela on 3 February 1916. Financial difficulties then plagued Lick, and Campbell hoped that Perrine would send an expedition and make the test "as you would probably be the only man of experience on the line of totality." Perrine did go, but not to attempt the Einstein test, since narrow finances prevented him from taking the necessary equipment.[105]

Campbell's next chance would come in the summer of 1918, in the United States. In January 1917, the month the United States entered the war, Robert Grant Aitken at Lick called attention to the Einstein test and the old Vulcan problem in relation to the coming eclipse. He pointed out that telescopes of the type previously employed to search for intramercurial planets could be used to look for light bending. He was still unaware that Einstein's 1916 theory had changed the predicted amount, which he estimated at 0.9 seconds of arc, as in Einstein's theory of 1911.[106] By 1918 many U.S. scientists had been mobilized for war work, and little money was available for attending eclipses. Nonetheless, by pooling equipment and manpower, more than a dozen parties were organized.[107] Vesto Melvin Slipher, director of Lowell Observatory and John A. Miller of Sproul Observatory in Swarthmore contemplated both the Vulcan and Einstein tests. On Campbell's published advice they rejected the former, but Slipher considered borrowing an 11-foot, 4-inch-aperture lens from Wilber A. Cogshall of Indiana University and adapting it for the Einstein test. In the end, he and Miller each opted for direct photography of the corona, leaving the search for light bending to others.[108] The Chamberlin and Alle-

gheny observatory expeditions attempted the Einstein test, but nothing
came of these efforts due to clouds.[109] Only Campbell and Curtis got results. They were made known in a climate charged by St. John's negative
results and by the international situation immediately after the Great War.

LICK ASTRONOMERS GO ECLIPSE HUNTING

When war precipitated the Lick party's hasty departure from Russia, they
left their instruments at the Pulkova Observatory. Russian colleagues had
promised to return them after the war, or as soon as transportation allowed. It was not until August 1917, under the Kerensky regime, that the
instruments began their long journey east to Vladivostok. They reached
the Pacific Ocean in December, weeks after the Bolsheviks seized state
power in western Russia. At Vladivostok, "the government, or the opposite of government," as Campbell wrote to a colleague, "put a business
boycott in force there and nothing could be moved."[110] The boycott ended
in April 1918, and the instruments eventually set sail for Kobe, Japan,
from where it was hoped a steamer would bring them to the west coast
of the United States.

The line of totality for the eclipse was to pass diagonally across the
United States from Washington in the Northwest to Florida in the South-
east. Campbell chose as his site the town of Goldendale, Washington (Fig.
4.3a,b), since it was the "westernmost" of all suitable stations, having
"quick connection with Pacific Ocean ports" and his traveling instruments. Campbell was not sanguine, however, that they would arrive in
time. To be safe, he cobbled together portable equipment from Lick and
borrowed some instruments from the Students' Observatory and the Department of Physics at Berkeley. "The expedition will be on a more modest scale than had previously been hoped, but the equipment will nevertheless be well worth while."[111]

The irony of Campbell's position, as one of the most fastidious eclipse
hunters, was not lost on Slipher. "It is very discouraging, I am sure," he
wrote to Miller, ". . . after he has for so many years been observing carefully solar eclipses going to distant parts of the Earth and then through
loss of his apparatus to be deprived of the opportunity to observe advantageously this one so near home."[112]

The equipment from Russia did not arrive in time. Campbell's improvised inventory included the following. For the Einstein test he used a 4.5-
inch photographic lens (focal length 15 feet), and a 3-inch Vulcan lens
(focal length 11 feet, 4 inches), both from the existing Lick equipment.
He borrowed two 4-inch, 15-foot focal length photographic lenses from

Figure 4.3a. Observers and guests assembled in front of the Einstein cameras at Goldendale, Washington, 1918. *From left to right*: Joseph H. Moore (Lick), his daughter Kathryn, A. H. Babcock (Southern Pacific Company), Warner Swasey, Douglas Campbell, Mrs. Campbell, E. P. Lewis (Berkeley), William H. Crocker, J. E. Hoover (Lick foreman), Mrs. Crocker, F. S. Bradley (San Francisco), W. W. Campbell, Samuel L. Boothroyd (University of Washington), Mrs. Plaskett, Heber Curtis (Lick), Estelle Glancey (Cordoba), J. S. Plaskett (Dominion Astrophysical Observatory, Victoria, B.C.), John Brashear, C. A. Young (Victoria), Mrs. Moore, Edward E. Fath (former Lick graduate), Mrs. Morgan (proprietor of lodgings at site). (Courtesy Mary Lea Shane Archives of the Lick Observatory, University Library, University of California–Santa Cruz.)

the Chabot Observatory in Oakland, California. He mounted these "Chabot lenses" on wooden tubes constructed at the eclipse station.[113]

Curtis was in charge of the Vulcan and Einstein cameras. After about a week at the site, he wrote Hale at Mount Wilson: "We have had a fair average of weather to date and are hoping for good luck:—the most disquieting feature noticed to date lies in the fact that we are running *thirteen* instruments and the house we have rented possesses *three* black cats and a dog named '*Shadow*'!!!"[114]

The night before the eclipse, clouds moved in and it looked like the black cats would have their way. Twice during eclipse day, the cloud cover thinned enough to secure time observations on the Sun, but the sky was

Figure 4.3b. Einstein cameras and spectrographs set up at Goldendale, Washington. (Courtesy Mary Lea Shane Archives of the Lick Observatory, University Library, University of California–Santa Cruz.)

not encouraging. At the crucial moment, however, a break in the clouds revealed the Sun. As Campbell described the drama: "By great good luck a small rift in the clouds formed exactly at the right place and right time. The clouds uncovered the Sun and its immediate surroundings less than a minute before totality became complete, and the clouds again covered the Sun less than one minute after the total phase had passed. The small clear area was very blue and the atmosphere was tranquil." Ethel Crocker, the wife of Regent Crocker, who paid for the expedition, witnessed the surprise clearing as a religious experience: "It *was* a miracle that little lake of blue sky in the center of which was the phenomenon we had all gathered to see—God is very good to people that believe in his power to perform miracles."[115]

The *New York Times* ran a story by Campbell. Under a subheading "Test of the Einstein Theory," Campbell explained the purpose favored by the weather: "It is hoped that the measured positions of the recorded stars will serve as a test of correctness or falsity of the so-called Einstein theory of relativity, a subject which has occupied a foremost position in the speculation of physicists and others during the last decade. . . . The test as an eclipse problem has never been made before, and it may be the only satisfactory test known to physicists, but whether our work will contribute evidence of value remains to be seen."[116] (Fig. 4.4.) The dramatic event of an eclipse, always of interest to the lay public, was thus linked in the public mind with a controversial theory called relativity and with the name of Einstein. It was probably the first time either was mentioned in the North American press.

Curtis had the job of measuring the plates. A hasty examination revealed that stars fainter than the eighth magnitude had been recorded, although clouds had interfered somewhat with the stars farthest from the Sun. Campbell reported these preliminary findings, but then all work on the plates ceased.[117] Curtis left Mount Hamilton for Berkeley to help train navigators for the navy. He had begun this work, under the supervision of Armin O. Leuschner of the Students' Observatory at Berkeley, in the summer of 1917.[118] His eldest son was a wireless operator in the navy, and Curtis wished to serve where he could. About a month before the eclipse, Campbell, who had three sons involved in the fighting, had offered to give him leave of absence from Lick should any opportunity arise for him "to get into technical scientific service relating to the war problems of our country."[119] After the eclipse was over, the opportunity opened at Berkeley. Curtis only had a few days at Mount Hamilton before he took up residence there. He did not want to give up the war work as long as the hostilities continued. As he explained to Hale: "[I] find it impossible to be content at astronomical work these days if I can be doing war work instead."[120] The measurement of the eclipse plates would have to wait.

There was also a technical reason for delay. Normally, the comparison plates of the region of the sky where the Sun was during the eclipse would have been taken at Mount Hamilton by night several months before the eclipse. Hoping for the arrival of instruments from Russia, Campbell and Curtis had not made comparison plates with the auxiliary equipment they finally used. Now they had to wait for the Sun to vacate the region of sky of interest so that they could photograph it at night with the cameras used at Goldendale. This bit of work would be possible in the late fall or early winter. In August, however, Curtis left the West Coast for war-related research work at the Bureau of Standards in Washington, D.C. Campbell made formal arrangements for Curtis's absence for the duration of the war. He decided that nothing would be done with the Einstein plates until the comparison plates could be taken the next winter.[121]

With the end of the war in November 1918, thoughts of getting back to normal lines of activity returned. Campbell broached the subject to Curtis, suggesting that he might wish to return by the middle of January, but leaving the choice to him. Curtis opted to stay until the summer of 1919. "I almost wish that you had ordered me back," he wrote, "as I feel

Figure 4.4. *New York Times* article, 10 June 1918, likely the first mention of Einstein's theory in an American newspaper. Lick newsclippings. (Courtesy Mary Lea Shane Archives of the Lick Observatory, University Library, University of California–Santa Cruz.)

a good deal as tho I were really 'playing hookey.' But I believe that I can now begin to get into several activities here which will be interesting and profitable." With the comparison plates yet to be taken, Campbell put the Einstein problem on ice, despite requests from colleagues like Samuel L. Boothroyd who was to give a paper on relativity and was "anxious to get the latest evidence regarding the Einstein effect."[122]

EINSTEIN LIBERATES FREUNDLICH

Back in Germany, with the war still raging, Einstein had been trying for a number of years to find a situation for Freundlich that would allow him to devote his time to observational tests of general relativity. Early in 1916 David Hilbert helped him try to get Freundlich a job at Göttingen. Johannes Franz Hartmann, professor of astronomy and the observatory director, refused "with invincible apathy" to mentor Freundlich on the astronomical side. Einstein would have to take on the responsibility of overseeing the research. He had second thoughts. Despite his desire to help the young astronomer, he had misgivings about working closely with him. "For such a half-marriage, not only is a certain respect for the other person necessary," he explained, "but also some of that personal congeniality which makes frequently repeated meetings pleasant and which sweetens shared disappointments. But in the present case this is decidedly lacking."[123]

Schwarzschild's untimely death later that year offered another possible route. If an amenable director were appointed to run the Potsdam Astrophysical Observatory, then Freundlich might obtain a position there. Karl Friedrich Küstner, professor of astronomy at the University of Bonn, was a candidate for the position and Freundlich went to visit him there. Einstein reported to Hilbert that Küstner "received him well and is prepared to support him in every way." Einstein was confident that if Küstner got the job "a conscientious examination of the experimental gravitation questions would then be safeguarded, and simultaneously Freundlich would be saved from a situation that has become positively heartrending." He wrote a letter to Naumann at the Education Ministry "to place my bit of weight on the scale toward the above."[124]

Einstein got himself onto the commission of the Prussian Academy that would recommend the new Potsdam director. He nominated Küstner and asked Willem de Sitter for an opinion. De Sitter wrote that Küstner "is the only man in Germany now who thoroughly grasps the major modern-day problems and who can still contribute much, particularly in the field of precision astrophysics, as well as toward the fusion of astrophysical and astronomical methods—where, in my view, the future lies." At the first commission meeting, Einstein learned that Hugo von Seeliger had written a letter supporting Gustav Müller, a senior observer at Potsdam, over Küstner. Struve, who was also on the commission, suggested that von Seeliger was acting from personal motives. Nonetheless, the commission moved that Müller and Küstner be ranked equally, but Einstein objected. He consulted de Sitter who replied that he hadn't mentioned Müller "because, frankly speaking, I consider him too old. . . . Küstner is the man who can be expected to make Potsdam move into the forefront: Müller would *perhaps* be able to keep it at the level it is." At the second commission meeting, Einstein presented de Sitter's recommendations and an oral one from Jacobus C. Kapteyn in favor of Küstner. The commission recommended Küstner first, and Müller second. The full Academy accepted the commission's advice and forwarded the recommendation to the Education Ministry. Despite all of Einstein's efforts, however, Müller got the appointment. "All who mean well in the matter are unhappy about it," he told de Sitter. "It is unclear what forces are to blame in this. There is talk of von Seeliger."[125]

Early in 1917 a unique opportunity presented itself, finally allowing Einstein to free Freundlich from his servitude at Struve's observatory. When Planck and Nernst had initially enticed Einstein to come to Berlin in 1913, their offer package included the directorship of the Kaiser Wilhelm Institute for Physical Research. The institute existed only on paper and was to be founded after Einstein's arrival. The outbreak of war diverted all government funding to finance the war effort. Plans for Einstein's institute were put on the back burner until the war was over. Einstein was unperturbed, as he did not need an institute, hated administration, and preferred to work on his own. However, private money from Leopold Koppel became available, so early in 1917 the institute went forward. In June 1917, Einstein asked Freundlich to prepare a detailed research plan in anticipation of his institute's creation. The industrialist and privy councilor Wilhelm von Siemens was appointed chairman of the board of trustees. Einstein chaired the board of directors, which included Planck, Nernst, Haber, Rubens, and Warburg as advisers. Einstein received an annual honorarium of 5,000 marks and held meetings of the board in his apartment. There was no building or equipment. Einstein hired his cousin Elsa's twenty-one-year-old daughter, Ilse, as part-time secretary for 50 marks a month. On 17 December 1917 the press announced the launch of the new Kaiser Wilhelm Institute for Physical Research. Freundlich received a three-year contract, renewable for two years, to conduct research on the experimental testing of general relativity. His freedom began officially on 1 January 1918.[126]

Freundlich developed an ambitious research program that had grown out of his long association with Einstein and their discussions and work on testing relativity.[127] The research fell into two main categories: the light-bending test and the gravitational redshift. For the light bending, Freundlich proposed three modes of attack: the solar eclipse method, daylight photography of the solar vicinity, and photography of stars at Jupiter's edge. On his aborted eclipse expedition to Russia in 1914, Freundlich had "devised a series of improvements toward obtaining sufficiently precise data" that he hoped to develop further in time "for the coming unusually favorable solar eclipse in 1919." He proposed the other two projects "to become independent of the rare moments of total solar eclipse in studying these problems." As early as 1913 Freundlich had suggested daylight photography of stars near the Sun but never pursued it "since neither funds nor instruments were made available to me." Frederick Lindemann had recently published results of tests done in England,[128] and Freundlich hoped to pursue this avenue of investigation. He was also "having a measuring instrument built by the Toepfer Company in Potsdam" to use in precise measurements of stars at the edge of Jupiter and comparing their positions when Jupiter was not nearby. "The development of this method thus requires a series of preparatory researches which are anyway of interest in astrophotography."[129]

For the gravitational redshift, Freundlich was still insisting that his earlier statistical analysis "demonstrated that for the B (Orion) stars, an effect of the expected type is indeed indicated."[130] He wanted to pursue this line of research further. He also wanted to take up a line of investigation that Einstein had suggested two years earlier. Some double star systems are so close that they cannot be resolved visually with the telescope. As the stars orbit each other, the spectrum of the receding star is redshifted and the approaching star is blueshifted. The spectra of the two stars shift back and forth, reflecting the orbital motion. Einstein had suggested that a detailed analysis of how the wavelengths in these "spectroscopic binaries" varied could detect gravitational effects superimposed onto the motion effects.[131] Now Freundlich wanted to use a new technology developed in 1913 by the Hamburg physicist Peter Paul Koch. The Koch photometer shines light through a stellar photograph onto photocells that convert the light into an electric current that moves a pen. As the plate is moved, the pen produces a line tracing of the spectrum. Positions of spectral lines and the energy distribution within the spectra can be measured "with an accuracy not even remotely achieved previously." Freundlich was having a Koch photometer built for this new test of the gravitational redshift.[132]

Even though Freundlich was finally free to pursue the relativity questions, there were problems with his new situation. In Germany, all employees of state-funded observatories were civil servants, with generous benefits and pensions. After his contract with the Kaiser Wilhelm Institute was finished, he would have to look for a job. He tried to have his position at the Royal Observatory held for him, obtaining a leave of absence for the duration of his contract with the institute. As an alternative, Einstein made the case for him to be hired at the Potsdam Astrophysical Observatory, where Schwarzschild had been director. The situation was awkward, however, because he had backed Küstner over Gustav Müller, who was now director. Einstein wrote to Hugo Krüss at the Ministry of Education, suggesting that Freundlich be hired at the Potsdam Astrophysical Observatory. Krüss was sympathetic, and suggested that Einstein write Müller directly "and persuade him a little." Müller did not have any openings on staff, but agreed that if the ministry would fund another assistant position, he would take on Freundlich. He asked Einstein to make the case directly to Krüss. Müller was willing to go further, in the event the ministry nixed the arrangement. "I am prepared to support the planned investigations by Dr. Freundlich toward testing the general theory of relativity, in that I am granting him permission to acquaint himself with the various observation methods employed at our Observatory, and in that I am willing to take care of his instruction and, as far as is possible, to place the Observatory's equipment at his disposal."[133]

Freundlich did not have experience in modern astrophysical methods, having worked with more classical observational techniques. So, Einstein had to make a case that Freundlich needed to work at an astrophysical observatory such as Potsdam. In presenting his argument to Krüss, he emphasized Freundlich's training and expertise in theory, and denigrated the astronomy profession's understanding of theoretical physics—a not-so subtle dig at the old guard.

> The advances in astronomy in recent decades have been based far more on the perfecting of precision in observational methods than on fundamental, theoretically based innovations. That is how it came about that purely practical observational skill was valued more highly than theoretical knowledge and expertise. The employees were chosen in keeping with this approach and eventually acceded to the leading positions at the observatories. The consequence of this is that collection of the most precise data possible became an end in itself and that, in general, leading astronomers have quite poor theoretical training and insight. Thus we see that, among astronomers, the most modern efforts in the area of gravitation theory is generally met with a complete lack of understanding and interest, even though the empirical investigation of gravitation must be seen as astronomy's most important task.

Einstein was quick to add that Schwarzschild had been a "striking exception" who had the requisite combination of "brilliant theoretical talent" and "sufficient practical skill in the observational methods." Among the younger astronomers, he argued, Freundlich "is the only one with a solid knowledge of mathematics, celestial mechanics, and gravitation theory. Although considerably less talented than Schwarzschild, he nevertheless recognized the importance of modern gravitation theories to astronomy many years before the latter did and has worked fervently toward verifying the theory along astronomical or astrophysical lines."[134]

It is not clear whether a new assistant position was created at Potsdam, but Müller was as good as his word.[135] Einstein was "very pleased that you [Freundlich] are being received so amicably at Potsdam. The ice seems finally to have broken."[136] By March 1918 Freundlich could write to his former professor, Felix Klein, that he had left the Royal Observatory and was working full-time on testing general relativity. "I am devoting myself entirely to the experimental testing of the theory now that that the Kaiser Wilhelm Institute for Physical Research has made me independent for some years, enabling me to leave my position at the Royal Observatory. At the moment I am working exclusively at the astrophysical institute in Potsdam and am developing a number of methods, first to determine the gravitational shift of spectral lines, should they exist, and then also to verify the deflection of light in a gravitational field."[137]

Freundlich was keen to organize an expedition to observe the eclipse of 29 May 1919—if the war ended in time. The Russians had confiscated his instruments when war broke out, aborting his plans to test Einstein's light bending prediction at the 1914 eclipse. By the autumn of 1915 the instruments had been transferred to the observatory in Odessa. Early in 1917, Einstein offered to try to get someone to retrieve them. "Perhaps it is better if I use my personal influence concerning that instrument, since no bellicose odium weighs on me." As a known internationalist and pacifist, he might fare better than a German astronomer.[138] When the German army occupied Odessa in March, Freundlich proposed to the Ministry of Education that the troops retrieve his instruments. Einstein recruited Wilhelm Schweydar, an observer at the Geodetic Institute in Potsdam, who was on a scientific mission in Romania, to go to Odessa if Freundlich's plan went through. Einstein also consulted Planck, who advised that Freundlich ask the Academy to make the arrangements. He also offered to support the proposal. Freundlich argued that the instruments were needed for an expedition to Brazil to observe the favorable eclipse of 29 May 1919. Planck also suggested that Freundlich consult Struve, who had initially sent out the 1914 expedition. Struve suggested another approach involving the observatories that had participated in the expedition, rather than the Academy. This strategy was approved and plans were set in motion to retrieve the instruments. Unfortunately, they did not arrive in Germany until 1923.[139]

As the end of the Great War approached, Einstein had finally succeeded in setting up Freundlich to take on experimental testing of general relativity. By this time, astronomers outside Germany were aware of his theory and rival testers were already actively looking for light bending and the gravitational redshift. The early lead that Freundlich had gained before the war had disappeared. The postwar period would dramatically change the nature of debates about the astronomical predictions of Einstein's theory, moving them into the glare of the public spotlight. The center of gravity also shifted from Europe to America.

Chapter Five

1919: A YEAR OF DRAMATIC ANNOUNCEMENT

THE END OF THE WAR saw scientists begin the difficult task of rebuilding the international cooperation that hostilities had destroyed. Some, like Max Planck who lost a son, had suffered personal tragedy. Others, like Campbell, were lucky: his three sons returned. Despite the horrendous toll on personal and professional lives, scientific research continued. Einstein had become famous within German-speaking Europe. His theory of relativity was controversial but highly acclaimed. It became almost a foregone conclusion that he would be awarded the Nobel Prize, but St. John's negative results concerning the gravitational redshift in the Sun proved to be an obstacle. Einstein received seven nominations for the 1919 prize for physics. Some were for his work on general relativity. A number of them were for his early work on Brownian motion. The committee decided that Einstein's contribution in relativity and quantum mechanics were more important. Nonetheless, they opted to wait for clarification of the redshift problem, and for results of the solar eclipse that would take place on 29 May 1919.[1]

That year turned out to be a dramatic one for relativity. By year's end, Einstein and his theory would vault to international fame. Astronomers were at the heart of these dramatic developments.

EVERSHED'S EARTH EFFECT VERSUS RELATIVITY

The year did not begin auspiciously for relativity. Evershed's observations of Venus during 1918 had been blessed with good weather "and the fortunate arrival of fast plates, which escaped the submarine." By January 1919 he was able to announce preliminary results.[2] He obtained four series of observations, with Venus well in front of the Sun and also behind it (relative to the Earth). He measured Venus and iron arc spectra, using sunlight and iron arc as a control. Evershed found that the wavelengths of the solar lines reflected off Venus gradually diminished as the planet moved toward the far side of the Sun, indicating that the redshifts were due to the side of the Sun facing the Earth. When Venus was at an angle of 135° from the Sun-Earth line, the reflected sunlight (from a hemisphere of the Sun turned 135° from the Earth) exhibited a *violet* shift compared with the terrestrial iron lines. Evershed could only conclude that the "sunlight of the Sun reflected by Venus differs from ordinary sunlight."[3] While he admitted that he had "been reluctant to accept the Venus results" because they implied that the Earth controls the motion of solar gases, he decided "the evidence now appears to me to be conclusive."[4]

Evershed did not mention that his colleague Narayana Aiyar had been measuring plates from an extensive series of observations made in the spring of 1918 comparing limb and center solar line displacements with the iron arc. This work was the latest in the Kodaikanal group's attempt to check St. John's measurements of cyanogen lines. The work was not complete, but the results were confirming his earlier assertion that St. John's small or zero shifts at the limb were wrong. Aiyar found positive shifts, in some cases close to Einstein's predicted amount.[5] However, the shifts differed for different substances, precluding a definitive confirmation of relativity. Furthermore, the unusual motion hypothesis could explain these results as well as the Venus observations. Although Evershed did not publish Aiyar's results until the following year, they gave him increased confidence in his Earth effect. He concluded that "whether we like it or not it seems necessary to admit that the Earth does affect the Sun, causing a movement of gases analogous to that taking place in a comet. Is it possible that this action controls to some extent the distribution of sun-spots and prominences, which seem also to betray an earth influence?"[6]

Evershed ruled out the relativity explanation as long as there was an Earth effect: "The bearing of these results on the 'Relativity' effect is obvious, for we now find that the shift towards red of the solar lines only occurs on the side of the Sun facing the Earth. It cannot, therefore, be a gravitational effect which would be constant all over the Sun."[7] At Mount Wilson, St. John and his colleague Seth Nicholson immediately began taking plates of Venus with the Snow telescope in search of "Evershed's mysterious earth effect."[8] Unfortunately, clarifying the redshift problem was proving to be elusive and difficult.

DELAYS AND TECHNICAL CHALLENGES AT LICK

With Curtis in Washington, Campbell did what he could to advance the work on the Einstein problem. Early in 1919 he consulted Curtis about taking the comparison plates. The star images on the eclipse plates were not clearly defined, and Curtis responded that for comparison purposes, the "double image and the little 'jump' will be very troublesome." At this early stage, neither astronomer knew that the poor definition of the star images on the Goldendale plates came largely from movement of the telescope mounting during exposure, which caused doubling and tailing of the stellar images. Curtis was inclined to blame the clocks and the drive for the poor images, and the lack of any guiding of the telescope during the exposure of the plate. "With our good clocks in Russia," he reminded Campbell, "all our clocks at Goldendale were 'seconds.' " In taking the comparison plates, he advised, an auxiliary lens should be used "to guide in R.A. so as to get a good plate the first time." Curtis felt that "we are expecting too much of the apparatus" to obtain good star images. He reminded Campbell that even the Crossley reflector, with its excellent drive, had little chance, "WITHOUT GUIDING, [of] turning out a sharp image. You will recall that in Russia I had planned to guide in R.A. on Regulus, which was advantageously placed." Curtis stated emphatically that if the problem were to be attacked again "it should either be tried with the best drive we can devise, or not at all."[9]

At the end of February, Campbell reported to Curtis that he and his colleagues had gotten comparison plates after a hard struggle. Three nights had started out beautifully, but each time the wind had made trouble for them. They had used a guiding telescope as Curtis had suggested, "and that helped immensely."[10] Campbell decided that "at the 1923 eclipse the Vulcan mounting, etc., must be designed anew, and to include a guiding telescope and other conveniences and necessities." The eclipse of 1923 would be visible from nearby southern California and Mexico. Evidently in February 1919 Campbell had no intention of observing the eclipse of the following May, nor that forecast for 1922, which would be visible from Australia.

With the comparison plates in hand, Campbell had put the wheels in motion for Curtis's return to Lick beginning 1 May. "I wish you were here to take up this work immediately," he admitted, and hoped to have the comparison apparatus ready for Curtis's return. He described a design for a differential measuring apparatus to Curtis:

> It seemed to me that we could construct, solidly, a framework to hold the positive of a Goldendale plate and the corresponding Mount Hamilton negative face to face vertically in front of a north or west window, with a large ground glass between them and the window; that a sort of double slide, of strong design and accurately constructed in wood, could hold a micrometer eyepiece in front of the plates, so that the eyepiece could be shifted by finite jumps over the whole area searching for any Goldendale objects not duplicated at Mount Hamilton, and measuring the distances between the corresponding images. The observer could stand, if he prefers, on a series of 1-inch boards, so as to adjust himself to the height of the eyepiece above the floor, or he could have a suitable stool constructed for his greater comfort.[11]

Curtis doubted whether the fainter stars would be made out as well on a positive as on the original negative, and preferred a procedure that

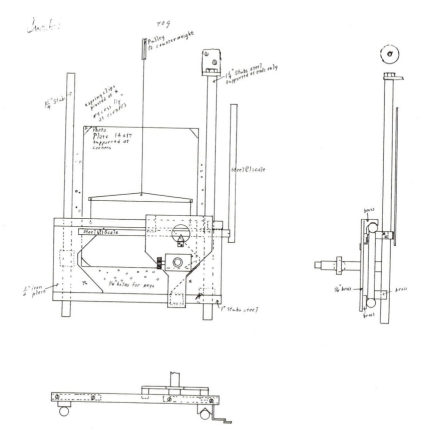

Figure 5.1. Curtis's sketch of March 1919 for constructing machine to measure the Goldendale eclipse plates. (Courtesy Mary Lea Shane Archives of the Lick Observatory, University Library, University of California–Santa Cruz.)

would not require positives at all. He wanted to have two glass scales, 16 inches long, ruled at the Bureau of Standards to use instead of a steel or wooden scale. Since these would be highly accurate, he hoped they would enable him to measure the negatives of both sets of plates (from Golden-dale and Mount Hamilton) absolutely in rectangular coordinates, "and get rid of scale and orientation error by actual solution over the entire plate." He conceded that the differential scheme "would probably work all right, though I am a little afraid of the difficulty of allowing adequately for scale and orientation differences in such a method." Curtis enclosed a sketch of the apparatus he had in mind,[12] shown in figure 5.1.

Campbell grew increasingly anxious about Curtis's return to Lick to work on the Einstein plates. He told Curtis that he had been asked to head the American delegation of astronomers attending the inaugural meeting of the International Astronomical Union in Europe and that he would be leaving some time in June. Campbell wanted Curtis back on Mount Hamilton before his departure. That suited Curtis, who had grown tired of peacetime Washington. When Campbell learned this, he regretted having set up a 1 May return instead of an earlier one.[13] What fired up Campbell was the news that the British intended to send two expeditions to observe the May eclipse, and that they would concentrate on the Einstein problem. Campbell wanted to be first in announcing definite results: "I am really anxious to have you get out of the Vulcan-Einstein photographs whatever may be in them very promptly, not only on general principles, but because our friends across the water are sending expeditions to Brazil and Africa for the eclipse of this year, devoting themselves intensively to the Einstein problem. It behooves us to get what there is coming to us, quite promptly. So I am deciding to count pretty heavily on your advice as to what comparing and measuring apparatus to get ready for your arrival."[14]

With the pressure on, Campbell deferred to his colleague's advice on the matter of the measuring machine. He authorized Curtis to order steel rods for use as traveling guides upon which the optical part of the measuring apparatus would ride. He urged Curtis to proceed as quickly as possible with the working drawings, as he would like to put the Lick shop men onto the project in early April. "That may rush you a little," he admitted, "but I am anxious to have the apparatus well along to completion when you arrive, so that you may have something to report pro and con at the Pasadena meeting of the Astronomical Society of the Pacific in connection with the Pacific Division of the 3 A.S. [AAAS] about June 19–20."[15] So Campbell was pushing for an announcement of preliminary results by the middle of June, two weeks after the May eclipse and well in advance of any likely substantive statement by the British expedition.

ENTER THE BRITISH

About the time that Campbell and Curtis were feverishly preparing for Curtis's return to Mount Hamilton, two teams of British astronomers set sail for Lisbon, on their way to different locations along the line of totality of the May 1919 eclipse. Charles Rundle Davidson and A.C.D. Crommelin were bound for Sobral in Brazil; Arthur Eddington and E. T. Cottingham set off for Principe, a small island off the west coast of Africa in the Gulf of Guinea. On the last evening before sailing, they bid farewell to the Astronomer Royal, Frank Dyson, in his Greenwich study. As the discussion turned to the amount of the light deflection, Eddington insisted that he fully expected to find the Einstein value, and not the Newtonian

half-value. When Cottingham asked what it would mean if they got double the Einstein deflection, Dyson evidently replied: "Then Eddington will
go mad and you will have to come home alone!"[16]

The British decided to enter the fray during the war. The May 1919
eclipse promised to be particularly auspicious, with over five minutes totality and the eclipsed Sun in the constellation Taurus, surrounded by
thirteen bright stars in the Hyades cluster.[17] Eddington was central to the
British effort. As the war approached its climax, however, it threatened
to intrude on the astronomer's plans. During the early part of the war,
Great Britain had relied on volunteers to enlist, but by March 1916 the
government required conscription to fill its military ranks. At the request
of Cambridge University, where Eddington was Plumian Professor of Astronomy and director of the Cambridge Observatory, the war office ex-
empted Eddington on the grounds that it was in the national interest that
he remain in his university post. Eddington was a member of the Society
of Friends, and had been prepared to claim exemption on grounds of
conscience, even entering the plea; but the occupational exemption had
been sufficient. By mid-1918 the Ministry of National Service was desper-
ate for fighting men. Eddington had been graded 2 by the medical board;
he was thirty-five years old and single. The ministry appealed his exemption. The appeal was granted at a hearing in Cambridge on 14 June. Eddington's exemption was to end on 1 August 1918. On 27 June he formally applied for exemption from military service on religious grounds.
The Astronomer Royal supported his application:

> I should like to bring to the notice of the Tribunal the great value of Professor
> Eddington's researches in astronomy which are in my opinion to be ranked
> as highly as the work of his predecessors at Cambridge—Darwin, Ball and
> Adams. They maintain the high tradition of British science at a time when it
> is very desirable that it should be upheld, particularly in view of the widely
> spread but erroneous notion that the most important scientific researches are
> carried out in Germany. . . . The Joint Permanent Eclipse Committee, of
> which I am Chairman, has received a grant of £1000 for the observation
> of a total eclipse of the Sun in May of next year, on account of its excep-
> tional importance. Under present conditions the eclipse will be observed by
> very few people. Professor Eddington is peculiarly qualified to make these
> observations and I hope the Tribunal will give him permission to undertake
> this task.[18]

At the hearing Eddington explained that during the coming eclipse he
could test Einstein's theory of relativity against the rich star background
of the Hyades. An equally favorable eclipse would probably not occur
again for centuries. The tribunal granted him a twelve-month exemption on condition that he continue his astronomical work, in particular

planning for the coming eclipse. Long before the exemption expired, the war ended.

Eddington and Cottingham arrived at Principe on 23 April 1919. They successfully obtained check plates on three different nights by the middle of May. But it was touch and go on eclipse day. Eddington reported in his notebook:

> On May 29 a tremendous rainstorm came on. The rain stopped about noon and about 1:30 when the partial phase was well advanced, we began to get a glimpse of the sun. We had to carry out our programme of photographs in faith. I did not see the eclipse, being too busy changing plates, except for one glance to make sure it had begun and another half-way through to see how much cloud there was. We took 16 photographs. They are all good of the Sun . . . but the cloud has interfered with the star images. The last six photographs show a few images which I hope will give us what we need."[19]

Crommelin and Davidson got good plates at Sobral, and the race was on for who would announce results first.

The Lick Verdict: "Einstein Is Wrong"

In late March, as the British set sail on their eclipse expeditions, Curtis was working intensively on the project. Although he agreed to try Campbell's differential method first, he was having the glass scales ruled up at the Bureau "as an 'anchor to windward' in case that scheme does not work." His plan was "to put the plates on in the same position, so that I shall use the same division of the glass scales for the eclipse plate and the comparison plate." Absolute measure was then the standard practice, the differential method being Campbell's innovation to suit the problem at hand. If it didn't work, Curtis remarked, "we shall have the real glass scales to fall back on in case we have to measure the plates in the ordinary way."[20]

The steel rods Curtis ordered were not perfectly straight. Rather than lose time in an uncertain attempt to obtain straighter ones, Campbell elected to use them, but their fault ruled out Curtis's absolute method. "I think we are going to be obliged to depend upon differential measures," Campbell decided. "We cannot hope for accuracy of the slides except in an instrument costing a thousand or two dollars, which would let us use absolute values of the coordinates."[21] Curtis ended up using differential measures, but with reference to the glass scales, rather than by using a micrometer to measure the actual distance between each eclipse star image and its companion image on the comparison plate. This hybrid technique would lead to problems.

By about the middle of April, all was ready for Curtis's return. Yet Campbell could hardly have been pleased with the situation. The Goldendale eclipse plates, taken with improvised equipment, were not of first quality; and rival Einstein testers in the form of two British expeditions were waiting at separate locations to photograph the star field around the Sun during an especially favorable eclipse. Moreover, the entry of the British raised the stakes of Einstein testing. Eddington had convinced British astronomers that the theory of relativity had fundamental implications not only for gravitation theory, but also for basic concepts of physics and astronomy. Campbell had been interested in the Einstein test partly as an instrumental challenge, and partly as a judgment about a prediction of a controversial theory. The British were looking into a possible revolution in science. Their excitement gave a new urgency to the project Campbell had been engaged in for over seven years.

While Curtis was in transit from Washington to Mount Hamilton, Campbell was in Philadelphia presenting a paper on eclipse problems, including the "Einstein effect." He reported on the Lick Goldendale project, the delay in measuring the plates due to war service, and announced that the plates would "receive attention" in May. He shared some hard-earned tips for astronomers planning to attempt the test at future eclipses: "In securing both sets of Einstein photographs, the driving clock should be reliable, and the observer should 'guide' in right ascension on a bright star in the immediate neighborhood of the sun. A guiding telescope of 3, 4, or 5 inches aperture and of focal length equal to that of the Einstein cameras and making an appropriate angle with the axes of the cameras, should be able to pick up the image of the selected bright guiding star a few seconds before contact II."[22] Campbell and Curtis would become sought-after consultants on this difficult procedure in subsequent eclipses.

Campbell's attitude toward relativity and the eclipse test around 1919 is clear in his correspondence with Arthur Hinks, formerly chief assistant at the University Observatory, Cambridge. He had resigned in 1914 rather than serve under Eddington. Hinks opened the exchange: "Now that Peace is in sight," he wrote, "I find my thoughts reverting to astronomy a little, and I hope eventually to finish off some things I had to leave incomplete in 1913." He did not expect to take up newer developments. "The statistical stuff with its integral equations was bad enough. But relativity is much further beyond the limits of my comprehension, and I shall find when I start to make up my two years arrears of reading that I am hopelessly outclassed." Campbell replied that "most astronomers could conscientiously make the same confession." As for himself, he said, "I have not yet made up my mind what to think of 'relativity' as applied to our subject. I have not attempted to go through the mathematics, but the applications have interested me very much in a general way. Eddington is rendering valuable service in keeping us posted on the applications and implications."[23]

The application of greatest interest to Campbell of course was Curtis's measurement of the 1918 eclipse plates. Writing Hinks on 2 June 1919, he anticipated results in three or four days. "We hope that a week of intensive computing may [then] give us at least a hint as to what the final results of his work will be." Campbell favored a negative result. "I must confess that I am still a skeptic as to the reality of the Einstein effect in question, but I would not be willing to undertake a technical defense of my skepticism. I am quite ready to welcome a positive result, though I am looking for a negative one."[24] Campbell retained this open-minded skepticism until working through results from the Australian eclipse of 1922.

Campbell left California for a meeting in Washington before Curtis had anything definite. But by the middle of June Curtis could speak, as he did to Charles Burkhalter of the Chabot Observatory. "You will be interested to know that the plates taken at the Goldendale station for the Einstein effect have given good results on measurement; these were taken with the two Chabot lenses which you loaned to us." As an extra line of evidence, Curtis had measured up plates taken at the eclipse of May 1900, in Georgia. The results of both sets of measures came out negative. "I expect to do some more measuring before publishing, but the conclusions from these plates and from the 40-foot plates taken in 1900 is very definite to the effect that the Einstein effect does not exist, and that there is no deflection of the light ray when passing through a strong gravitational field." The additional measures Curtis contemplated included ones on plates taken by Burckhalter with the Chabot lenses. "You have some valuable material, as it turns out, bearing on this same problem, taken with these same lenses at the 1900 eclipse." Curtis thought it "probable that these may prove considerably more valuable than the Goldendale plates, because of favourable arrangement of relatively bright stars."[25]

Curtis prepared an extensive paper on the results of his measures of the Goldendale plates and the 1900 Vulcan plates for the Astronomical Society of the Pacific meeting in Pasadena. He kept on, measuring and remeasuring, always with the same result. "Recent measures only corroborate the results sent in my paper, [he wrote Hale] namely, that there is no Einstein effect."[26]

In his talk for the Pasadena meeting (which, as he did not attend, was read for him),[27] Curtis began with a brief recap of relativity's development from Einstein's original 1905 theory to the present. He noted that since Einstein's first "epoch-making paper on this subject in 1904 [sic]," the literature on the subject had reached "enormous proportions." Over this period the "original hypothesis itself has first been modified, and later rejected." His evaluation of the resulting general theory was: "The more modern form of the theory, a four dimensional time-space concept, is far different, and certainly far less simple, than the original form." In discussing the three astronomical tests of relativity, Curtis pointed out that "St. John has found no evidence of . . . a displacement in the lines of the solar spectrum."[28] He did indicate, however, that a gravitational redshift might explain the residual radial velocity for the B stars that Campbell had discovered in 1911. "The B Class stars, after correcting for the effect of the Sun 's motion, show a puzzling positive residual velocity of about 4 km. per second. This may be due to an actual expansion in space of this type of stars, a theory which is difficult to accept. It may likewise be due to causes inherent in the mode of excitation of the spectral lines of these stars, and a relativity shift of the spectral lines to the red would offer a possible explanation."[29]

In 1915 Freundlich had run afoul of von Seeliger in Germany claiming that relativity explained this effect quantitatively. Curtis was unaware of this work because German periodicals were unavailable.[30] Other astronomers in Europe had discussed this idea, in particular de Sitter, whose suggestion was picked up by Eddington.[31] Several astronomers entertained the idea of a relativity shift for the massive B stars throughout the 1920s. The discussion was mixed up with considerations of a cosmological redshift derived by de Sitter's solution of Einstein's field equations for the universe as a whole in his third *Monthly Notices* paper in 1917. By the 1930s, evidence had accumulated that favored a relativity shift for the even more massive O stars.[32]

Curtis considered both values for light bending that Einstein had discussed. He observed that the 1911 value applied "quite apart from any theory of relativity, if light is acted upon by a gravitational field in the same way as ordinary matter. The problem may then be described [in Eddington's phrase] as an attempt to weight the light ray."[33] As for the full general relativistic value, Curtis merely stated it without interpretation: "In Einstein's later theory, as published in 1915, he postulates a relativity deflection of twice the former amount, or 1″.75 for a star at the sun's limb."[34] Curtis considered his measurements as arbiters not only "on the validity of the various theories of relativity which have been propounded," but also among "the numerous physical theories bearing on the nature of light, the ether, and the structure of matter."[35] The arbitration had proceeded as follows. Curtis had used two cameras at Goldendale, with "some second-rate driving clocks" that proved slightly defective. While he admitted that the resulting star images were not of first quality, showing a slight elongation, he claimed they were "fairly satisfactory for measurement."[36] A total of fifty-five stars appeared on the plates, some too faint to examine quantitatively; forty-three to forty-nine images on each plate were measureable. The region immediately around the Sun was poorly represented, with stars 40' to 2°30' from the Sun being picked up.

Curtis used Campbell's differential method to compare the stellar images so that "slight errors in the straightness of the ground steel ways, departure from rectangularity, scale error, etc., might have no effect."[37] For each star, he measured the displacement of the star in the eclipse field relative to the same star on the comparison plate. Even if there were no displacement due to gravity, the star positions would shift due to a variety of factors. The scale of the plates will change because of the different conditions when the photographs are taken months apart in different locations with slightly different focus. It is also impossible to place the eclipse plate and the comparison plate in exactly the same position in the telescope. So there will be a displacement due to slight differences in plate orientation relative to the optical axis. If the telescope points at the stars at a different angle during the eclipse and comparison observations, the light from the stars passes through the Earth's atmosphere along different path lengths. Differences in refraction will cause displacements as well. Curtis used a standard procedure to determine all these factors. X, Y are the measured coordinates of a star, and n_1 the difference between the X coordinates for the eclipse and the corresponding comparison plate, and n_2 the difference between the Y coordinates for the same pair. According to a method of reduction devised by Herbert Hall Turner,

$$\left. \begin{aligned} aX + bY + c = n_1 \\ dX + eY + f = n_2 \end{aligned} \right\} \tag{2}$$

where a, b, c, d, e, f are constants due to the effects of orientation of the plates, scale difference, refraction, and so on. Curtis got a set of equations for each star, measuring the plates both direct and reversed, and solved the equations by least squares.* He determined star positions by reference to the glass scales he had had made at the Bureau of Standards.

Curtis minimized the effect of refraction by taking the comparison plates at such an hour angle** that the refraction, estimated under Mount Hamilton winter conditions, should be the same as for the eclipsed region photographed at Goldendale. In the reduction, Curtis pointed out, "we have to consider in the method employed only the difference in the differential refraction. As the refraction for the originals and the comparison

---

* A mathematical procedure for finding the best-fitting curve to a given set of points. The amount that each point lies off the curve ("residual") is measured, and the best curve is the one where the sum of the squares of all the residuals is a minimum.
** Angular distance from the meridian.

TABLE 5.1.
Curtis's Preliminary Results (Goldendale, 1918)

	Inner Group 20 Stars Aver. Dist. 68ʺ.9	Outer Group 23 to 29 Stars Aver. Dist. 124ʺ.6	Excess of Inner Group (+ = Expans.)
Plate #2, direct	−0ʺ.03	−0ʺ.22	+0ʺ.19
Plate #2, reversed	+0.09	−0.09	+0.18
Plate #3, direct	−0.09	+0.01	−0.10
Plate #3, reversed	+0.04	+0.17	−0.08
Average expansion of inner group of 20 stars on four plates			+0ʺ.05
Predicted expansion from Einstein's later theory (1ʺ.75)			+0ʺ.18
Predicted expansion from Einstein's first theory (0ʺ.87)			+0ʺ.09

plates was practically identical, this second differential is linear over the entire plate far within the limit of accuracy required."[38] Earlier Curtis had worried that "the increase of altitude above sea level of the instrument . . . will completely change the refraction conditions, and the effects will have to be determined separately from the plates themselves for each set."[39] He now felt that the refraction conditions had not differed as radically as he had expected. He estimated that the probable error of a single star place ranged from 0ʺ.4 to 0ʺ.6, and thought that further analysis might reduce it further.

In the final procedure, Curtis corrected the differential displacements between stars on the eclipse and comparison plates, or "residuals" as he called them, by Turner's method. He then "projected [each residual] on the line joining the center of the Sun and the star."[40] In this way, Curtis obtained the radial components of the "corrected residuals" for each star in the eclipse field. He could then present his results in two ways. For one, he arranged the stars in order of distance from the Sun's center. If the Einstein effect was real, then there should have been a systematic decrease of displacement from the nearer stars out to the farther ones. Curtis reported that "no regular run of differences can be made out between residuals of the closer and more distant stars." Alternatively, Curtis divided the stars into two sets: an inner group of twenty stars, mean distance 68ʺ.9, and an outer group of from twenty-three to twenty-nine stars at a mean distance of 124ʺ.6. The results, projected as a slide at the meeting, are given in table 5.1.

The results from plate 2 were in excellent agreement with the prediction based on general relativity. Curtis only considered the mean from the two plates, however, which was proper scientific procedure. He concluded that the results "indicate no expansion of the inner group of stars such as

TABLE 5.2.
Curtis's Results from the 1900 Eclipse Plates

Results, 1900 Georgia eclipse; 40-foot telescope.
1mm = 16″.9

	LATER THEORY (1″.75)	EARLIER THEORY (0″.87)
Average deflection predicted by theory for outer group, stars A, D, E, F	+0″.54	+0″.27
Average deflection predicted by theory for inner group, stars B and C	+0.84	+0.42
Excess of B, C, by theories	+0″.30	+0″.15
Sum of deflections, from measures, A, D, E, F	+0″.059	
〃　〃　〃　〃　〃　B, C	+0.054	
Average deflection, from measures, A, D, E, F	+0″.015	
〃　〃　〃　〃　B, C	+0.027	
Excess of B, C, by measures	+0″.012	

is called for in Einstein's later theory, and, less definitely, pronounce against the smaller value predicted from his earlier theory."[41]

Curtis confirmed this outcome by analysis of the plates from the 1900 eclipse. Six bright stars on the 16-second and 8-second exposures had better images than those obtained at Goldendale. The stars on the 8-second exposure plates were "particularly fine with small, round sharp images susceptible of very accurate measurement." They were also closer to the Sun than any on the Goldendale plates. The 40-foot focal length camera had increased the accuracy manyfold, "as on these 40-foot telescope plates 1mm. = 16″.9." By comparison, the Chabot lenses were less than 15-foot focal length each, and the scale value on the Einstein plates was approximately 1 mm = 45′. Of course, there were no comparison plates for the 1900 eclipse. Curtis therefore measured the six stars and compared the measures "with the rectilinear coordinates given for the region in the Paris zone of the Carte Photographique du Ciel."[42] He was forced in this case to use the absolute method of measuring corrected by Turner's method.

Curtis divided the six stars into two groups: farthest (stars A, D, E, F) and closest (B, C) to the Sun, and gave their displacements (table 5.2). The displacements were an order of magnitude less than what Einstein's theory predicted. In sum: "It is not believed that the proposed further working over of the Lick material will change the conclusion derived from this present investigation, namely, that there is no deflection of the light ray produced when the ray passes through a strong gravitational field, and that the Einstein effect is non-existent."[43]

Attendance at the Pasadena meeting was fairly light because postwar conditions minimized travel and many men were still abroad or winding down war work. Two circumstances, however, ensured that many astronomers heard about Curtis's results. First, the meeting took place at Mount Wilson, and Hale, who had decided to send others to the International Astronomical Union in Brussels, attended. He was very pleased with Curtis's results: "Accept my hearty congratulations on the results of your eclipse work," he wrote. "I confess that I am much pleased to hear that you find no evidence of the existence of the Einstein effect. I listened to your paper at the A.S.P. meeting with a great deal of interest, and was really delighted with the results obtained."[44] Hale was the nerve-center of much of American astronomy and science, and he was sure to spread such excellent news, particularly as the negative results concurred with St. John's.

Campbell cabled Curtis from Washington asking for results suitable to present at a special meeting at the Royal Astronomical Society (RAS) in London. He asked for "Number, limiting magnitudes stars, probable errors," and added a caution, "careful about proper motion 40-foot plates." Curtis had compared the positions of the stars on the Georgia eclipse plates from 1900 with their positions in the *Carte de ciel*, which had been determined from photographs taken three and a half years earlier on 30 November 1896. Campbell wanted to be sure that the stars had not moved in the intervening years. Curtis telegraphed that the proper motions would be negligible; he reported a probable error of $0\rlap{.}{''}4$ to $0\rlap{.}{''}6$ for the Goldendale plates and $0\rlap{.}{''}05$ for the Georgia plates. Curtis assured Campbell: "I am confident that there is no Einstein effect whatsoever; have gone over results again and you can make it as strong as you like."[45]

Campbell sailed for Europe on 30 June in company with seven colleagues: Walter Sydney Adams, Frederick H. Seares, and Charles Edward St. John of Mount Wilson; Lewis Boss of Dudley Observatory; Frank Schlesinger of Allegheny Observatory; Samuel Alfred Mitchell of Leander-McCormick Observatory; and Joel Stebbins of the University of Illinois. On board ship, these leaders of the American astronomical community heard firsthand from Campbell about Curtis's results.[46]

A month before Campbell set sail for Europe, Eddington began work on his eclipse plates. He and Dyson wanted some preliminary indication of the results as soon as possible. So rather than making the long voyage back to England, Eddington had set up a dark room and measuring equipment at the site. For six nights after the eclipse, he and Cottingham developed the plates, two each night, and Eddington "spent the whole day measuring." The weather had adversely affected the plates, but by 3 June Eddington had some evidence that Einstein was right: "The cloudy weather upset my plans and I had to treat the measures in a different way from what I intended, consequently I have not been able to make any preliminary announcement of the result. But the one plate that I measured gave a result agreeing with Einstein."[47]

The special meeting of the Royal Astronomical Society took place in London on 11 July, four days after the Americans arrived. All of them spoke at the meeting.[48] As chairman, Campbell was asked to speak first, and he presented Curtis's results on the Einstein effect. After summarizing the main elements of the work, including the troubles that had beset the enterprise from the beginning, he reported that Curtis had arranged the corrected differences of position for each star in order of distance from the Sun, and that he "was not able to say that there was anything systematic about these differences, which showed no change of the order required by Einstein's second hypothesis."[49] Campbell noted that the error was "regrettably large," and that a telescope of long focal length would have been a great help. "For the one we used," he explained, "the stars were too faint and in the long exposure required we suffered from the increased extent of coronal structure." He then described Curtis's method of dividing the stars into an inner and outer group, and noted that Curtis obtained a differential displacement between the two groups of $0.''05$, whereas it "should have been $0.''08$ or $0.''15$, according to which of Einstein's hypotheses was adopted." In discussing Curtis's examination of the 40-foot Lick plates from the 1900 eclipse, Campbell noted that it would be useless to take a duplicate photograph now "owing to uncertainty in the values of the proper motions." He reported that reference had been made to the Paris plates in the *Carte du ciel*, "but Curtis was unable to say from the comparison that the innermost star showed a displacement due to the Einstein effect." Campbell concluded: "It is my own opinion that Dr. Curtis' results preclude the larger Einstein effect, but not the smaller amount expected according to the original Einstein hypothesis."[50] The Americans voted no to Einstein.

After Campbell had spoken, the president of the RAS, Alfred Fowler, asked the Astronomer Royal, Frank Dyson, to relate news from Eddington's eclipse expedition. "Prof. Campbell could not have chosen a more interesting subject just now than the problem of relativity," Dyson remarked. "It is an extremely difficult question to settle." He had received a letter from Eddington two days earlier, expressing great disappointment over his results at the recent eclipse. Out of sixteen photographs secured, only the last six showed any stars on them, the first ten being ruined by cloud. No more than five images appeared on any of the six plates, and on none were the images well distributed. Nonetheless, Dyson reported that Eddington was hoping "to get good enough measures to determine the displacement definitely. . . . From his best plate, however, he has some evidence of deflection in the Einstein sense, but the plate errors have yet to be fully determined."[51] Should Eddington vote "yes" to Einstein, there would be a hung jury.

Meanwhile, back on the mountain, Curtis was pressing on, remeasuring the plates and revising his Pasadena paper. Early in July, a letter addressed to Campbell arrived from T.J.J. See, whose piece in the *Observatory* two years earlier had elicited a comment by Curtis in the Pacific Astronomical Society's *Publications*. Knowing that Campbell had looked for the Einstein effect, See asked him "whether the Einstein calculations point to a refraction of the ray which throws the star apparently farther from the Sun than it actually is at the time of observation? That is, is the Einstein effect like that of a slight atmosphere about the sun?" See claimed that "in my own researches I have reached some very remarkable results, and I may be able to throw light on this problem. Hence I wish to be sure of the Einstein conclusion."[52] Curtis replied for Campbell, mentioning Einstein's two values for light bending. He attributed the earlier prediction to relativity and gravity, the other to "a four-dimensional time-space manifold." He did not hint why or how the mechanism for the bending changed from the earlier to the later relativistic prediction. Neither, in any case agreed, with Curtis's observations. "So far as I have gone in the work of measurement, the conclusion seems very definite that there is no marked deflection of the light ray when passing through a strong gravitational field, and that the Einstein effect is non-existent. I do not believe that the proposed further working over of the Lick Observatory material will change this conclusion. I can, however, speak more definitely on this point a month from now, by which time I hope to have additional measures made."[53]

The British Declare, "Einstein Is Right"

The summer of 1919 was not a good one for Einstein's relativity. Both the light-bending and the gravitational redshift tests from America were negative or inconclusive. No one accepted Freundich's positive interpretation of the stellar gravitational redshift in B stars. The only positive evidence was Mercury's perihelion motion. Einstein's jury was not favorably inclined toward his case.

The situation started to turn around in July. Still in Europe, Campbell continued to discuss Curtis's results with colleagues. He became increasingly concerned about the "regrettably large" probable errors in his colleague's measurements. The Goldendale errors were of the order of $0.''5$, about an order of magnitude larger than routine parallax measurements. Five days after he had spoken at the Royal Astronomical Society, Campbell sent a cable to Lick: "Both Curtis Einstein Results Small Weight Errors Large Use Cautiously" (fig. 5.2a).[54] Curtis wrote immediately to ask Robert Aitken, editor of the *Publications* of the Astronomical Society of the Pacific, to withhold his Pasadena paper from the coming issue. He told Aitken about Campbell's cable: "This coincides with my own view on the matter, although I think an effect of the size of the value predicted by Einstein would have shown up. Then the 40-foot plates and Burckhalter's depend still on the Carte du Ciel in the one case, and a combination of Boss and Paris in the other. These two must wait for comparison plates taken with the same lenses, to be as certain as we can make them, and these plates can not be taken till August 15."[55]

Curtis proposed to take comparison photographs later in August to increase accuracy for the Lick 40-foot plates and the Burckhalter plates. Until this was done, he took the lead from Campbell: "It is far better to 'be safe than to be sorry,'" he told Aitken, "So I am going to ask you to hold this paper out, and send back to me. Will have all the data worked up as well as we can do from the material in time for the next number." In case the printer had already set the type, making withdrawal impossible, Curtis would have provided a final paragraph "calling attention to the fact that, while the plates, as treated, indicated no marked expansion, still the probable error is large, and final decision must await the taking of additional comparison plates for the 40-foot and Burckhalter plates."[56]

Curtis still believed that he had ruled out the Einstein effect. He had already made some improvements in the measuring engine since the first measures, and he told Aitken that the probable errors had been reduced, though they were "still rather large." The six closest and easiest stars to measure for the Goldendale eclipse "show no expansion, but a contraction of 0″.9." So the results still seemed to be negative. In the end, there was time to withdraw the paper, and a substitute never appeared. Another cable arrived, this time from Brussels. It read: "Delay Publishing Einstein Results. Campbell" (fig. 5.2b).[57]

Campbell sailed from Europe in the middle of August and arrived at Mount Hamilton at the beginning of September.[58] News of his equivocal results reached Freundlich in Germany, who wrote to Einstein that Campbell "reportedly could *not* detect the effect of light deflection at the solar eclipse of 1918." He assured Einstein that Campbell's "result carries no great weight" but worried that the astronomers at Potsdam would not be favorably inclined to give him an observing position to pursue relativity testing. "I believe I am not mistaken when I say that all the gentlemen at Potsdam, even Director Müller, want to cover their rear on the point of the gen. theory of relativity and not advocate its verification any more than by allowing you informally to grant me the opportunity to work independently at their institute." Einstein "did not know anything" about Campbell's result. He agreed that general relativity "must win ac-

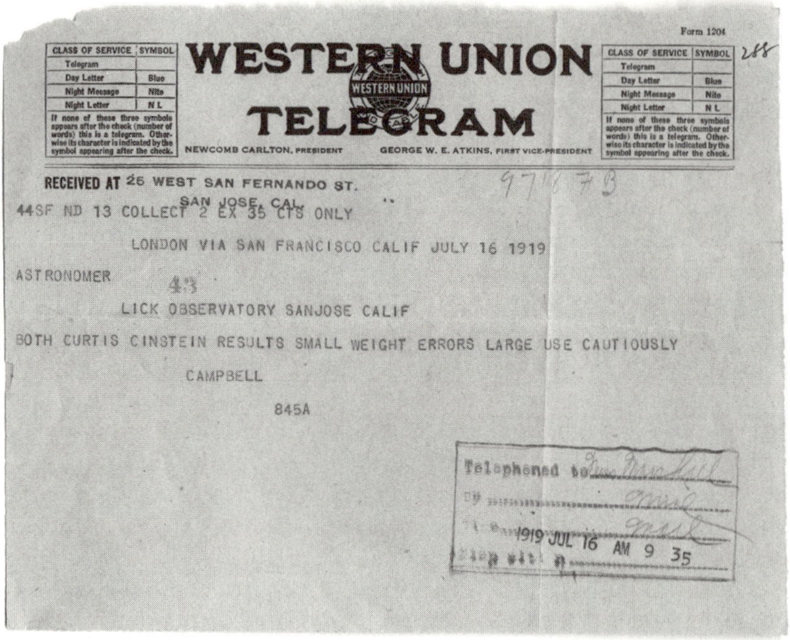

Form 1204

CLASS OF SERVICE	SYMBOL
Telegram	
Day Letter	Blue
Night Message	Nite
Night Letter	N L

If none of these three symbols appears after the check (number of words) this is a telegram. Otherwise its character is indicated by the symbol appearing after the check.

WESTERN UNION TELEGRAM

NEWCOMB CARLTON, PRESIDENT GEORGE W. E. ATKINS, FIRST VICE-PRESIDENT

RECEIVED AT 26 WEST SAN FERNANDO ST. SAN JOSE, CAL.

44SF ND 13 COLLECT 2 EX 35 CTS ONLY

LONDON VIA SAN FRANCISCO CALIF JULY 16 1919

ASTRONOMER

LICK OBSERVATORY SANJOSE CALIF

BOTH CURTIS CINSTEIN RESULTS SMALL WEIGHT ERRORS LARGE USE CAUTIOUSLY

CAMPBELL

845A

Telephoned to

By

Time 1919 JUL 16 AM 9 35

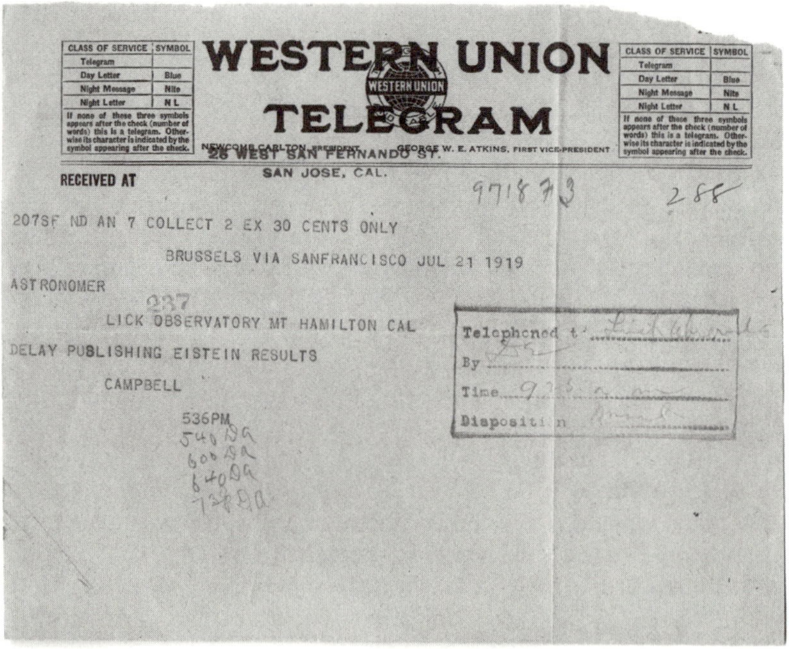

CLASS OF SERVICE	SYMBOL
Telegram	
Day Letter	Blue
Night Message	Nite
Night Letter	N L

If none of these three symbols appears after the check (number of words) this is a telegram. Otherwise its character is indicated by the symbol appearing after the check.

WESTERN UNION TELEGRAM

NEWCOMB CARLTON, PRESIDENT GEORGE W. E. ATKINS, FIRST VICE-PRESIDENT

26 WEST SAN FERNANDO ST. SAN JOSE, CAL.

RECEIVED AT

207SF ND AN 7 COLLECT 2 EX 30 CENTS ONLY

BRUSSELS VIA SANFRANCISCO JUL 21 1919

ASTRONOMER

LICK OBSERVATORY MT HAMILTON CAL

DELAY PUBLISHING EISTEIN RESULTS

CAMPBELL

536PM

Telephoned to

By

Time

Disposition

Figure 5.2a,b. Campbell's two cables from Europe in July 1919 highlighting errors in Curtis's Einstein results and halting their publication. (Courtesy Mary Lea Shane Archives of the Lick Observatory, University Library, University of California–Santa Cruz.)

ceptance among astronomers" before trying to get Freundlich a position at Potsdam.[59]

Meanwhile, Eddington had obtained comparison plates at Oxford and was busy measuring his plates. In September he gave a lecture at a meeting of the British Association for the Advancement of Science in Bournesmouth. He dropped a cautious hint that his results indicated a light bending somewhere between the Newtonian and Einsteinian values. The accuracy was not good enough to make a choice yet, but Eddington hoped the Sobral expedition results would settle the matter. A Dutch physicist (van der Pol) was at the meeting. He relayed the news to Lorentz in Leiden. Lorentz sent a cable to Einstein: "Eddington found star dislocation at solar rim provisional magnitude between nine tenths second and double—Lorentz." Einstein sent a jubilant note to his mother: "Today joyous news. H. A. Lorentz cabled me that the English expeditions have really proved the deflection of light." To colleagues he admitted that the accuracy of 0.9–1.8 seconds was "slight." Nonetheless, the word went out in Germany that the British had confirmed Einstein. Planck effused, "The close bond between beauty, truth, and reality has once again proved effective. You yourself have frequently observed that you had no doubt about the result, but it is a good thing that this fact has now been established beyond doubt also for others." The major daily in Berlin trumpeted the light-bending result, "possible only if Einstein's fundamental framework, the general theory of relativity, represents the true constitution of the universe." Einstein immediately published a cautionary note in *Naturwissenschaften*: "The provisionally established value lies between 0.9 and 1.8 seconds of arc. The theory demands 1.7."[60]

That the British astronomers had found a deflection greater than the Newtonian prediction made an impression on scientists, despite the possibility that it might not be as large as Einstein had predicted. Lorentz related to Einstein what his colleague had told him about reactions to Eddington's announcement in Bournemouth. "Van der Pol told me . . . that a dicussion took place (I would have liked to have been present) in which Sir Oliver Lodge expressed his congratulations to you and to Eddingon for the obtained result." Lorentz shared "a little calculation" he had done, showing that "an extremely low gas density could produce a deflection of 1″. Luckily, this deflection would then diminish very rapidly if the [light] ray . . . were situated farther away from the Sun. . . . Consequently it will be easy, provided many stars appear on the plates, to distinguish your effect from this refraction in a solar atmosphere." Lorentz felt sure that "this refraction is not involved at all and your effect alone has been observed. This is certainly one of the finest results that science has ever accomplished, and we may be very pleased about it."[61] The decrease of light deflection with increasing distance from the Sun would figure prominently in debates during the 1920s about whether refraction or Einstein's theory provides the correct explanation for the light bending. The law of decrease would prove more important than the actual limb deflection.

Meanwhile, the British team was having problems with the data. Eddington's plates were not very good, but he had been able to photograph a check field of stars at night at the eclipse station and at Oxford. This gave him greater confidence in his results, because it allowed him to determine the scale without relying on the eclipse stars. His plates gave a limb deflection of 0.61 seconds. Davidson and Crommelin had used two instruments—an astrograph from Greenwich Observatory similar to the one that Eddington had used, and a 4-inch refractor. The astrograph images were blurred, due to heating of the coelostat mirror. They gave a deflection of 0.9 seconds. The refractor had the best plates with clear images and more stars than the others. They yielded 1.98 seconds of arc with a small probable error, a value greater than Einstein's prediction. Faced with three values, Eddington could have averaged the three to get a value somewhere between the Newtonian and Einsteinian values. Instead, he decided to throw out the Sobral astrograph results. The average of the two remaining results gave 1.75 seconds. Einstein was right.[62] Einstein heard the news from his Dutch friends during a visit to Leiden late in October. The world heard about it two weeks later.

On 6 November 1919, Eddington and his colleagues presented their results at a joint meeting of the Royal Society and the Royal Astronomical Society. The room was packed. Alfred North Whitehead captured the tension and apprehension in an account published years later: "The whole atmosphere of tense interest was exactly that of a Greek drama. We were the chorus, commenting on the decree of destiny in the unfolding development of a supreme incident. There was dramatic quality in the very staging:—the traditional ceremonial, and in the background the picture of Newton to remind us that the greatest of scientific generalizations was now, after more than two centuries, to receive its first modification. Nor was the personal interest wanting: a great adventure in thought had at length come to safe shore."[63] The assembled astronomers and scientists learned that the British eclipse observers had found the full deflection of stellar rays by the gravitational field of the Sun, as predicted by Einstein's general theory of relativity.

The startling news was picked up by the newspapers,[64] from which T.J.J. See learned of the British claim. He found room for doubt: "Just how they could be sure of 0″87 [!] on plates covering several degrees of space, and thus being a quantity of the order 150000 I do not see, but not wishing to form hasty opinions I have written to the Astronomer Royal for light." See asked Campbell if he could send any further data on the Lick effort. "My *New Theory of the Aether* is now being arranged for the printer," he announced, "and I have so many new results that I am drawing no conclusions as to what may exist. My work throws definite light on the field about the sun, just as my work on the Lunar Fluctuations illuminated that subject."[65]

Having received no reply by early December, See wrote Campbell again: "I want the Lick view, independently of that sent out from London, of which I am somewhat skeptical, especially as Curtis wrote in June that there was no trace of Einstein Effect. So far as I can judge the quantity sought (0.″87) is likely to be one part in 10000 of angular space, and thus I am aware of the difficulty of experimental detection, even by the most perfect super position of plates. But I am not judging hastily, as I have proof of a new law of Density of the Aether about the Sun." Campbell answered promptly this time, calling the British results "especially interesting" and noting that the published details were being awaited "with great interest." As to the Lick results, he wrote that Curtis was still measuring his plates and reducing the data and "is not yet ready to say what conclusions will proceed from these plates." He elaborated all the difficulties they had faced with the expedition and the delays; but Curtis had mentioned all these details to See months earlier, and yet had still seemed so definite about a negative results. Campbell concluded that Curtis "is now in the midst of the rather extensive computations."[66]

See's anxious request for information regarding the Lick 1918 eclipse results upon hearing of the news from England was a harbinger of what was to come. As an ardent critic of the new-fangled theory from Europe, See had taken heart at Curtis's negative result. Now that scientists from abroad were claiming an opposite verdict, the first place he turned for counterevidence was Lick. Campbell, already exasperated about the delay and the inconclusiveness of the result, brushed the request aside, and dug in his heels to await further information from Britain. Meanwhile, a storm broke on the international scientific and popular scene that would forever change the atmosphere for research on this problem.

Chapter Six

MEN OF SCIENCE AGOG

REACTIONS TO THE BRITISH ECLIPSE RESULTS

If there was some interest in the Einstein theory before November 1919, there was a delirious fascination afterwards. News of the British verification flew around the world in the public press. U.S. scientists picked it up from the publicists, where they read that the eclipse result was "of such fundamental importance that confirmation is obviously most desirable, and . . . British astronomers . . . are already giving consideration to favorable eclipses which will occur in the next two or three years."[1]

Eddington wrote Einstein that since the 6 November announcement, "all England has been talking about your theory. It has made a tremendous sensation."[2] He hoped that the excitement might help to heal postwar tensions.

> It is the best possible thing that could have happened for scientific relations between England and Germany. I do not anticipate rapid progress towards official reunion, but there is a big advance towards a more reasonable frame of mind among scientific men, and that is even more important than the renewal of formal associations. . . . Although it seems unfair that Dr. Freundlich, who was first in the field, should not have had the satisfaction of accomplishing the experimental test of your theory, one feels that things have turned out very fortunately in giving this object-lesson of the solidarity of German and British science even in time of war.

Einstein congratulated Eddington on his successful eclipse results. "Considering the great interest you have taken in the theory of relativity even in earlier days I think I can assume that we are indebted primarily to your initiative for the fact that these expeditions could take place." Einstein expressed amazement at the British interest taken in the theory "in spite of its difficulty." He stressed the importance of the gravitational redshift test: "According to my persuasion, the redshift of the spectral lines is an absolutely compelling consequence of relativity theory. If it were proved that this effect does not exist in nature, then the whole theory would have to be abandoned."[3]

Frank Dyson at Greenwich Observatory, the prime mover of the British eclipse expeditions, accepted the results as verification of Einstein's prediction. In December 1919, he wrote Hale that at Greenwich they "were very satisfied with the eclipse results, as they seemed definitive." He admitted that he personally had been "a sceptic" and had expected a different outcome. "Now I am trying to understand the principle of relativity & am gradually getting to think I do." With the light bending prediction verified, Dyson emphasized the importance of the gravitational redshift: "Einstein says that the displacement of the spectral lines is an essential point. [Joseph] Larmor however has been questioning this. But Eddington won't have Larmor's explanation. I am glad St. John is going to pursue the matter further. There is no doubt that his word will be the final one as to the observational facts."[4] Dyson was aware that many astronomers might question the results of the British eclipse observers, and he sent Hale a print of one of the Sobral photos. "Naturally our opinion depends on whether the observational material was good," he explained, "& so I am going to distribute a few similar copies to show the goodness of the images." He also sent a print to Campbell, "so that you can see what the star images look like," and to Frank Schlesinger of the Allegheny Observatory.[5]

Hale agreed that the star images were good and admitted that his understanding of Einstein was bad. "I congratulate you again on the splendid results you have obtained, though I confess that the complications of the theory of relativity are altogether too much for my comprehension. If I were a good mathematician I might have some hope of forming a feeble conception of the principle, but as it is I fear it will always remain beyond my grasp. . . . However this does not decrease my interest in the problem, to which we will try to contribute to the best of our ability." Hale advised Dyson that it was not clear whether St. John's results could be applied as a test, but promised that his colleagues would work toward "leaving no doubt whatever regarding the position of the solar lines."[6]

As a specialist in stellar photography, Frank Schlesinger was ready to accept provisionally that "the photograph [sent by Dyson] leaves little doubt that the deflection is close to the total amount," although he wanted the test repeated. "I trust you are planning to observe the 1922 eclipse in the same way," he wrote, "as I am sure you will be among the first to agree that results of such importance should be thoroughly confirmed before we accept them as establishing Einstein's theory." On the theoretical side, however, Schlesinger hesitated. "I have a feeling that some explanation for the deflection will be found without resorting to non-Euclidean space."[7]

Dyson's reply indicates that he had worked a little more carefully with the details of the theory. Emphasizing that "I put a great deal of reliance on St. John," he explained that "for myself am only prepared to say that the law of gravitation is $\delta \int ds$ is stationary when $ds^2 = (1 - 2m/r)dr^2 - r^2 d\theta^2 - r^2\sin^2\theta dr^2 + (1 - 2m/r)dt^2$ with suitable units." Dyson also mentioned future expeditions: one to Christmas Island in 1922, and possibly another to the Maldives. He hoped that an American expedition would be sent "in view of the importance of having the point thoroughly settled. . . . I have tried to understand the Relativity business, & it is certainly very *comprehensive*, though elusive and difficult."[8]

Many astronomers refused to accept the results as definitive. For some, St. John's negative verdict on the gravitational redshift was enough to decide against Einstein. In the discussion that took place after the initial announcement at the joint meeting of the Royal Society and the Royal Astronomical Society, the spectroscopist Alfred Fowler urged that further work be done repeating all the spectroscopic tests. The theoretician Ludwig Silberstein took the strong position that the negative results of St. John and Evershed excluded the possibility of interpreting the eclipse results as verifying Einstein's theory.[9] Silberstein later attacked the eclipse data, arguing that relativity predicted a purely radial displacement of stars. The eclipse plates showed some nonradial displacements, and Silberstein insisted that some refraction effect could provide a better explanation.[10]

The Princeton astronomer Henry Norris Russell neutralized Silberstein's criticism, publishing a detailed analysis of the causes for nonradiality of the displacements on one of the Sobral sets of eclipse plates. He was able to trace the departure from pure radiality to heating of the coelostat mirrors.[11] In the spring of 1920 Russell gave a special lecture on relativity in which he downplayed the significance of the "still uncertain" question of the gravitational redshift for judging Einstein's theory. He felt "that Einstein's theory could be modified in such a manner as to account for the other effects already observed without demanding the existence of this one. Hence this can hardly be called at the present time a failure of the Einstein theory." Like Eddington, Russell liked relativity because it "reduces what previously appeared to be disconnected things to manifestations of a single underlying unity of principle."[12]

Others were not so ready to embrace Einstein's theory. Hale's old friend Hugh Frank Newall, director of the Solar Physics Observatory at Cambridge and professor of astrophysics, pursued a refraction explanation of the British eclipse observations. "I think it is only natural for one who has devoted his time to the consideration of the surroundings of the Sun to cry 'Pause' in the interpretations of the observations," he explained at a meeting of the RAS dedicated to discussing relativity. He insisted that "a solar physicist would have expected a deflection without the Einstein theory, and I am not yet ready to admit that the whole of the observed displacement is to go to explain the larger Einstein effect."[13] Newall tried to develop a theory of an extended atmosphere around the Sun, with refractive properties capable of deflecting the light from the stars in the

Sun's vicinity. He articulated his position at RAS meetings and in print for months after the eclipse results were announced. He had the pain of being systematically refuted by "relativists" like Dyson and Frederick A. Lindemann, who pointed out that comets do not appear to slow down near the Sun, as they ought to do should a significantly refractive gas exist there.[14] When Newall shared his doubts about relativity with Hale, he could not extract informed sympathy. "I have shared your hesitations regarding relativity," replied Hale, "though the evidence as presented to us certainly seems to be strong. . . . I cannot pretend, however, to have the smallest comprehension of the theory of relativity, and probably I shall never reach a much higher state." Hale told Newall that St. John was "making a most careful investigation of the displacement of the solar lines" and that "[Harold] Babcock is checking his results by Fabry's interferometer method."[15]

When Campbell learned more about the British data and how they had interpreted it, he was surprised. Having suspended publication of his own results due to large probable errors, he was not impressed with the British results. He eventually put his thoughts into print: "Professor Eddington was inclined to assign considerable weight to the African determination, but, as the few images on his small number of astrographic plates were not so good as those on the astrographic plates secured in Brazil, and the results from the latter were given almost negligible weight, the logic of the situation does not seem entirely clear."[16]

Robert Grant Aitken, associate director of Lick and editor of the Astronomical Society of the Pacific *Publications*, published an assessment of the situation for western United States astronomers. He emphasized that the British verification of a theory "which is assumed—whether correctly or not—to overthrow the accepted concepts of time and space as well as the Newtonian law of gravitation" had produced bewilderment. Citing Einstein from comments published in the *London Times*, Aitken outlined the three astronomical tests of general relativity and quoted Einstein's assertion that "if any deduction from it [the theory] should prove untenable, it must be given up." Then he considered each of the three tests, urging a cautious approach. He mentioned that von Seeliger's dust hypothesis "may be competent to produce the observed perturbation" of Mercury's perihelion. He quoted Hugh Frank Newall's suggestion that an extended atmosphere around the Sun might explain the British light-bending result, and repeated Newall's caution about accepting the relativistic interpretation of the eclipse results too readily. Finally, he related that St. John and Evershed had obtained negative results regarding the gravitational redshift in the solar spectrum. Aitken summarized: "While two of the tests proposed by Einstein give favourable results, each of them *may* also be accounted for on the old Newtonian theory," and that the third test "gives an unfavourable answer." He concluded: "Recalling once more Einstein's words, 'If any deduction from it should prove untenable, it must be given up,' we may concur in the opinion recently expressed by a number of prominent American physicists that the theory of relativity has not yet been established."[17]

Professional astronomers' skepticism and ignorance helped keep up public interest in a theory that appeared to elude the grasp of most scientists. The physicist Ernest Rutherford wrote to Hale from Cambridge: "The interest of the general public in this work is most remarkable and almost unexampled. I think it is due to the fact that no one can give an intelligent explanation of the same to the average man and this excites his curiosity."[18] Rutherford, who had met Einstein on two occasions at the Solvay Conferences of theoretical physicists in 1911 and 1913, was an experimental physicist, not a theoretician. When the German physicist Wilhelm Wien commented to Rutherford that no Anglo-Saxon could understand relativity, Rutherford readily agreed: "They have too much common sense."[19] His practical bent allowed him to accept the British observational verification, but he worried to Hale about the larger implications: "While I personally have not much doubt about the accuracy of Einstein's conclusions and consider it a great piece of work, I am a little afraid it will have the tendency to ruin many scientific men in drawing them away from the field of experiment to the broad road of metaphysical conceptions. We already have plenty of that type in this country and we do not want to have many more if Science is to go ahead."[20]

Another aspect of the publicity that worried some astronomers was its quantity. The press was so enthusiastic in England that Americans unfriendly to relativity believed mistakenly that the British had launched a publicity campaign to bolster the controversial and difficult theory. "I suppose we have all been worrying lately over the 'Einstein effect," complained William Hammond Wright of Lick Observatory to Hale, "chiefly as a result of the publicity which has been accorded to its so-called confirmation by the English Eclipse observations."[21] Yale astronomer Ernest Brown visited Cambridge in 1920 and confronted his British colleagues about their role in generating the enormous public interest. "Eddington told me that the booming of the relativity business was entirely the work of the London Times," he wrote Frank Schlesinger, who also believed the British had mounted a campaign. "Neither the Royal Soc. Nor the R.A.S. was responsible for the big advertisement it got. The Times reporter is always present at the meetings and having some knowledge of science he saw that a big scoop could be made out of it and he acted accordingly." Brown concluded that "our criticisms of the apparent advertising methods of the R.S. fall to the ground."[22]

Nonetheless, many American astronomers remained wary of their British confrères. Samuel L. Boothroyd, an astronomer at the University of Washington, encountered this attitude when he decided to organize a joint symposium on relativity for the Mathematical Society, Physical Society, and the Astronomical Society of the Pacific. Robert Aitken of Lick, chairman of the organizing committee for the astronomers, advised against the relativity symposium. Vesto Melvin Slipher, director of Lowell Observatory in Flagstaff, Arizona, wrote Boothroyd: "My guess is that at this time the average Pacific coast astronomer is not as enthusiastic over the astronomical standing of the theory as are our English friends, and naturally, too." Nonetheless, Boothroyd went ahead with his plan.[23]

Skepticism reigned in the eastern United States as well. The renowned physicist Robert Millikan noted at a meeting of the National Academy of Sciences in Washington, D.C., that Einstein's theory of relativity was not conclusive and that the bending of light near the Sun could be due to refraction.[24] Charles Lane Poor, a specialist in celestial mechanics at Columbia University, publicly blasted the theory, claiming that social unrest and creeping Bolshevism had invaded science, leading people to "throw aside the well-tested theories upon which have been built the entire structure of modern science and mechanical development in favor of psychological speculations and fantastic dreams about the universe." He refuted the relativistic explanation of Mercury's perihelion advance on the grounds that the calculations assume the Sun is a sphere when it is not, and appealed to refraction to explain the light bending near the Sun. Poor's remarks elicited a derisive editorial from the *New York Times* attacking the British astronomers for having "regarded their own field as of somewhat more consequence than it really is."[25] The *Times* eventually backed away from its critical position, but Poor did not. During the 1920s, Poor mounted a campaign to discredit Einstein, relativity, and observations verifying Einstein's predictions.[26]

Even scientists who took the Einstein theory seriously encountered prejudice against relativity from colleagues. Early in 1920, William F. Meggers, spectroscopist at the Bureau of Standards in Washington, proposed to Schlesinger a plan to use infrared photography to measure the displacement of red stars near the Sun during the daytime. Schlesinger thought that although the chances for success were probably less than half, the scheme was worth trying.[27] Meggers had met several astronomers who had served at the Bureau of Standards during the war, including Curtis from Lick and two specialists in spectroscopy, Keivin Burns and Paul Merrill. After the war, Burns had gone to Lick temporarily and Merrill had joined the staff at Mount Wilson. Burns responded to Meggers's suggestion: "I haven't tried to photograph the Sun in the long waves, nor stars in the day time, mainly for lack of a proper screen. I don't think much would come of it. Of course no one at Lick believes in the Einstein effect, it being contrary to philosophy, judgement and horse sense. But since so much is being said on the subject it is necessary to be interested. It may take a long while to show the error of the ways of the English astronomers." Burns admitted that the general problem of daylight photography of the Sun should be explored, and if a suitable screen were found, it would be well worth trying at Lick or Mount Wilson. But, he thought, "the Einstein effect is the result of some medium sized minds trying to lift their intelligence over a mental obstacle by their intellectual bootstraps. It corresponds to nothing objective."[28]

Merrill took a similar line: "Nobody here is very enthusiastic about the Einstein stuff. I have heard it referred to as an 'accursed theory.' Some of the coolest Englishmen even have no use for it. There is not much violent opposition to it here—the attitude is one of watchful waiting. There are dozens of other things which are more worth working on." Like Burns, Merrill was intrigued with the technical problem. "A first class wide angle lens would be the thing," he advised. "This might be used in fundamental star places, or in other problems such as that of disposing of the Einstein foolishness. . . . If I can get hold of a suitable screen I might get one of the direct photography men to shoot at stars in the daytime, but don't stop any experiments that occur to you and that you have the opportunity of making."[29]

Meggers dropped plans for the project. About a year later, Orley H. Truman, then at the Lowell Observatory in Flagstaff, asked him for information about the infrared procedure. Meggers replied that he was glad to hear of this interest: "Nearly a year ago, we suggested that either the Lick or Mount Wilson Observatories might undertake this experiment, but they seem to be prejudiced against the Einstein doctrine and so far as we know have done nothing in testing the deflection by daylight photography of stars near the Sun."[30]

If the Mount Wilson and Lick astronomers were not very enthusiastic about relativity, both centers of research were committed to trying empirical tests of the theory. In 1920 William H. Wright at Lick asked Hale if he had considered using "one of your big reflectors" to look for light bending around Jupiter. He had checked star catalogues to see what star fields would be available in the near future, but felt that the problem would be difficult and the "only answer seems to be Mount Wilson."[31] Hale checked the idea with his colleagues but received mixed reactions. Frederick H. Seares "thinks the experiment is perhaps worth trying," but Adrian van Maanen "does not think the chances very good unless Jupiter enters a crowded region or a cluster." Van Maanen tried some exposures using an occulting screen and Hale tried applying the recently developed stellar interferometer to the problem, but nothing came of these experiments.[32]

Despite the resistance evident in America, leading members of the European scientific community were highly supportive of Einstein and his theoretical contributions. The Nobel committee received a number of recommendations to give Einstein the award for 1920. A strong letter came from the Netherlands, signed by H. A. Lorentz, Willem Julius, Pieter Zeeman, and Heike Kamerlingh Onnes. They focused on Einstein's theory of gravitation, which had successfully explained the motion of Mercury's perihelion and predicted the bending of light. They considered Einstein to be "in the first rank of physicists of all time." The committee asked the physicist Svante Arrhenius to prepare a statement on the consequences of general relativity. Arrhenius noted that observations had still not confirmed the gravitational redshift. He pointed out that various criticisms of the British eclipse measures of the light bending had been made. He also mentioned Ernst Gehrcke's specious criticism of general relativity and his claim to have come up with an alternative explanation for the perihelion motion. The committee decided that relativity could not yet be the basis for the award, so Einstein was passed over again.[33]

During the 1920s American astronomers played a critical role in testing general relativity. The most important test that Mount Wilson attempted was the search for a gravitational redshift in the Sun, and Lick continued to pursue the eclipse test for light bending.

PRESSURE FROM THE PRESS

Einstein was already famous in European scientific circles when the British announced their eclipse results verifying his relativity theory, but he was virtually unknown elsewhere. The English-speaking world picked up the story from London with an alacrity and intensity that launched Einstein and his relativity theory to international fame. The announcement came not long after the war had ended and Europe was still reeling in its aftermath. The United States was watching from afar, sending aid to countries ravaged by lack of supplies, currency crises, and political revolutions. In this charged atmosphere, the news acquired a political flavor that captured attention. The *London Times* quoted Sir J. J. Thompson, president of the Royal Society, claiming that relativity was "one of the most momentous, if not the most momentous, pronouncements of human thought," although no one had yet succeeded in stating in clear language what this theory of Einstein's really was. The Cambridge physicist Sir Oliver Lodge, who was also Member of Parliament for his district, was "besieged by inquiries as to whether Newton had been cast down and Cambridge 'done in.' " Lodge wrote an editorial in the *Times* cautioning against a hasty rejection of the ether concept and against "a strengthening of great and complicated generalizations concerning space and time."[34]

As the public endeavored to extract explanations from their scientific fellow citizens, the scientists became increasingly and painfully aware of their own inadequacy. Herbert Hall Turner wryly remarked: "The vain attempts of the reporters to apprehend exactly in what the revolution consists have been amusing, and would have been more so but for our own similar difficulties."[35] He quoted at length from an article, "Newton Put in the Shade" from the *Evening Standard*, written by a reporter who had attended the joint meeting where the eclipse results had been announced:

> But what is Einstein's theory? At the meeting yesterday two noted scientists got up and begged the speakers to explain it in easy language, and—they couldn't do it.
>
> ### SCIENTISTS CAUGHT OUT
>
> A representative of the *Evening Standard* went to the Royal Society this morning and asked for a popular explanation. . . . The Secretary rubbed a hand over a dome-like brow, and frankly admitted he was beaten. The theory is down in black and white, with plenty of $x = 0$, but compared with it the Rosetta stone in the British Museum is a child's rag alphabet.
>
> ### SORROWFUL CONFESSIONS
>
> A distinguished scientist was next seen. 'I don't understand it at all,' he said, wearily. 'Don't mention my name.'
>
> Another equally distinguished scientist said:— 'Einstein says, I think, that the qualities of space are relative to their circumstances. Is that what you want?' 'No. Can you put the theory in terms of plums and apples?' 'It can't be done. The theory is terribly involved and mathematical, and—don't mention my name—I don't understand it.'
>
> ### ONE MAN WHO KNOWS
>
> That is how the matter stands. Einstein's theory is no doubt tremendously important and frightfully interesting, if only he would consent to come down from Euclid to terms of the earth, earthy. At the present time his meaning, like light, is seriously deflected.
>
> The writer went up to [the] Library of the Royal Society and read through the theory three times and was led out sobbing.

The article concluded that Eddington claimed to understand the theory, but that "until he consents to put it in schoolroom prose—Gott strafe Einstein [God punish Einstein]."[36]

Journalists had an opportunity to hear Eddington in December 1919, when he opened a relativity discussion at the Royal Astronomical Society's monthly meeting. Eddington chose "simply to explain, as far as I can, what are the conceptions of space and time and force which the relativity theory introduces." He described nonmathematically how time and space measurements were observer dependent, and explained almost in a "plum and apples" way about geodesics—the shortest distance in a curved space. He showed how observers whose "world-lines" deviate from geodesics experience a field of force around them.[37] The *Times* reported on the meeting, devoting almost an entire article to Eddington's remarks, leaving out contributions by others (Jeans, Dyson, Larmor, Silberstein, Lindemann, and Jeffreys). Only Sir Oliver Lodge was quoted at the end of the article, expressing reserve in accepting the whole of the "theory on time and space," and astonishment at the fact that Eddington thought he understood it.[38]

Eddington was one of the few British scientists who demonstrated intimate familiarity with the theory, and he expounded upon it in an unambiguously favorable light.[39] Lodge was a renowned physicist from an earlier generation, whose career had flourished in the heyday of the ether and Newtonian mechanics. He could not tolerate the "metaphysical" notions of relativity. As various articles on relativity came and went in the pages of the press, with little or no clarity being gained, Lodge emerged for a short time as the symbol of the scientific community that was floundering in the face of the new and revolutionary theory.[40] It became increasingly clear to the public that most of the scientists did not understand relativity well and that those, like Eddington, who did, had worked hard to gain their understanding.

In America, the editors of the *New York Times* played on the fact that no one could explain the theory. "Men of science more or less agog over results of eclipse observations," declared an early article reporting on the story. Reporters were quickly dispatched to interview local astronomy professors to explain what was going on. They were not very helpful. Clinton H. Currier, professor of astronomy at Brown University, told a reporter: "It was not until 1915 that the four-dimensional theory of the universe, with time as a fourth dimension, was definitely conceived. This was contained in Einstein's famous relativity theory." He claimed that Newton's theory of gravitation did not predict any bending of light, electromagnetic theory subsequently predicted bending of light, and Einstein's theory predicted a deflection of twice that amount. Caroline E. Furness, professor of astronomy and director of the observatory at Vassar College, candidly admitted that "Einstein's theory is one of the most difficult parts of mathematical physics. As yet I have not followed strictly its application to astronomy. Its results are remarkable and are such that it must be accepted." She remarked that the course of a star may be deflected many times, according to relativity, "and that the true positions of stars will be confused for a while." John M. Poor, professor of astronomy at Dartmouth College, declared that if, "as reported in the daily papers," Einstein's theory had really been confirmed, then Newtonian mechanics would need modification: "That will be a matter which . . . will concern the student in mathematics and pure science."[41]

The editors reacted strongly to the lack of explanation coming from scientists abroad and at home: "As all common folk are suavely informed by the President of the Royal Society that Dr. Einstein's deductions from the behaviour of light as observed during an eclipse cannot be put in language comprehensible to them, they are under no obligation to worry their heads, already tired by contemplation of so many other hard problems, about this addition to the number." They remarked that the "masters" would probably explain more if they could, but that to have them decide in advance for "the rest of us" to give it up is "well, just a little irritating."[42] When Robert Millikan and Charles Lane Poor later proclaimed that other explanations for the British results would be found, the editors ridiculed British scientists for having been "seized with something like an intellectual panic," concluding that although "these gentlemen may be great astronomers . . . they are sad logicians."[43] However, within days, the *New York Times* retracted. Having been reminded by "a reader of *The Times* who possesses a well-trained scientific mind," the editors told readers that those unwilling or unable to get the necessary training in mathematics were advised to "accept the expert's conclusion on the authority of its maker, supported by the acceptance of the few others like him." In Einstein's case the "few others" numbered "a minority of twelve."[44]

The intense interest of the public put tremendous pressure on those astronomers who had taken on the task of making the difficult observations to test Einstein's astronomical predictions. Their inability to understand or explain the theory made their position even more tenuous. At Mount Wilson, Hale worked diligently for years to strengthen theoretical capabilities in physics at the observatory. He continually invited European theorists to lecture and tried hard to attract theoreticians to take positions on the staff. In 1922, Hale succeeded in bringing H. A. Lorentz to California to give a two-month course of lectures on topics in mathematical physics. Paul Merrill was impressed: "Lorentz is great. He makes everything seem very simple. His treatment of special relativity was of course extremely valuable, (while I am not converted to it as yet, I certainly feel that I have a much more satisfactory knowledge of it than ever before.)"[45]

During the decade following the announcement of the British eclipse results, astronomers at Mount Wilson and Lick, as well as elsewhere, concentrated on making new observations to test relativity further.

THE ROLE OF ARTHUR EDDINGTON

Eddington had been instrumental in bringing Einstein's full-fledged theory of relativity to the attention of English-speaking astronomers during the war. He continued to play a significant role in elaborating the theory to colleagues and the general public. In December 1919, less than a month after the first announcement of the British eclipse results, he wrote Einstein: "I have been kept very busy lecturing and writing on your theory. My Report on Relativity is sold out and is being reprinted. That shows the zeal for knowledge on the subject, because it is not an easy book to tackle. I had a huge audience at the Cambridge Philosophical Society a few days ago, and hundreds were turned away unable to get near the room."[46] In 1920 Eddington published a popular exposition, giving a history of the theory's development. The publisher sold 1,886 copies in the first year and 1,789 in 1921.[47] In 1923, Eddington published a technical treatise on relativity that sold 999 copies in the first year and over 2,000 in 1928, becoming a standard reference work for scientists all over the world.[48]

Eddington had immense influence among astronomers, but in certain ways his earlier expositions misled more than enlightened. His use of analogies such as refraction to explain the physics of what was going on reinforced classical ways of thinking about the problem. His first published remarks on the light-bending prediction appeared in February 1915, before Einstein announced his final theory. Drawing on Einstein's earlier prediction of a deflection near the Sun of $0''.87$, Eddington explained "a wave of light will travel more slowly when it enters an intense gravitational field of force. This must lead to a refraction of waves of light passing near a massive body."[49] Two years later Eddington discussed light bending again, after he was familiar with Einstein's later theory. Here he appealed to the notion that "electromagnetic energy wherever found must possess mass" and hence "be subject to gravitation."[50] In discussing the light-bending prediction, he first treated light "like a stream of material particles" and calculated Einstein's 1911 deflection amount. Then he simply stated with no further explanation: "This mode of treating the problem is, however, too crude, and the actual deflection given by the complete theory is twice as great."[51]

In 1918 Eddington derived the full general relativistic bending of light in his "Report" for the Royal Society. Although he obtained an expression for the velocity of light from Einstein's metric tensor, he used two classical analogies—refraction and Newtonian motion of a material particle—in deriving the light bending. "The course of the ray will . . . depend only on the variation of velocity, and will be the same as in a Euclidean space filled with material of suitable refractive index. . . . We thus see that the gravitational field round a particle will act like a converging lens." Using his "refractive index" formula, he obtained two expressions representing angular momentum and energy "for the Newtonian motion of a particle with velocity μ." He obtained the full Einstein deflection of 1″.75 but did not explain it: "It is curious to notice the occurrence of the factor 2 . . . in the dynamical analogy. The deflection is twice what we should obtain on the Newtonian theory for a particle moving through the gravitational field with the velocity of light." He then redid the calculation using Einstein's special theory to obtain an equivalent mass of E/c^2 for the light ray to get "only half the deflection." Eddington concluded that "the experimental amount of the deflection should thus provide a crucial test" between Einstein's full-fledged relativity theory and the simpler notion that light has mass and is subject to gravitation.[52]

In March 1919, just before he sailed to Africa on the eclipse expedition, Eddington discussed the coming event in the pages of *Observatory*.[53] He noted that "apart from the theory of relativity" there were reasons to anticipate the smaller deflection originally predicted by Einstein's equivalence hypothesis, and expected from the equivalence of mass and energy:

> If gravitation acts on light, the momentum of a ray will gradually change direction when acted on by a transverse field of force, just as that of a material projectile does. . . . The effect is that light is deviated just as a particle moving with the same speed would be deviated according to Newtonian dynamics . . . and it is easy to calculate that the total deviation of such a body on passing the Sun, if it grazed the surface, would be 0″.87, or half the Einstein deflection. It may happen that the ratio of weight to mass for light is not the same as for matter. If so, the deflection will be altered in the same proportion. The problem of the coming eclipse may, therefore, be described as that of *weighing* light.[54]

When it came to Einstein's 1915 prediction, Eddington emphasized that "we have to take a somewhat different point of view, because the theory is geometrical and it goes behind the conceptions of force and inertia which are fundamental in Newtonian dynamics." He fell back on his refraction analogy to try to explain the general relativistic prediction:

> We can imitate the effect of the Sun's gravitational field by filling the space with a medium of suitable refractive index, increasing as we approach the Sun—forming, in fact, a converging lens. It is thus seen why the ray should be deflected. Evidently, the effect is qualitatively the same as that due to refraction by coronal matter, but it is presumably much more intense. The effect may also be compared to the bending of the sea-waves in a bay as they approach the shallow shore; the wave-front of the sea-waves and of the light gradually turns, because one end is moving slower than the other.[55]

As source material for later brushing up on relativity by other astronomers, Eddington's early articles may have been more confusing than helpful. His use of refraction or particle analogies to illustrate relativistic effects were similar to alternative theoretical explanations that others offered against the new and complicated Einstein theory. Eddington remained clear about what was an alternative and what was merely an analogy. When M. Jonckheere suggested that "a condensation of the aether around the Sun would produce a similar effect," Eddington allowed that this picture would explain the changing of the velocity of light that caused the bending. However, since that was just what one was looking for, the ether condensation suggestion "amounts to a hypothetical explanation or illustration of the Einstein effect, and is not to be regarded as an alternative to it."[56]

Not until the 1920s, when interest in general relativity grew to enormous proportions, did Eddington begin to elaborate the uniquely geometrical way of looking at general relativity and its astronomical predictions. As the decade progressed and Eddington's writings increasingly presented general relativity more cogently, astronomers benefited from his ability to popularize the difficult theory. Referring to his popular book *Space, Time and Gravitation*, Walter Adams of Mount Wilson remarked in 1924: "The aptness of illustration and the lightness of touch which we find throughout the pages of this book lead us to suspect that the author found nearly as much pleasure in writing these delightful chapters as the reader finds in returning to them after long hours among the tensors and vectors of the more formal treatises."[57] Nonetheless, others continued to find the theory difficult, even with Eddington's help. When his popular book first came out in 1920, Ernest W. Brown told a colleague: "Eddington has got out a new book which I've bought & read—supposed to be explanation for the beginner—but I'm more mixed up than ever. And I can't get much out of Eddington by conversation either. I feel as though I'd lost all my apologies for a brain."[58]

Critics and supporters alike tried to understand general relativity in terms of Newtonian theory. It was so new that they needed something familiar to help them. Eddington himself noted that fundamental physical concepts such as mass, time, and distance "are all ambiguously defined in Newtonian dynamics" and that there was "some freedom of choice" in defining them for general relativity. To help limit this choice, it was best to make sure "our definition agrees with the Newtonian definition in the limiting case of a vanishing field of force." Looking back half a century later, the relativist John L. Synge remarked: "In the days when relativity had to win credence in an incredulous world, it was natural to give it respectability by explaining it as far as possible in terms of the old concepts. But this led to getting the concepts mixed up."[59]

During the 1920s, the "curious" factor of 2 difference between Einstein's early and later light-bending predictions caused much confusion, and critics pushed refraction explanations for a long time.

Einstein's life in Berlin changed forever after the British confirmed his theory. "Ever since the announcement of the deflection of light a cult has been practiced with me, so that I feel like a graven image," he complained to his friend Heinrich Zangger. "But that, too, will pass with God's help."[60] It did not pass.

The enormous attention Einstein and his theory attracted after the announcement of the British eclipse results caused consternation in some quarters in Germany. Weeks after the news broke, the government received demands that it should "make available the necessary means to enable Germany to cooperate successfully with other nations on the development of the fundamental discoveries of Albert Einstein and to facilitate his own further research." Not to be outdone by Britain and other nations, the Prussian minister of culture informed Einstein that a special fund of 150,000 marks would be made available for his research. Postwar Germany was suffering a financial crisis, and Einstein had misgivings. "Would not a decision like this justly arouse bitter feelings among the public?" Remembering the lack of support Freundlich had been receiving, he added that "if the observatories and astronomers of this country were to place a part of their equipment and their energy at the service of this cause" there might not be any need for special financing from the state.[61]

Freundlich saw an opportunity in the government's haste to capitalize on Einstein's international fame. After his mentor had liberated him from under Struve's thumb at the Royal Observatory, he had initiated plans to build a tower telescope similar to the type that Hale had developed at Mount Wilson. His idea was to create his own research institute to undertake solar research and test Einstein's gravitational redshift prediction. Gustav Müller, the new Potsdam director that Einstein had initially opposed, turned out to be accommodating. He offered Freundlich a plot of land on the observatory grounds for the building. Freundlich recruited his architect friend, Erich Mendelsohn, to draw up the plans (fig. 6.1). He drafted a grant proposal, which Müller submitted to the ministry on 16 August 1918. Yet it stalled in postwar bureaucratic paralysis. When Einstein and relativity rocketed to international stardom a year later, Freundlich leaped into action. In December 1919 he drafted an "Appeal for the Einstein Donation Fund," asking for contributions to the Astrophysical Observatory to get into the business of Einstein testing. His pitch was a

Figure 6.1. An early sketch for Freundlich's tower telescope by Erich Mendelsohn. (Courtesy Astrophysikalisches Institut Potsdam.)

nationalistic one: "The academies in England, America and France have recently set up a commission, which excludes Germany, to establish actively experimental bases for the general theory of relativity. It is an obligation of honor to those who are concerned about Germany's cultural standing to come up with whatever funds they can afford in order to enable at least *one* German observatory to work directly with its creator in testing the theory."[62]

All the important physicists and astronomers of the Prussian Academy signed the appeal. Haber and Nernst assisted with fund-raising. The campaign raised 350,000 marks from private individuals and industry, including contributions of optics and equipment at cost by Zeiss and Schott in Jena. By May 1920, the Prussian government had still not approved the 150,000 marks it had offered to Einstein. Freundlich prodded the bureaucrats with a plea "to prevent German science's being excluded from the further development of this important field of research—especially considering that not only is the conception of the new ideas the intellectual property of a German, but also, it was in Germany that the first attempts to test it through experiment were made and its consequences pursued."[63] Inflation was rampant at the time, and the government had to move quickly to avoid erosion of the private money that was committed. In September 1920 the ministry finally contributed 200,000 marks.[64] Freundlich became the scientific director of the observatory. At last, he could pursue research on relativity testing. Einstein laid the foundation stone of the new observatory, confident that the search for the gravitational redshift in the Sun would "eventually provide a brilliant confirmation of the theory; I never had a second's doubt of that."[65] The building was completed in 1922. For years people called it the "Einstein Tower" (fig. 6.2a, b).

Einstein did not like his name being used for nationalistic purposes. In June 1918, he had received copies of a lecture on theories of gravitation, including general relativity, by Adolf Kneser, a professor of mathematics

Figure 6.2a. The *Einsteinturm* (Einstein Tower) for astrophysical research on the testing of Einstein's theory of relativity, ca. 1940. (Courtesy Astrophysikalisches Institut Potsdam.)

Figure 6.2b. Einstein posing in the *Einsteinturm*. (Courtesy Astrophysikalisches Institut Potsdam.)

at the University of Breslau. Kneser had proudly emphasized that this work was done by Germans as a "by-product of war" and constitutes the "work of peace which continues behind the front with the full pulsating energy of our nations." While Einstein found the lectures to be "original and inspired" there was "*one* thing" that pained him: "It hurts me when my name and my work are abused for chauvinistic propaganda, as has been happening frequently in recent time. This is out of place even objectively. I am by heritage a Jew, by citizenship a Swiss, and by mentality a human being, and *only* a human being, without any special attachment to any state or national entity whatsoever. If only I could have said this to you before you held your lectures!"

Kneser respectfully replied: "Your brilliant discoveries were made in Germany during the war, where you are being granted protection and leisure for scientific research. So you must submit to having your researches credited to Germany and ranked as part of the German peace effort behind the frontline." Kneser was pleased that "you did not take part in the exodus of many Swiss scholars, who left Germany as if abandoning a ship thought to be sinking."[66]

While there were many in Germany who were proud of Einstein's achievements and wanted to claim him as a national treasure, others felt differently. In the difficult months after Germany's defeat, attacks on Jews became rampant. Einstein noticed it. "There is strong anti-Semitism here and raging reaction," he wrote to his sister in the spring of 1919, "even among the 'educated.' "[67] When his name became a household word after the British results were announced, he became a target for some of these attacks. Einstein had a premonition that this would happen. When the *London Times* solicited an article from him shortly after the British announcement, he ended with a wry comment, now well known, in reaction to the *Times* reference to him in an earlier report as a "Swiss Jew." "The description of me and my circumstances in *The Times* shows an amusing feat of imagination on the part of the writer. By an application of the theory of relativity to the taste of readers, today in Germany I am called a German man of science, and in England I am represented as a Swiss Jew. If I come to be regarded as a bête noire, the descriptions will be reversed, and I shall become a Swiss Jew for the Germans and a German man of science for the English!"[68]

It did not take long to happen. A shady character, Paul Weyland, raised money from anonymous donors to form the anti-Semitic "Working Party of German Scientists for the Preservation of Pure Science." He offered large sums of money to scientists who would speak publicly against Einstein and his theory. He attacked Einstein and relativity in print and elicited outraged replies from some of the leading physicists of the day, who jumped to Einstein's defense. Weyland rented the Berlin Philharmonic where he organized a public meeting on 26 August 1920. He opened the proceedings with an anti-Semitic rant. Ernst Gehrcke, who had already attacked Einstein in the scientific journals in 1916, followed him. Among other things, they accused Einstein of plagiarizing from the nineteenth-century German physicist Johann von Solder. For sale in the foyer were reprints of a scientific article by Nobel laureate Philipp Lenard, in which he had taken exception to Einstein's general theory of relativity. Einstein, who attended the event, wrote a devastating and angry response, which was printed in the daily newspapers. He attacked Gehrcke, but also made disparaging comments about Lenard, whom Weyland was advertising as speaker for a follow-up event. It turned out that Lenard had not given his permission, and was incensed at Einstein's attack. Yet Lenard also threw himself into the anti-Einstein campaign.[69]

Einstein was shaken by the experience and decided to leave Germany. His friends in Zurich and Leiden had standing offers for him. With rampant inflation in Germany, and a family to support in Switzerland, he could also improve his financial situation. On 27 August 1920 the newspapers reported:

Albert Einstein, disgusted with the Pan-Germanic bickering and the pseudo-scientific methods of his opponents, wants to turn his back on the capital city and on Germany. In the year 1920 it has thus become a question of the cultural spirit of Berlin! A German scientist of world acclaim, whom the Dutch have appointed in Leiden as an honorary professor . . . whose work on *relativity theory* was one of the first German books to appear in English after the war, such a man is forced by disgust out of a city that holds itself to be the center of German intellectual and cultural life. A scandal![70]

A week later, a worried chargé d'affaires in London wrote to his superiors in Berlin: "The attacks on Prof. Einstein and the agitation against the well-known scientist are making a very bad impression over here. At the present moment in particular Prof. Einstein is a cultural factor of the first rank, as Einstein's name is known in the broadest circles. We should not drive out of Germany a man with whom we could make real cultural propaganda."[71]

Einstein's physics colleagues were mortified. Max von Laue appealed to Arnold Sommerfeld, chairman of the German Physical Society, to bring about a resolution from the society against Weyland's group. "If anything more is needed to stimulate your zeal, it would certainly be the news that in view of the persecution Einstein and his wife appear to have definitely decided to leave Berlin and Germany at the next opportunity. Should this come to pass, we would experience, in addition to all our other misfortunes, the fact that those who want to be 'national' have forced out a man of whom Germany could be proud, as it could be of few others. Sometimes one has the feeling of living in a madhouse."[72] In the end, Einstein decided to stay in Berlin. But Gehrcke's and in particular Lenard's participation would make an impression on less well-informed scientists and nonspecialists. Weyland's campaign eventually came to nothing. Gehrcke and Lenard acknowledged to each other that he was a "dubious type" and "turned out to be a crook." Yet they would both return to the nationalist, anti-Semitic and anti-Einstein theme a decade later with the rise of national socialism.[73]

HALE REALIZES HIS VISION

While Germany was struggling to hold on to Einstein, Hale's plans to create conditions for cooperative research in physics, chemistry, astronomy, and astrophysics were coming to fruition. The 100-inch telescope was installed and in operation. The neighboring California Institute of Technology (Caltech) emerged in full flower with munificent funding. There was a steady flow of visitors and research associates to Mount Wilson. Hale's goal was no less than to establish his observatory and the Pasadena area as a research center with the very highest reputation for breaking new ground in physical research of the most fundamental nature. It would also be the place to provide definitive results on well-known problems requiring the best in modern technology and expertise.

Attracting European theoreticians of international reputation was an integral part of Hale's overall strategy. He used the lure of contact with Europe's best and brightest to convince Robert Millikan to leave Chicago and take charge of the Caltech physics laboratory. The plans were laid during the dark years of the war. In 1916, Hale saw American involvement in the war coming. He convinced President Wilson that the National Academy of Sciences should coordinate research efforts for the war effort under the auspices of a newly created National Research Council. He recruited the chemist Arthur Noyes and Millikan to oversee their respective fields of research. Hale used the war to highlight the importance of scientific research for the nation, in wartime or peacetime. He leveraged this heightened awareness among political and financial leaders to raise funds necessary to build a physics institute, with Millikan at its head. Plans were suspended when the United States entered the war on 31 January 1917.[74]

After the war, Hale pushed forward with his dreams. Millikan was one of the major planks in Hale's arsenal to attract renowned scientists of great ability in mathematics and theory to the vicinity. His international reputation would easily draw capable men to the area. A January entry in his 1921 diary reveals his thinking: "Bring to Pasadena: Majorana, Lorentz, Epstein, Ehrenfest, Fowler, Fabry, Perrine, Jeans, Eddington, Rutherford, Silberstein, Mees, Langevin." Several months later: "Additional inducements to M[illikan] 2 Obsy Res. Assocs. would lecture each year at Inst. for $1000 extra. Fowler, Eddington & others on const[itution] of matter from ast[ronomical] side." Millikan finally accepted in June 1921. By the fall he was established in Pasadena.[75]

Millikan carried out Hale's vision. Even before he was officially on board and was still negotiating terms with the Caltech board, he was recruiting visiting associates and institute staff. On a visit to Leiden in the spring of 1921 he convinced H. A. Lorentz to give a series of lectures at Caltech the following year. He also recruited Paul Epstein to cover theoretical physics at Caltech. Epstein's name had been on Hale's shopping list. A European-born and trained theoretical physicist, he was familiar with the new quantum theory. He was working with Lorentz in Leiden at the time but wanted to come to America. Although Millikan expressed some doubts, because Epstein was a Jew, he invited the young physicist to come to Caltech for a year to teach theoretical physics. Epstein joined the Caltech staff on this basis in 1921, but within several months Millikan asked him to stay on permanently.[76]

Lorentz's visit was a major triumph for Hale. Several years earlier he had invited the famous Dutch physicist to be a research associate of the observatory, "but he could not spare the time to come over here." Now his Caltech strategy to create an additional magnet for eminent physicists was paying off. He put aside $1,000 from his research associate fund for 1921 to go toward Lorentz's expenses.[77] By the end of 1921, Hale could write happily to his friend, the Cambridge physicist Joseph Larmor: "Our physics colony here is now greatly strengthened by the permanent addition of Millikan, Epstein, and other good men to the faculty of the California Institute of Technology." He also announced that Richard Tolman, an American physical chemist and physicist, was coming to Caltech, and that Lorentz would be lecturing at the Institute in January and February of the coming year. Hale was delighted. "All of this is of course a great help to the Observatory, which has been so weak in mathematical physics."[78]

A glimpse of Hale's achievement emerges from his 1921 annual report. Hale singled out three "outstanding events" of the year: the publication by W. S. Adams and his associates of the absolute magnitudes and parallaxes of 1,646 stars, and their development of the theory of spectroscopic parallaxes; the application of the interferometer by Albert A. Michelson and Francis Pease to the measurement of star diameters; and the establishment in Pasadena of the Norman Bridge Physical Laboratory of the California Institute of Technology and the acceptance by Dr. Robert A. Millikan of its directorship.[79] The "third event" underscored Hale's research strategy. The physical laboratory "has proved so necessary to the interpretation of solar and stellar phenomena that it is now being advanced from a secondary to a primary place in the scheme of the Observatory."[80] Two great laboratories, one chemical and the other physical, at neighboring Caltech now complemented the observatory installation. Hale described the intense collaboration between Caltech and Mount Wilson:

> As during the past year, the members of the Observatory staff will meet weekly with the investigators of the Bridge and Gates laboratories to hear reports on current research and discuss problems of common interest. They will also be invited to attend the courses of lectures to be given at the Institute by eminent men of science, who will include for the coming year Professor H. A. Lorentz of Haarlem and Professor Paul Epstein, formerly of Leiden, now a member of the faculty of the institute. Furthermore, a joint study of the constitution of matter and the nature of radiation will be orga-

nized, in which the astronomical, physical, and chemical aspects of these problems will be attacked by the members of the three groups immediately concerned.[81]

It was this cooperative research that Hale hoped would provide an antidote to the lack of theoretical expertise in his own establishment, and in American physics and astronomy.

PART THREE

1920–1925
ASTRONOMERS PUT EINSTEIN TO THE TEST

Chapter Seven

TACKLING THE SOLAR REDSHIFT PROBLEM

EVERSHED AND ST. JOHN DECLARE THE CASE UNRESOLVED

With public and scientific attention focused on Einstein's theory, Evershed and St. John felt obliged to report on the status of their respective searches for a gravitational redshift in the Sun. In 1920, reviews by both men appeared back-to-back in *Observatory* summarizing current results and how they might be interpreted if the relativity prediction were in fact true.

Evershed made it clear that the recent "brilliant confirmation" of the light deflection motivated his present review of the situation. He related in detail his early work disproving the role of pressure in producing the solar redshifts, and how these studies had led him to the hypothesis of an Earth effect. "It was with a view to verifying or otherwise [*sic*] this extraordinary movement, implying repulsion by the Earth, that the observations of Venus were instituted," he explained, "and it must be admitted that, contrary to expectation, the Venus spectra have, so far, given almost unqualified support to this hypothesis."[1] Evershed urged that "from this defiance of Einstein's theory and the favouring of a very incredible hypothesis . . . the most careful confirmation is needed—preferably by independent investigators."[2]

"Assuming for the moment that the Venus measures are affected by some undiscovered source of error," Evershed allowed, "we will see how far it may be possible to bring the direct measures of the displacements into line with Einstein's prediction." He compared his and St. John's results on the limb and center solar line-displacements relative to the carbon arc. His colleague Narayan Aiyar had completed measuring all the plates from his series of observations of 1918, and his results "unfortunately do not bear out St. John's conclusion" of a zero shift at the solar limb.[3] The Kodaikanal observers found that the mean shift of "the most characteristic triplet-bands" in a part of the cyanogen series "closely approximates to Einstein's prediction." Other measures of band-lines and metallic lines showed "that there is a general displacement of the lines at the Sun's limb, which, if not in exact agreement with Einstein's prediction, is of the right sign and the right order of magnitude." Evershed emphasized that this general displacement "cannot be explained by pressure, nor by motion, unless we admit an Earth effect."[4]

Nonetheless, there were complications. Evershed cautioned that the shifts differed for different substances, and for different lines in the same substance, "so that if Einstein's hypothesis be true there is some unknown modifying influence at work." He noted that there were no such difficulties with the motion hypothesis except having to believe in the Earth's influence. "This being the present position of the problem," he concluded, "it is evident that the most pressing need is to obtain further confirmation of the Venus measures, because it is these which offer the most stubborn opposition to Einstein's theory."[5]

St. John weighed in with a more pessimistic view on a possible relativity explanation for solar line displacements. In view of the eclipse result, he intended to pursue various lines of inquiry in order to decide "the now critical question," but he emphasized: "Such investigation is not so easily carried to a definitive result as those without experience might at first think."[6] St. John insisted that the problem required stable equipment, simultaneous observation of the Sun and the arc, high resolving power, large solar image, and extreme care in selecting lines.

The present state of affairs looked bad for relativity. St. John presented sample measures of Sun-arc shifts for the solar center: five different means from twenty lines observed on five different days were consistently smaller than the Einstein prediction. Reviewing his early cyanogen work, he admitted that these lines were problematic because they were so closely packed. This difficulty might account "for the differences between the results obtained by different observers employing spectrographs of quite different dispersion and resolving power." Nonetheless, he felt that he had adequately accounted for these factors. He presented his figures again, indicating that the discrepancy from the Einstein prediction was "very large." While he had not carried out any further work on the problem, he reported, "one or two by-products of other investigations have a bearing upon the question and are of interest."[7] Studies of the wavelengths of magnesium lines indicated shifts much smaller than the Einstein amount, and Sun-arc displacements of iron lines in the blue region were an order of magnitude less than the relativity effect. St. John attached great importance to the magnesium result, since it depended upon work done independently by himself and by Walter Adams.

St. John summed up by comparing the status of the two relativity effects. He noted that the "two lines of observation in their bearing upon the equivalence hypothesis seem to lead to opposite conclusions." In the case of the solar redshifts he allowed that conceivably "unrecognized causes are at work counteracting the displacement to the red required by the equivalence hypothesis." He reiterated "the characteristic differences for different lines and elements complicate the situation." In the case of the light bending, St. John was not ready to abandon Newton: "The point of view that the deflection of light in passing the Sun may, after consideration has been given to co-operating sources, be found to be in harmony with the Newtonian law seems not yet excluded."[8]

Further work on the redshift problem at Kodaikanal and Mount Wilson took two directions: (1) continuing to investigate limb and center shifts for different lines and trying to determine the causes, including a possible gravitational effect, and (2) observing the spectrum of Venus in an attempt to test Evershed's Earth effect. As to the theoretical side of the question, the Mount Wilson astronomers relied on European theorists for edification, who in turn watched with interest for empirical results emanating from their American astronomical friends.[9]

"EINSTEIN'S THIRD VICTORY"

For some influential Germans, Einstein's international fame provided a valuable boost to Germany's reputation in the postwar environment. Hence it was of considerable interest when two young spectroscopists, Leonhard Grebe and Albert Bachem, working in the laboratories of Heinrich Kayser in Bonn, published results in 1919 and 1920 that seemed to favor relativity's gravitational redshift prediction.[10] Not only did they find the right amount of line displacement, but they also explained why other observers had failed to find the Einstein effect.

Einstein had arranged to lend Freundlich's new microphotometer to the Bonn spectroscopists in order to measure their spectra as accurately as possible.[11] He was thrilled when he received news of their results. He wrote Eddington that Grebe and Bachem had obtained results agreeing with relativity, using the same cyanogen bands as had Evershed, Schwarzschild, and St. John. Eddington replied that the Bonn results "look convincing although I am scarcely qualified to judge the questions involved." He noted that "St. John has been making further researches, with magnesium and other lines, still getting a zero result; so I expect that for some time to come spectroscopists will be divided as to what the result really is."[12]

An American reviewer of a paper by Grebe and Bachem declared that the Germans' result "seems to remove the last practical objection to Einstein's theory of gravitation."[13] The British journal *Nature* quoted Einstein's enthusiastic response: "Two young physicists in Bonn have now proved with certainty the redshift of the solar spectral lines and cleared up the reasons for former failures."[14] The *New York Times* picked up the story, publishing an article under the headline: "Einstein's Third Victory: Red-Displacement of Spectral Lines Regarded as Completing Proof of Relativity Theory." The author, Robert Daniel Carmichael, enthused:

"Perhaps no other theory in science ever had three so remarkable and diverse confirmations in so short a time."[15]

The Mount Wilson team was busily engaged in research to unravel the complexities of the problem, and they did not feel the Bonn work was rigorous enough nor its results valid. St. John took pains to point this out in detail in the pages of *Observatory*. He gave four reasons why the Bonn data were questionable: the spectrograph's dispersion was "too low for work of this exacting character in a region of the solar spectrum where the Fraunhhofer lines are so closely packed"; no provision was made to ensure that the slit of the spectrograph was parallel to the solar axis;[16] the small size of the solar image required "extreme care" in guiding and "there seems to have been no provision for accurate guiding"; and finally, the Bonn collaborators had used the standard procedure of obtaining the comparison spectrum half before and half after the solar measures were made. Mount Wilson astronomers had pioneered a technique using mirrors to obtain the comparison and solar spectra simultaneously, a method that Evershed had adopted at Kodaikanal. St. John emphasized that the standard practice employed by the Bonn observers "is not sufficient, as it does not eliminate the possibilities of spurious displacements, large in respect to the precision sought. It has been shown at Mount Wilson Observatory that such errors are not prevented by simply stabilizing the apparatus."[17]

Grebe and Bachem had also supposedly "cleared up the reasons for previous failures" to find the gravitational redshift in the Sun. They argued that the emission lines in the arc had asymmetries that the solar vapor removed in the corresponding solar absorption line. When they compared arc and solar lines, the asymmetry in the arc lines made the apparent displacement of the solar line appear smaller than it really was. In 1914, Karl Schwarzschild had obtained consistently smaller values for the redshifts than relativity had predicted; Evershed's limb displacement had been too small, and St. John's had been essentially zero. Grebe and Bachem originally obtained small values for the shifts as well. They felt that their explanation would bring all the previous investigations in line with relativity.

St. John attacked their contention that absorption in the Sun removed asymmetries in the spectral lines. Radiation from the solar interior passes through an absorbing layer of gas on its way to the surface. When the light emerges, certain spectral lines are darker, because the gas absorbs selective frequencies. Grebe and Bachem assumed that the absorbing layer is thick, hence removing asymmetries in the absorption lines. St. John countered that such a thick absorbing layer of vapor would radiate light itself, "the very light that the spectroheliograph utilizes." Since the light from the absorbing vapor would come from different depths and thus different temperatures, the intensity across the width of the absorption line would vary. This would result in an unsymmetrical absorption line—the opposite of what Grebe and Bachem had postulated. St. John also attacked how the Bonn physicists determined the asymmetries in the arc emission lines. He pointed out that over-exposure of the arc "is likely to bring out apparent asymmetries, depending upon ordinarily unimportant grating errors."[18] He showed how Grebe's and Bachem's explanation for previous low values of the redshift would give different results for the iron or carbon arc. St. John concluded that the case for or against relativity was still undecided: "The purpose of this note is not to show that the predicted Einstein displacement is not present in the solar spectrum, but to call attention to some considerations that, in the minds of solar physicists, make it doubtful that the proof in its favour is as complete, and the explanation of previous failures to find it is as convincing, as the brilliant author of the equivalence hypothesis finds them."[19] As far as the prestigious Mount Wilson group was concerned, the jury was still out on Einstein.[20]

The persistent uncertainty surrounding the redshift problem continued to affect the annual deliberations of the Nobel committee. Einstein received over a dozen nominations for the 1921 prize for physics, most of them for his work on relativity. The committee commissioned reports from two of its members: one on relativity from the brilliant ophthalmologist, Allvar Gullstrand, a Nobel laureate from 1911; and a report on the photoelectric effect from Arrhenius. Gullstrand was highly critical of relativity and turned in a damning report, which included a statement that it was still not certain that Einstein's theory could account for Mercury's perihelion motion. Arrhenius was not keen to give a prize for quantum physics so soon after one for quantum theory had been given (Planck, 1918). So the committee decided not to give a physics prize during that year.[21]

<h3 style="text-align:center">UNRAVELING COMPLEXITIES—EVERSHED VERSUS ST. JOHN</h3>

By the middle of 1920, Evershed was beginning to find discordant results with his Venus measures.[22] Using a positive-on-negative method of measuring the plates, he was able to obtain a high degree of accuracy in measuring the spectra, but when comparing different plates, "we are apt to encounter rather serious discrepancies . . . even when the definition of the lines is the finest possible." To check his 1918 results favoring an Earth effect, he had taken a new series of Venus observations in 1920: "Although very special precautions were taken to guard against every possible source of error, this latest series of plates shows much larger discrepancies from plate to plate than was the case in the earlier series. The mean results do, nevertheless, confirm the earlier series in showing smaller wavelengths in the Venus light as compared with direct Sunlight."[23]

About this time St. John and his colleague Seth Nicholson were coming to the conclusion that a systematic error was causing Evershed's Venus results, not his hypothetical Earth effect. They used the Snow telescope at Mount Wilson to obtain spectrograms when Venus was west of the Sun in 1919 and west of the Sun in 1919–1920.[24] As Venus changed positions in its orbit, the planet's altitude when they took photographs would change. Further experiments indicated that a low altitude tended on average to reduce the wavelength of reflected light from Venus due to dispersion in the Earth's atmosphere. They suggested the change in altitude was causing the systematic decrease in wavelength that Evershed had observed and presented results to this effect at a scientific meeting in Seattle in June 1920. Evershed ultimately rejected their explanation.[25]

Evershed also reported "a considerable amount of work" on the cyanogen bands, which now seemed to favor relativity. "The shifts we get for the stronger bands are nearly in accordance with the predicted shift of Einstein," he reported, "and are not in agreement with St. John, whose results depend mainly on the fainter lines of the series."[26]

The Kodaikanal team had obtained this latest series of observations during March and April 1918, expressly to clear up the discrepancy with the Mount Wilson limb shifts. Evershed took great care in selecting lines that he believed were unaffected by superposition of other lines. For the plates of the solar center, he used a negative-on-negative method of measurement, by which duplicate negatives are superimposed film to film, but not reversed end over end, so that the arc lines of one coincide with the solar lines of the other. Much higher accuracy is attainable using this technique. For the limb measures, Evershed used the normal method because duplicate negatives were not available. Evershed found that the third-order spectra with lower dispersion and greater contrast were "more satisfactory" to measure than the higher dispersion fourth-order plates, due to the widening of all lines near the limb. Preliminary results, based on mean shifts for ten measurable bands, seemed to indicate an Einstein shift (table 7.1).[27]

Evershed emphasized the preliminary nature of his results since the individual plates both of limb and center showed "somewhat discordant values," and it would be necessary to measure a large number of plates before a definite conclusion could be reached. Nonetheless, his tentative opinion favored Einstein: "The result for the south limb plates is remarkably close to the shift corresponding to 0.634 Km/sec, predicted by Einstein and taken by themselves these results must be considered distinctly favourable to the relativity effect. The smaller shift at the center of the disc is readily

TABLE 7.1.
Evershed's Preliminary Redshift Results for the Solar Limb
and Center

	In Angstroms	In Km/Sec
North limb spectra	+0.0057	0.44
South limb spectra	+0.0080	0.62
Centre of disc	+0.0037	0.29
Einstein prediction		0.634

explained by an outward radial movement of the solar gases which might produce a partially compensating shift towards violet at the centre." He emphasized that the results "appear to be in serious disagreement with those of St. John," noting that St. John's limb measures were based on the mean shift of seventeen lines of small intensity in a different region of the cyanogen spectrum from the one Evershed used.[28]

The editors of *Observatory* reported "a curious feeling that the situation with regard to the displacements of stellar images in eclipse photographs is reversed in the matter of the spectroscopic displacements. In the former case the best evidence is completely favourable, the inferior evidence generally unfavourable to the full displacement. In the latter case the best evidence is completely unfavourable to the existence of the effect, but a large amount of inferior evidence supports it." They were quick to point out that "the inferiority of the other evidence is largely a question of instrumental equipment, with necessary consequences in the selection of material to be studied." St. John's solar image was larger and his dispersion "one-third as large again as that of Evershed." Evershed had found the lower dispersion more satisfactory for the limb plates, "and it is only in the limb plates that he gets the full Einstein displacement." The editors concluded on the side of the naysayers: "The weight of the evidence is still with the skeptics, or with those who hold that Einstein's generalized relativity theory does not involve the shift of the spectral lines."[29]

Evershed was quick to defend himself. Before designing the spectrograph at Kodaikanal, he had spent a month at Mount Wilson, where the spectrograph at the time was "in essentials similar to" the one St. John used in his relativity shift research. Evershed's experience with the Mount Wilson instrument "led me to start work at Kodaikanal on entirely different lines in the design of our spectrograph." He cited his own discovery of the radial movement in sunspots (the Evershed effect) before its detection at Mount Wilson as evidence of the advantages of his setup. He argued that there was nothing to be gained at the limb of the Sun by using large dispersion, and that using a smaller solar image had decided advantages in this work. "The difficulty of deciding for or against the Einstein shift in the Sun lies in the conflicting nature of the evidence itself," he insisted; "it is not a question of weighing the relative accuracy of the measures made at Mount Wilson and at Kodaikanal."[30]

Evershed emphasized that the cyanogen bands, even if they showed an Einstein shift at the limb, could not decide the matter, since there were anomalies in the lines of other substances that needed to be resolved. He reminded readers that he had ruled out pressure shifts and that others had confirmed his conclusions. He called for more work on iron lines, "instead of wasting more time on the much more difficult cyanogen band-lines," and he hoped that "by co-operation with Mount Wilson" the problem might be cleared up and a definite conclusion reached. He concluded that recent work with iron lines "is not altogether unfavourable to an Einstein effect, superposed on a motion shift which has its greatest effect at the center of the disc." However, he noted that his Venus results were still an obstacle unless it could be shown that they "are affected by some as yet undetected source of error." He found it "very difficult to believe" St. John's suggestion that unequal illumination of the slit had contributed to his results.[31]

By the end of 1921 the situation remained unresolved. "Because of numerous fragmentary attacks upon this question," Hale reported, "the situation is becoming more and more involved and unsatisfactory, as the following brief summary by Mr. St. John shows."[32] By that time, St. John had come to the same conclusion that Evershed had, namely that the cyanogen lines could not resolve the issue. He showed that with completely different assumptions for motion of gases in the solar atmosphere and for pressure shifts, different investigators working with the same cyanogen lines were reporting displacements equal to the relativity amount. At Mount Wilson, studies of these lines by Harold D. Babcock failed to confirm some of the results. Furthermore, Arthur S. King had discovered that lines of the cyanogen band vary in relative intensity with change in the furnace temperature. Raymond T. Birge at Berkeley had recently found extensive overlapping of lines of the different series for cyanogen. St. John concluded: "In view of this superposition of lines, of the changes in relative intensity with temperature, and of the line-density in the solar spectrum, it appears that the cyanogen band is not well adapted for a definitive test of the theory."[33]

As for studies using lines of other elements, the problem was similar. St. John criticized recent publications by Alfred Perot and by Charles Fabry and Henri Buisson (both announcing displacements roughly equal to the relativity prediction) largely on grounds of unwarranted assumptions regarding pressure and motion effects. He decided that the only way to approach the problem was to launch an extensive campaign to study all the causes of solar redshifts: "The problem must be envisaged as a whole and not in detached portions and a consistent and probable rôle found for the gravitational effect if the theory of relativity is to find confirmation in the displacement of Fraunhofer lines."[34] Hale put the matter more strongly, noting that neither St. John nor Evershed had found the desired displacement. He reported that St. John and Babcock "have renewed their attack with improved apparatus, involving many refinements of procedure overlooked by less careful spectroscopists, some of whom have found no difficulty in confirming Einstein's prediction."[35]

The controversy over the third test of Einstein's theory continued to dog those who wanted to proclaim relativity verified and gave hope to the critics who preferred to see the theory disproved. The British astronomy establishment looked to the superior technology at Mount Wilson to resolve the issue once and for all, while placing less confidence in Evershed, much to his chagrin. Evershed continued his work with dogged determination and ultimately concluded that there is indeed a gravitational redshift in the Sun.

EVERSHED VOTES FOR EINSTEIN

In 1923 Evershed received strong vindication of his stand against significant pressure in the solar atmosphere from the Indian astrophysicist Megh Nad Saha. Saha proposed a theory of ionization that allowed Ralph Howard Fowler and Edward Arthur Milne in Cambridge, England, to calculate that pressure in the Sun's reversing layer must be extremely low.[36] Evershed immediately realized that he could now use metallic lines subject to pressure effects in the arc to give measures of solar redshifts just as accurate as for the cyanogen bands, as long as he corrected the corresponding arc spectra for the pressure shift.[37]

In the same year, Evershed reported a series of results that were distinctly favorable to relativity. His measures for strong iron lines, assumed to originate high in the solar atmosphere, had shown shifts in excess of the Einstein prediction, more so at the limb than at the center. Comparative measures of plates taken in 1914, 1921, 1922, and 1923 indicated that these lines were not constant over time, especially in the limb spectra. Evershed suggested that the excess redshift might be due to some sort of instability. For weaker lines originating at lower levels in the solar atmosphere "the results are in much better agreement with prediction, and the wave-lengths appear to be quite constant throughout the Sunspot period."[38] Also, the change of redshift with frequency was in agreement with the relativity effect. For ultraviolet and red lines, the corresponding Doppler shift was 0.75 kilometers per second at the limb and 0.46 at the center, compared with a predicted 0.634 kilometers per second. For eleven medium-intensity lines at an intermediate frequency, the mean limb shift was 0.71 kilometers per second, and at the center, 0.45 kilometers per second. Evershed explained the lower center shifts again by a superimposed ascending motion, which would not show up at the limb (being tangential to the line of sight). He noted that St. John's previously reported shifts for the magnesium triplet, which had been too small for the Einstein shift, had included "no allowance for the pressure in the arc."[39] When Evershed applied the correction, he obtained results very close to the Einstein value. He also presented new measures on the D-lines of sodium, which yielded redshifts agreeing very well with relativity. Taken as a whole, his results for individual lines were impressive.

Evershed included a brief report on his Venus work as well. While he persisted in his claim that the results "cannot have been due to any effect of atmospheric dispersion, owing to the low altitude of the planet," he admitted "that these plates give unreliable results."[40] Using a new prism spectrograph "built especially for the Venus work," Evershed had taken a series of twelve spectra photographed in November and December 1921. With Venus illuminated by the hemisphere of the Sun facing away from the Earth, these measures "gave shifts in close agreement with the control plates of direct Sun-light." Evershed took six more plates in April and June 1922, when Venus was an evening star, obtaining the same result. "These results I therefore take to be final," Evershed concluded, "in proving that the shift to the red is found in light coming from any part of the Sun."

Having disposed of the Earth effect, and finding shifts of the order of Einstein's prediction in limb shifts of metallic lines, Evershed was ready to vote in favor of relativity:

> Reviewing the evidence as a whole, there seems to me to be very little doubt that the Einstein effect is present in the solar spectrum. The observed shifts over the entire face of the Sun, and in the unseen hemisphere, seem impossible to explain by motion, pressure, or anomalous dispersion. Assuming the gravitational effect to be the principal factor, there now remains to be explained the considerable excess of shift shown by the high-level lines in the ultra-violet, especially at the Sun's limb, and the large differences of shift in individual lines observed throughout the spectrum.[41]

By about this time, St. John's detailed investigations were beginning to point in a direction similar to Evershed's. He and Seth Nicholson had resolved the Venus question to their own satisfaction with further observations after 1920. They derived and tested an empirical formula for the redshift of deflected Sunlight from Venus as a function of altitude of the planet and size of the image. As Venus went around in its orbit, its altitude in the sky changed, with low altitudes occurring when the Venus-Sun-

Earth angle was large. The size of the planet's disk also appeared smaller, since it was farther away. According to St. John and Nicholson, the low altitude would scatter the high frequency light away from the spectrograph slit, combining with the small image to produce an unsymmetrical illumination of the slit. The Mount Wilson astronomers attributed the changed line displacements for the large Venus-Sun-Earth angles to this altitude effect.[42] Though Evershed never accepted their explanation, their careful work helped to remove the unpopular Earth effect from the redshift discussion, until Evershed's later measures did away with his suggestion for good.

St. John's parallel work on the solar line displacements proved to be as systematic as he had promised in 1920. He attacked the problem on three fronts:

1. An accurate determination of terrestrial wave-lengths.
2. An accurate determination of solar wavelengths.
3. An extensive study of the causes giving rise to displacements of lines in the Sun, such as general and local convection, lateral drifts, pressure and possible effects from density distribution, and irregular refraction and dispersion.[43]

The first investigation was tied in with an ongoing international effort to establish universal standards from the iron arc spectrum for use by spectroscopists worldwide. The data from the second effort would provide the basis for a new table of standard solar wavelengths, as well as material for the gravitational redshift discussion. As part of the third project, St. John and Harold Babcock examined the whole question of limb-center shifts in the solar lines, independent of comparisons with the arc. They needed to explain the excess redshift and general broadening of lines at the limb compared to those at the center, referred to as the "limb effect." St. John worked with the spectrographs while Babcock did independent work with an interferometer. By the fall of 1923, both investigations had confirmed that the limb displacement increased with wavelength and depended upon the intensity of the line.[44]

In May 1923, St. John began to feel that his long labors on the problem might be coalescing into a consistent hypothesis for the cause of the solar line displacements. He wrote to Hale that it looked like the relativity effect was one of the contributing factors:

I am driving away on the gravitational shift and for the first time I am striking something that looks like that effect. To get free from pressure I am trying to use the high level lines, such as the Mg triplets and in the green and violet the Al lines at 3900, Cu 4226 and the D lines. It is difficult to obtain accurate solar and terrestrial measures. It looks as though these lines might show a displacement at the center of about the right order and practically the same shift at the limb as at the center. If this turns out to be the case I do not see any other explanation of such a behaviour.

Yet there were problems. St. John still felt that "the great body of solar lines give, on the face of the returns, great difficulty on the relativity view." Assuming zero pressure, St. John told Hale that lines of medium intensity "give not large displacements and become more troublesome still when limb shift is taken into account." For low intensity lines "far too small" displacements were obtained at the center, and "not much limb shift." Nonetheless, St. John now had a working hypothesis:

> There appear to me to be at least three things acting, and I am trying to reconcile the observations with some working hypothesis which at present is something like this, an Einstein shift for all lines; a Doppler effect for low-level lines decreasing the Einstein effect. This disappears at the limb showing as a limb-center displacement. For the great majority of lines of medium intensity, no Doppler effect but a limb-center shift due to anomalous refractions, which according to [Willem] Julius is small for weak and very strong lines and largest for lines of medium intensity. As yet this is only a working hypothesis but it has the virtue of directing investigation.

St. John had been working on the problem for almost ten years, and he remarked wistfully: "I hope to live long enough to be able to satisfy my own mind at least as to the effective causes in the relative wave-lengths in solar and terrestrial sources."[45] Hale assured his colleague and friend that he was "sure you will ultimately be rewarded for your long and careful work." Hale was in England at the time, and though he admitted that "I have had to avoid discussions here" due to chronic fatigue, "I must tell [James] Jeans what you are doing and see if he has any new views as to line-shifts in the Sun."[46]

As the Mount Wilson astronomers pushed ahead with their solar redshift work, their colleagues at Lick continued to work on the eclipse problem. Campbell had decided to suppress publication of Curtis's results from the 1918 eclipse until his measures and calculations could be thoroughly checked. The Lick astronomers worked on this problem at the same time that they were preparing for the next eclipse, which would be visible from Australia. The difficulties they encountered in resolving the Goldendale project gave them invaluable experience in bringing the second one to a successful outcome.

Chapter Eight

MORE ECLIPSE TESTING

PERSONNEL CHANGES AT LICK

During the first months of 1920, Curtis worked steadily at the Goldendale problem, but he did not complete the task. Despite the pleasant life at Lick, an offer of the directorship at the Allegheny Observatory as replacement for Schlesinger, who was going to Yale, and a salary of $6,000 drew him east. Curtis tendered his official resignation on 16 April, setting July as his departure from Lick.[1] His move to Allegheny came at an awkward time for Campbell, who was left holding the bag on the Einstein problem. His usual response to requests for information on the Goldendale results around this time was to defer to Curtis, as he did when approached by Charles E. Adams, director of Hector Observatory in New Zealand. Adams was preparing to observe the Australian eclipse and wanted to know if Lick's Goldendale plates had "confirmed the British results or not." Campbell's stock answer was that Curtis was still working on them: "Thus far he has not been able to satisfy himself that our 1918 eclipse plates confirm the Einstein effect. This comment, however, should not be put in to print."[2]

Campbell soon found someone to hand the Einstein problem to: Robert Trumpler, who (to complete the circle) had recently come to Lick from the Allegheny Observatory. Born in Zurich, Trumpler had studied there and in Göttingen, where he obtained his Ph.D. in 1910 under the astronomer Leopold Ambronn. His arrival in Göttingen coincided with Hermann Minkowski's presentation of his four-dimensional formulation of relativity in 1908. Trumpler came to America in 1915, as an assistant at Allegheny Observatory. His work there was primarily observational, conducting a detailed study of the Pleiades star cluster.[3]

Trumpler went to Lick as a Martin Kellog Fellow for the academic year 1919–1920. He intended to finish the Pleiades work as quickly as possible, and return to Switzerland if he were offered a good position there. When Campbell learned that Curtis would be leaving, he offered Trumpler a position as assistant astronomer, at a salary of $1,800.[4] Trumpler stayed at Lick for fifteen years. His training in precise stellar photography with the Pleiades cluster was perfectly suited to measuring the star displacements on the eclipse plates. And there was an added bonus. Trumpler emerged as the only astronomer in America capable of treating the theoretical aspects of relativity. Shortly after the news of the British eclipse results had come out, the astronomer Paul Biefeld wrote to Trumpler at Lick, "I know you have the theory [of relativity] so well in hand that you could give me the essentials." Five years later the Lick astronomer William Wright boasted: "We have here at the Observatory a specialist on Relativity (Dr. Trumpler)."[5] In 1920, however, Campbell had not yet realized Trumpler's potential in this regard, and he regretted Curtis's departure keenly.

Before Curtis left, Campbell discovered that Curtis's procedure for the measurement of the eclipse plates was the source of the large probable errors in his work. As a remedy, Campbell devised an intermediate plate with pairs of short diamond scratches ruled at right angles, intersecting at points corresponding to the positions of the star images on the day and night plates. In his revised measuring procedure, the eclipse and intermediate plates were put face to face with the emulsion side of the eclipse plate in contact with the ruled surface of the intermediate plate. The microscope with micrometer was readily moved over each star image, and the intervals between star and the intersection point of the two rulings quickly and accurately measured. The night comparison plates were measured in the same way, relative to the intermediate plate. "Curtis blessed me for hitting upon this simple and accurate device for comparing the night and day Einstein plates differentially and accurately," Campbell wrote Schlesinger. "He immediately ruled the intermediate plates in this manner and the corresponding night plates in June of that year 1920, before going to Allegheny. The results were a vast improvement upon his plan of absolute measurement."[6]

Joseph Haines Moore mentioned the improved measures publicly at the joint symposium on relativity that Samuel Boothroyd organized for the Mathematical Society, Physical Society, and Astronomical Society of the Pacific, against the advice of Slipher at Lowell Observatory and Aitken at Lick. Boothroyd had lined up Charles St. John to talk on "The Astronomical Bearing for the Theory of Generalized Relativity."[7] The meeting took place in June 1920. Moore, Campbell's emissary, thus reported the strange turn of events there:

> Dr. St. John was unfortunately not present and Professor [Edward P.] Lewis and I to our horror were called upon with a few minutes notice to try and fill out his end of it. Fortunately I had familiarized myself with the English astronomers work, taking of course the opportunity of expressing my admiration for their fine piece of work, and a statement of their results. In this connection I spoke of the L.O. [Lick Observatory] expedition, explaining about the difficulties under which we had worked, on account of our apparatus not arriving in time from Russia, and the delay of the measurement and

discussion of the plates on account of the war, with a statement that the plates were being remeasured by a method which we believed possessed considerable advantage over the method which had been used in the previous measures of our plates, the results of which had been presented by Curtis last year at Pasadena.

Moore hinted that the measurements made by the new technique were incompatible with Einstein's theory. "The various points of which we had spoken," he wrote Campbell, "especially the one concerning the usual statement 'that displacements of 1.75 are easily measured,' which is true but we are not measuring displacements of this size, struck several rather forcibly."[8]

Conflicting Announcements on the Goldendale Results

Curtis left Lick before obtaining definitive results with the new technique. Campbell soon sent him the first results of computations done by Adelaide Hobe based on the remeasures (table 8.1). Because the probable errors were still large, Campbell wrote that he was "provisionally thinking of asking Dr. Trumpler to look over the computations with me in the next two or three days, to see if we can find any chance for improvement."[9] This remark was the first indication that Trumpler might become involved in the Einstein work.

It turned out that Curtis's haste in his last weeks at Lick had contributed to the large probable error. "Am sorry that the results seem indeterminate," Curtis apologized to Campbell, "and have larger p.e.'s than my former solutions." Among other difficulties, the intermediate plate had obscured one star image that he had used in his previous "final" solution. To compensate he had added four others. "Perhaps this was a mistake," he admitted, "and it might pay to run through the solution with the same stars as I used before, cutting out the four I added, and of course, the one I could not get." Curtis was beginning to suspect, however, that little more could be squeezed from the data, and he hoped results would soon be available for announcement at a coming astronomical meeting in September.[10]

Campbell responded quickly, asking Curtis to keep the Einstein results confidential for awhile, as potentially serious errors were showing up in the current work "partly through errors in transferring data from note books to computation sheets."[11] Curtis acknowledged the likelihood that he had made mistakes in measurement, transcription, and computation "as I was working under tremendous pressure the last two months on the hill, and that *never* pays." Curtis recommended that Trumpler go

TABLE 8.1.
New Goldendale Results by Miss Hobe

Plate 2:	in Dec.	$+0\rlap{.}{''}11 \pm 0\rlap{.}{''}16$
	in R.A.	$-0\rlap{.}{''}35 \pm 0\rlap{.}{''}27$
Plate 3:	in Dec.	$+0\rlap{.}{''}40 \pm 0\rlap{.}{''}16$
	in R.A.	$+0\rlap{.}{''}18 \pm 0\rlap{.}{''}45$

over everything, and advised Campbell that certain pencil marks on the eclipse plates were wrong "evidently my error, made in the dark room at Goldendale." He also admitted to errors in deriving a correction factor for the computations: "Recall doing it twice, once just the day before leaving; thought it checked, and must have had the same quirk in my brain each time." He backed off from his efforts to get Campbell to prepare material for a formal announcement of results. "No need for the lantern slides for a long time, I can see."[12] Campbell himself remeasured the better of the two Goldendale plates using the intermediate plate method; but when he came to do the comparison plates, "I gave it up when I saw how much the chart stars are elongated in declination. I had not realized that the wind had played such havoc with the image." He decided to take a new set of comparison plates. "Just recently Dr. Moore gave me the valuable suggestion that we mount the Chabot lenses in a suitable wooden tube on the side of the Crossley reflector and use that instrument and its excellent clock for guiding, with all its advantages. This I decided within ten seconds to do. I wonder why nobody thought of this before. Hoover will be making the camera tube when the first storm comes, and we shall plan to get the chart plates in the first week of November."[13] But Curtis was not convinced that further refinement could improve the outcome. "It appears to me that the limit is here set by the character of the eclipse plates."[14]

The first opportunity to get new chart plates came in November between storms and under poor conditions. The plates were exposed, or rather overexposed, for three minutes. The trouble with the previous comparison plates had not only been the wind, as Campbell had assumed. Aberration from the lenses had elongated the star images away from the center of the field. Campbell decided to try again, with one-minute exposures, but he now doubted that the Chabot lenses could "answer the Einstein question either positively or negatively."[15] If the new exposures corroborated this judgment, he would be ready to abandon the project as hopeless, and publish an appropriate statement describing the failed effort.

"Looks like a year of work gone to [hell]," Curtis commiserated. He agreed that a definite disclosure should be made promptly, describing all the troubles incurred in the 1918 eclipse. He advised Campbell to "go light on the matter of the non-return of our eclipse lenses from Russia" as he felt that the Chabot lenses and the 3-inch Vulcan lenses "are really on an equally bad footing" and that no two-piece lens would be appropriate for the Einstein problem. Curtis hoped that, if possible, Campbell would also present some results. "The matter is entirely in your hands. It may be best, as you suggest, simply to state that, as a result of extensive measurements and tests, 'we are forced to the conclusion that no sufficiently definite results can be secured from the Goldendale plates to warrant their publication as a trustworthy authority either for or against the existence of a deflection effect.' My own strong preference, however, would be to append some such statement as that given above to a brief description of the plates, the measurements and the results."

Curtis sent along a table including values of the deflection calculated for the two Goldendale plates and the six Chabot Observatory plates (1900 eclipse), with the number of stars for each measure and their relative reliability, the Goldendale having the least merit. As a conclusion he suggested giving the mean result for gravitational deflection at the Sun's limb as 0.″87, but emphasizing:

> From the data given earlier as to the character of the plates employed, from the probable errors of the separate plates, and from the serious lack of agreement in the individual results, we do not believe that these results permit a decision for or against the Einstein or other deflection hypothesis, and these indecisive results are published simply as a matter of record, etc., etc.
>
> A simple, frank statement, as above, of indecisive results secured will, so far from hurting the L.O. increase its already great reputation for sanity and conservatism, and for not announcing theories till it can deliver the goods. When the Einstein theory goes into the discard, as I prophesy it will go within ten years, these negative or indecisive results will be more highly regarded than at present.[16]

Yet Campbell decided to continue to work on the problem. Moore helped him "in the struggle to obtain with the Crossley reflector the best possible photograph of the 1918 field,"[17] and by early March 1921, still another measurement of the two Goldendale plates was well under way. "We had a long struggle with the weather to get satisfactory plates for comparison with your two Einstein plates of 1918," Campbell reported to Curtis. "Good weather finally came in the middle of February after a wait of two and a half months, and we found that the correct exposure times were twenty seconds."[18] But as we know, the eclipse plates had a "very inferior" definition, which looked all the worse in comparison with the latest check plates, and Campbell was "not very hopeful" that the probable errors would be reduced in the eventual solution. Sparing no effort, he had invented another way to measure the plates. He illuminated "the images and diamond rulings" by an electric lamp placed several feet behind the star image to be measured and lined up with the axis of the micrometer telescope. "Not only are the star images on the eclipse plate much better defined, but the diamond scratches are perfect."[19]

Still, the definitive announcement was delayed. By early April Campbell was almost finished with the plate taken with the shorter focus lens (Plate No. 2), and would "soon have something to communicate as to the results. We are applying a few more checks before daring to make any announcements of results."[20] As for its companion, "Dr. Moore has finished the measures of the other eclipse plate and the night plate, and measures are half way through on that pair. We have no idea what the least squares solutions may reveal as to them."[21] And to extend this agonizing reappraisal further, Campbell now wanted to rework the 1900 Georgia eclipse plates that Curtis had borrowed more than a year earlier from Charles Burckhalter of the Chabot Observatory in Oakland.[22]

Campbell was under immense pressure around this time. Except for his administrative duties at Lick and at the university in Berkeley, he had been devoting his time fully to the Einstein problem since the previous July. The arrival of spring meant annual meetings of national scientific societies, with concurrent invitations to give lectures at eastern universities. Campbell decided not to go east at all. As he plunged into the task of redoing the Georgia eclipse plates, he confided to colleagues that the work had been giving him "serious anxieties and regrets" and "has caused me to lose a good deal of sleep" for over a year.[23]

About the middle of May, Campbell wrote Curtis that Trumpler had just begun "what I think is the last section of computational work."[24] Curtis agreed with Campbell's view that irregularities in the clock could explain only a part of the appearance of the Goldendale images. The wooden mounting and tubes had not been sufficiently rigid, he now believed, and could "give" a little as the instrument moved through different hour angles. Curtis calculated that it would need only about 1/250 of an inch movement for the tubes as a whole to cause a deflection of 5 inches on the plate, "and this is not a great deal for a wooden framework." He went on to consider the rigidity requirement for the mounting of the new quadruple lenses that were being made by Brashear for the Lick expedition to Australia. He described a lattice design for the tube, which he thought could be made rigid enough to give no deflection as great as 0.001 inch between zenith and horizontal. To be sure of getting "perfect, round images . . . under the temporary conditions of an eclipse camp," he emphasized the need "to go the limit in rigidity of tubes and axis and excellence of clock."[25]

Campbell and Curtis continued to correspond on matters related to the 1922 eclipse plans, as well as preparations for the 1923 eclipse; but the results from the 1918 and 1900 eclipses that Campbell had promised to tell Curtis about never materialized. Though the two men discussed many things of mutual interest over the next year, besides eclipses, not a word was mentioned about these results.

As cautious as Campbell was about presenting the fruits of his labors, he did make an announcement at the astronomical session of the Pacific Division of the AAAS, held in Berkeley in August 1921. The only printed report of his negative conclusion appeared in the *New York American*, written by Edgar Lucien Larkin, director of Lowe Observatory in California, under the title "Einstein's Theory is Not Proved." The article ran:

> A striking feature of the recent session at Berkeley was a paper, illustrated by lantern slides, in which Astronomer William Wallace Campbell gave his observations of the Sun at instant of total eclipse at Goldendale Wash. Station, June 8, 1918. . . . Months passed in the work [of measurement]. The first result of all this toil is that the Lick Observatory expedition secured measures of bending of rays of light from the immensely distant star of one second of arc, slightly more than half of the deflection predicted by Einstein. . . . At the conclusion of this valuable paper of Dr. Campbell, he said that this bending of light had not settled the question. . . . For fear that I had not heard, I asked him at the close of his lecture if I had. His reply was: "Yes, the Einstein theory is unsettled." Now these words count, for they were spoken by an astronomer second to none, one who had toiled for more than a year making ready, and still more to make final reductions.
>
> [Then] Astronomer and Astrophysicist George Ellery Hale, planner of the largest observatory in the world, inventor of that powerful instrument the spectroheliographic telescope, used before the mighty 100-inch mirror was mounted at Mount Wilson, arose and made a speech commenting on Campbell's now historic work, and said in conclusion: "I still hold Einstein's theory in abeyance."

Larkin concluded that "in my humble opinion all of Einstein's other hypotheses may be held in 'abeyance' and considered 'unsettled.' . . . The bending aside may have been caused by the refraction of the light energy waves by the rare gases of the corona of the Sun, through which they had to pass to reach the earth."[26] (Fig. 8.1.)

It appears that Campbell intended to follow his announcement at Berkeley with the long-awaited definitive report. We know his plans and distress about the Einstein business as of November 1921, from a letter to Dyson:

> I have been very unhappy the past two years and more over the results of Dr. Curtis's efforts to test the Einstein hypothesis at the eclipse of 1918. Following Dr. Curtis's departure, I have done considerable work on the 1918

Figure 8.1. Newsclipping (August 1921) of an article reporting on Campbell's announcement of his revised Goldendale results showing a light deflection less than Einstein's prediction. (Courtesy Mary Lea Shane Archives of the Lick Observatory, University Library, University of California–Santa Cruz.)

and the 1900 plates myself, in company with Dr. Moore, and I plan to publish the indecisive results very soon. The fact is that we should not have attempted any observations on that subject with the imperfect and untested lenses which we borrowed only one month before the date of the 1918 eclipse, when it became apparent that our own eclipse equipment, already nine months out from Russia on its home journey, would not arrive in time. Having put our hand to the plough, and certain preliminary announcements of results having been made, I am going to see the thing through to the final publication, letting astronomers judge pretty much for themselves as the weights and values of our conclusions.[27]

No publication appeared.

Why did Campbell decide in the end not to publish his results? No doubt he hoped to get better data from the Australian eclipse of 1922 and could afford to discard the troublesome material of 1918. This course of action probably was clinched by an error-ridden account of Campbell's results in the *New York Times* that came to his attention shortly after he had presented his negative results. The news story involved Henry Norris Russell, who had visited Campbell during the summer of 1921 and discussed his eclipse results. Early in December 1921 Campbell received a letter from Edward E. Slosson, the editor of the newly established Science Service.[28] Campbell was well acquainted with the fledgling organization, which was the brainchild of E. W. Scripps of San Diego. Scripps wanted to create a communication network for the popular dissemination of scientific news. He recruited a number of eminent scientists to help him set up a news service for science. Scientists across the nation could submit articles on recent scientific developments to a central bureau in Washington, D.C., where they were edited for newspaper readers and sent out to the press. Campbell was a member of the preliminary committee of five that Scripps invited to implement the service; but he eventually felt obliged to relinquish his duties. Nonetheless, he remained sympathetic to the effort. When Science Service was formally in place, he urged his colleagues to contribute to it regularly.[29] Slosson was initially interested in the Lick plans for the coming Australian eclipse, and Campbell promised to send a report. But Slosson's attention was soon diverted to Campbell's past eclipse work. He sent Campbell "a clipping from today's *Times* . . . about your confirmation of the Einstein Theory through early photographs." He asked the Lick director if he could "give me a brief, non-technical article on these results, explaining their bearing upon the Einstein Theory? Such a statement from you would tend to prevent the spread of unauthorized rumours and exaggerations."[30]

The article, entitled "Einstein Theory Again Is Verified," carried the subtitle, "Prof. Campbell of Lick Observatory Confirms Calculations on Sun's Curvature of Light." The anti-Einstein campaign being carried on by Ernst Gehrcke and Philipp Lenard in Germany had percolated across the Atlantic and had been picked up by the American press. Reporters uncritically repeated the story that in 1801 a German astronomer, Johann von Soldner, had predicted that the Sun's gravitational field would bend light. The *New York Times* reported: "Soldner's work had been forgotten until recently it was discovered by German scientists, who have been using it to show the great advances made in astronomical science in the last century."[31] In fact, that was not the purpose of resurrecting Soldner. Lenard had published a fragmentary reprint of von Soldner's paper with extensive comments of his own in order to discredit Einstein. Lenard belonged to the German nationalist organization called Study Group of German Natural Philosophers. He trumpeted von Soldner as the precursor of Einstein to support his organization's claim that whatever was valuable in Einstein's work had previously been discovered by "Aryans."[32] Russell was aware of the real motives behind Lenard's story and he defended Einstein to the reporter. Russell mentioned that general relativity predicted a result different from von Soldner's and from Einstein's estimate of 1911. He pointed to Campbell's work as corroborating the general relativistic result, and the *Times* seized upon this story for its headline. The article emphasized that Campbell's "confirmation of the correctness of Einstein's calculations . . . is made doubly strong by the fact that it is based on observations extending over a number of years, instead of on one eclipse as in the case of the 1919 expeditions."[33]

Campbell ignored Slosson, but the news did not take long to travel west. The managing editor of the *San Francisco Journal* sent Campbell the same *New York Times* article, suggesting that "a more explanatory and extensive treatment of the subject, up to say 2500 or 3000 words, would make a most interesting feature for the magazine pages of our Sunday issue, particularly if this confirmation of Professor Einstein's conclusions antedated that attributed to British astronomers." Campbell replied promptly: "I have as yet published nothing on this subject, and am not ready to do so." He elaborated the misunderstanding: "When Professor Russell was here in August I described my work to him, as far as it had then gone, and it is a matter of surprise to me that he has passed it on, somewhat incorrectly, to the public press without asking or securing my permission. In time I shall hope to say something on the subject."[34]

Campbell felt obliged to relate the incident to Curtis, who had been waiting for years to see some concrete results published:

> There has been an article going the rounds of the newspapers in this country saying that my observations of the past twenty years have confirmed the predicted Einstein eclipse-star displacement. I have published nothing on that subject except a guarded statement made before the Royal Astronomical Society two and one half years ago, and I believe I am under obligations to Professor Henry Norris Russell for the newspaper activity referred to. Unfortunately, his apparent conclusion from our work on the subject here in the past year does not agree with mine. I thought this explanation was due you.[35]

Slosson persisted, despite Campbell's silence, in apparent ignorance of the mistake that Russell had perpetrated: "We are very anxious to secure such material in advance because as soon as the news 'breaks' in any way there will be a revival, in an intensified form, of the popular interest in Einstein of two years ago, and unless we have prepared ready as soon as possible articles on the subject, the field will be occupied by sensational or uninformed articles as before. The theory on which I am working is that of an agriculturist who aims to keep down weeds mostly by the growth of good crops."[36] Campbell shut up even tighter. This attitude annoyed Curtis, still waiting for some statement of the outcome of the labor poured into the eclipse projects. The following from Campbell gave him opportunity to express his feelings:

> Moore and I discovered, from looking over your measures, and especially the computations, for the Einstein effect, that computing is not your strong point. Examination of your computations made me extremely sorry that I had not insisted upon your following my advice, twice offered, that Miss Hobe check your computations. The sheets contained so many errors that we were led to regard your final results as fairly representative of your original measures, because the computational errors were so numerous, as to be themselves subject to the law of accidental errors! These comments apply not only to the Goldendale work, but also to the Burckhalter plates. I am sending you this letter in a perfectly friendly spirit, and I hope you will forgive me for telling you what I really think is to your advantage to know. I would not advise you to trust your computations, present or future, until you check them absolutely independently yourself or have an independent check from another source.[37]

"With regard to the errors you mention," Curtis replied, "[I] can only say, as I did when you wrote me before to this effect, that I am greatly chagrined that they should have occurred, and feel that I am thoroughly cured as to the adequacy of anything else than absolutely independent checking, *always*!" But Curtis had something else to say:

> It is now two years since I left. I have felt a bit hurt at times that you have never written me a line as to the *results* of the improved methods of measuring used by you and Miss Hobe, with more carefully checked computations. I have figured that you were perhaps saving these till after the coming eclipse, but you ought to know me well enough to realize that I would keep any such figures confidential, if you wished it so. I put considerable energy in on that proposition, enough, even if it is now regarded as valueless, to earn the right to know how things came out when no error was made. It impresses me as not quite fair.[38]

Perhaps the most important legacy from the efforts that had brought so much annoyance, disappointment, and misunderstanding was awareness of the many precautions and the high standards that would be required to collect eclipse data suitable for testing Einstein's theory. Unhappy experience with the Chabot lenses, the clock drive, and the telescope mounting taught Campbell and his colleagues how to prepare for the few minutes of measuring that would be theirs in Australia.

The continued negative conclusions that the Lick observers announced at various meetings in the first years after the British results were announced served to keep alive the skepticism toward the theory. Russell's rumor that Campbell had verified the British result, plus his own favorable opinion of the theory, somewhat mitigated the negative results, but the general opinion among astronomers was in favor of retesting the effect at the 1922 eclipse. When Slipher at Lowell Observatory received a query about a theory of gravity due to compression of the ether, he recommended T.J.J. See's "A New Theory of the Aether," which had recently appeared.[39] "The much mooted Einstein theory also has much to say on the subject, and the various theories brought forward against the seemingly satisfactory one of the great Newton, need thorough confirmation—which is not as yet forthcoming."[40]

The persistent controversy around relativity affected the final decision of the Nobel committee charged with proposing the 1921 (retroactively) and the 1922 prizes for physics. The committee faced a groundswell in favor of Einstein—sixteen letters of recommendation. "Imagine for a moment what the general opinion will be fifty years from now," wrote the French physicist M. Brillouin, "if the name Einstein does not appear on the list of Nobel laureates." Most of the nominators wanted Einstein to be awarded the prize for relativity, although some cited his work in quantum theory, in particular his discovery of the photoelectric effect. The committee commissioned two reports, one from Gullstrand again on relativity and one from Carl Wilhelm Oseen, professor of physics at the University of Uppsala, on the photoelectric effect. Gullstrand remained stubbornly critical of relativity, while Oseen was eloquently favorable about Einstein's quantum work. The committee finally capitulated and awarded Einstein the 1921 Nobel Prize; but it chose not to grant the award for relativity, singling out Einstein's work on the photoelectric effect instead.[41]

PREPARATIONS FOR THE AUSTRALIAN ECLIPSE

Frank Dyson decided to send an expedition to Australia to confirm the British 1919 results and silence the critics. He assigned Arthur Hinks the task of investigating the geographical conditions for the eclipse. On 12 March 1920 Hinks informed the Royal Astronomical Society that the eclipse path would begin on the east coast of Africa and cross the Indian Ocean, passing the Maldives and Christmas Island reaching Australia in

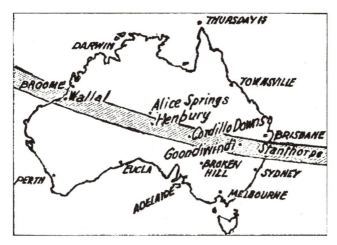

Figure 8.2. Map of the eclipse path across Australia. (Courtesy Mary Lea Shane Archives of the Lick Observatory, University Library, University of California–Santa Cruz.)

the early afternoon. Its landfall was not enticing. "The eclipse track reaches Australia at Ninety-Mile Beach, a hopeless part of the coast, and strikes into the great desert. There are no facilities for landing. . . . The desert is inaccessible, except to camels. There are no railways within hundreds of miles, and motor cars are out of the question."[42] Hinks judged that the first feasible Australian site was Cunnamulla in South Queensland, the terminus of the railway from Brisbane. The eclipse would arrive there around 4 P.M. Hinks also recommended two island sites. Dyson decided on Christmas Island, where a British company had built serviceable facilities to work the phosphate deposits. He sent an expedition under Harold Spencer Jones.

Campbell collected additional information about Ninety-Mile Beach from the New Zealand astronomer Charles Edward Adams, a former Lick fellow. Curtis judged from the material Adams sent that the site "appears to beat them all, if only one could get to it."[43] The eclipse path would enter Australia close to a combined post and telegraph station called Wallal (fig. 8.2). The Sun would be quite high (58°), and the eclipse would last 5 minutes, 18 seconds, the longest at any of the possible sites. The dryness of the desert region insured little chance of rain. Campbell wrote Adams that he disagreed with Hinks that the Ninety-Mile Beach was "hopeless." He asked whether "the Australian Government would be disposed to dispatch a small Government steamer from Perth or Port Darwin at the proper time to carry an eclipse expedition to the vicinity of Wollal [*sic*] and back to the point of starting."[44] He wrote Curtis that though the location "would be expensive in time, money, and comfort . . . the astro-nomical advantages are so evident that I do not want to give it up."[45]

Campbell had already started planning to order equipment. Thomas Cooke and Sons Ltd in York, England, set the standard in lens making, but Campbell was in a rush to get started. "I would correspond with Cookes on the subject," he told Curtis, "except for the great loss of time resulting from the distance." He asked Curtis's advice about optical speci-fications that should be met by the Brashear Company in making two quadruple eclipse lenses, aperture 5 inches, focal length 15 feet, that "will be equal in quality to the best work of the Cookes in England." He wanted to be sure that the star images would be good over a large field, and he wanted time to practice: "If I order the lenses I want them soon, so that Trumpler or someone else here may set the completed apparatus up at least a year before the eclipse date and determine by actual experiment what we may expect in the way of exposure times, driving clock control, probable errors of measured star positions, etc."[46]

Regent Crocker informed Campbell in October 1920 that he would fund an expedition and authorized the purchase of the two Einstein lenses. Curtis took on the responsibility of supervising and testing the lenses from Brashear, whose plant was near him in Pittsburgh. Although he was disappointed not to be able to go himself ("I doubt if I shall try to get the funds to go to Australia," he told Campbell; "we would need $5,000 to do it well, as we would have to build everything"), he gener-ously helped with Campbell's plans. This time the thing would be done right. "During the next ten years we are going to have available, with luck, only about eighteen minutes for further tests on the deflection effect, and it would seem as though the Australian eclipse should be well ob-served, as it is one of the best of the lot."[47]

By the end of 1920 Campbell had matured his plan for observing the eclipse in Australia, to Wallal if at all possible, otherwise to Queensland. Two parallel cameras would be erected on one equatorial mounting, each with quadruplet objective lenses 5 inches in aperture and 15 feet in focal length. The British had employed coelostats, flat mirrors driven by a clock, to track the Sun and to reflect the solar image into the fixed main telescopes, in order to avoid building clockwork mounts for their tele-scopes. Campbell had always used clock-driven mountings at eclipses and saw no reason to change his ways since the coelostats had a habit of heating up and troubling the images.[48] He decided to put Trumpler in charge of the observations, despite Trumpler's junior position on the Lick staff: "The Einstein problem, in my opinion, provides the limit of diffi-culty in the determination of accurate star positions. It would not be fair to assign this work to Dr. Moore or other astronomers whose training has been exclusively in spectroscopy: it would be the plunge of a non-swimmer into deep water without previous trial in shallow water. The work should be done by someone who is thoroughly informed and experienced in the photographic star position problem. Dr. Trumpler's experience conforms to these requirements." Campbell wanted to construct the mounting as soon as possible after the lenses arrived and perform tests on the Pleaides star cluster "to determine what may be expected from the lenses, and to estimate the unforeseen difficulties and get rid of them as far as possible." He estimated the tests would "consume three solid months" and he believed that "Trumpler is the man to make them."[49]

By early February, Curtis had tested the first lens from Brashear. "It is so far ahead of a two piece lens that there is no comparison." Campbell and Trumpler agreed, and the lens was shipped to Lick in March.[50] Several months later Curtis had a chance to the see the British 1919 plates. "They are certainly good plates [he wrote Campbell]. The farthest star shows a little coma on one or two, but the rest are fine images. But still I am not converted. I think that no two-piece lens is adequate for this problem, and feel that your four-piece lenses will settle it."[51]

As the new lenses for the Australian eclipse were being made and tested, a new feature of the program emerged. Campbell's original program was to use two cameras with the Brashear quadruplet lenses of 15-foot focal length each to photograph the immediate surroundings of the Sun, a region approximately 5 degrees by 5 degrees, providing a large scale on the photographic plate (45 seconds of arc in the sky to 1 millimeter on the plate). To get a larger field around the Sun, with more stars to measure, a lens with shorter focal length would be needed; but that would result in a smaller scale on the plate and probably also distortion near the plate edges from lens aberration. The design of the Brashear lenses was the best compromise between large linear scale and field of view. In the spring, Campbell heard from Frank E. Ross of Eastman Kodak about a new short-focus lens that made possible photographing a large region around the Sun without distortion.[52] Ross urged that one of his new lenses be tried on the Einstein problem and on another related problem, named after Leo Courvoisier of the Berlin Observatory. The Courvoisier effect was a yearly refraction cycle Courvoisier reported in 1913 from daylight observations with the meridian circle. No one had succeeded in verifying the effect. In 1920, Courvoisier suggested it could explain the 1919 British eclipse results.[53] Campbell decided that it would be very desirable to use a pair of Ross 60-inch (5-foot) focus lenses in addition to the pair of 15-foot focus quadruplets. He ordered them from Brashear.[54] The Lick program now included two pairs of cameras, one pair of long focus (15 foot) to get a magnified view of the immediate solar vicinity, and one pair of short focus (5 foot) to get a larger field of stars, which would be useful for studying the detailed decrease of stellar displacement for increasing angular distance from the Sun.

In the fall of 1921 Campbell received notification that the Australian navy would provide transport for eclipse observers directly between the Wallal site and Fremantle, cutting out the necessity of traveling to Broome by commercial steamer. Australian generosity allowed Campbell to increase the size of the Lick expedition, which until then was to consist solely of Trumpler and himself. Campbell's wife and Lick spectrosopist Joseph Haines Moore would join the party. Campbell had chosen Trumpler over the more senior Moore for the Einstein program, and now Moore could come along and get other observations. Campbell told Chant of the University of Toronto the welcome news and suggested that he might also avail himself of navy transport to Wallal. "The more the merrier." Campbell also urged Dyson to send a British party there.[55]

The expedition was promising to become an outing, but still a prolonged stay on inhospitable Ninety-Mile Beach did not appear attractive; so Campbell invented a more pleasant way to take the comparison night plates. Three months before the eclipse Trumpler would take the Einstein instruments to Tahiti, "which is nearly the same latitude as Wallal." Trumpler would take comparison plates of the eclipse field at Tahiti and develop the plates there. On the same plates, he would photograph another region of the night sky which could be observed at both Tahiti and Wallal. In reducing the data, Campbell and Trumpler would use this "auxiliary night field" to determine the scale of the plates. This was the technique that Eddington had used in Principe. At the eclipse, Campbell planned to take two plates with each camera. His plan would maximize the length of each exposure to ensure that many stars would be on each plate. It would also reduce the number of plate changes to one, allowing ample time to let the vibrations die down before beginning the second exposure. "We plan for only two exposures with each camera, exposing the first one of the two to the auxiliary night field the night before the eclipse and exposing the second eclipse plate to the auxiliary night region on the night following the eclipse. This is a rather ambitious program and the observers will have to pay strict attention to meeting all requirements."[56]

Eddington had taken his comparison plates at Oxford before the 1919 eclipse. He photographed a check field (Campbell used the term "auxiliary night field") to ensure some control over differences between conditions at Oxford and Principe. He had originally been wary to use the check field to determine scale, because of the different temperatures during day and night conditions. In the end, the temperature at Principe had been uniform due to clouds, so Eddington had confidence in the method for the Principe results.[57] The Berkeley astronomer Charles Donald Shane proposed a further refinement to Campbell: obtain the check field during the eclipse itself, thus providing an independent set of data for calculating effects of scale and plate orientation without using the eclipse-field stars. One would first photograph the check field containing moderately bright stars some 10 degrees away from the eclipse field in right ascension on the comparison plates. At the eclipse, one would take an exposure of the eclipse field, and then, "by rotating the cameras against suitably placed stops, take an exposure of 10 or 20 seconds on the auxiliary field 10° away."[58] Campbell rejected Shane's innovation because of his plan to take long exposures at the eclipse. He wished to get as many stars as possible on each exposure and to change plate holders as infrequently as possible to ensure steadiness of the apparatus.

Curtis described Shane's idea to the British on a visit to London in the summer of 1922. Charles Davidson noted the advantage of taking the check field exposure "at practically the same time" rather than under day and night conditions as with Eddington's method.[59] Frank Dyson wrote to Harold Spencer Jones, who was already on Christmas Island, advising him to adopt the new procedure. Jones was enthusiastic. He planned many short exposures of the eclipse field (10, 20, and 30 seconds) between exposures of the check field. He allowed 15 seconds for movement of the telescope between eclipse and check field, and 12 seconds for each plate change.[60] By contrast, Trumpler's two exposures took 2 minutes each, and he allowed 50 seconds to change the plate holders and steady the telescope from the resulting vibrations.

By February 1922 Campbell was feeling well prepared, writing to Samuel Alfred Mitchell: "The pair of cameras, lenses quadruplet aperture 5 inches focal length 15 feet, and another pair of cameras lenses quadruplet aperture 4 inches focal length 5 feet, including the steel and cast iron tubes, aluminum plate holders 17 x 17 inches, focusing and adjusting arrangements, polar axes, driving arms, everything complete except the driving clocks are new, and for the most part in accordance with my designs and sketches." He had learned from Goldendale. "Trumpler and Moore are going to Australia, but their eclipse experience has been small, and I have not found any way of shifting the responsibility for securing workable designs with minimum chances of something going wrong. If I could only have done the same thing for the eclipse of 1918 I would be very much happier than I am today."[61]

Trumpler left for Tahiti on 30 March 1922, planning to arrive on the 10th of April. Reaching his destination on schedule, he established "a splendid observing site" in the "garden of an American resident, with valued advantages of work rooms and other conveniences immediately on hand."[62] In June he wrote that he had successfully taken all the comparison plates, including ones for the University of Toronto expedition, as well as one long-exposure plate to use for the intermediate plate in the measuring process. Campbell and the rest of the Lick party were to sail from San Francisco in the middle of July, meeting Trumpler in Perth, where he would have previously begun to measure the positions of the stars on the night comparison plates relative to the intermediate plate.[63] Campbell had worked out a plan so that measures of one of the eclipse plates would be made while in Australia, allowing him to make a preliminary announcement of the results. In this way he hoped to avoid the pressure that would attend a delayed announcement. Again, the specter of Goldendale spurred him to make these detailed provisions. "If our plans as to the Einstein tests go through successfully, I am going to be happier next year than I have been this year and last."[64]

The 1922 Eclipse: All Eyes on Lick

Seven expeditions set out to measure the light-bending effect around the eclipsed Sun of 1922. Three represented serious competition to Campbell in the race to test Einstein once again. Spencer Jones's expedition, sent by Greenwich Observatory, was the British sequel to the 1919 expeditions. It offered Greenwich astronomers the chance to duplicate the previous measures, vindicating or disproving their original dramatic results. The British went to Christmas Island as originally planned, since the alternative at Wallal did not materialize until preparations for Christmas Island had gone too far to be altered.[65] Near the British camped a German-Dutch expedition headed by Erwin Freundlich. This would be Freundlich's first shot at the Einstein problem since his thwarted attempt in 1914. John Evershed, sent by the Indian government, intended to set up in the Maldives. When he encountered transportation problems, Campbell invited him to locate at Wallal.[66] Dyson had arranged for Evershed to borrow a 16-inch coelostat from the British Joint Permanent Eclipse Committee (JPEC) to use with his Einstein camera. He hoped to vindicate the method against the criticism of Campbell and others.

The Adelaide Observatory sent a party to Cordillo Downs, in the extreme northeastern corner of South Australia. Curtis loaned the Australians a quadruplet Einstein camera, and Campbell provided the polar axis mounting, driving clock, and driving arm. He also loaned them a 40-foot coronal camera, a duplicate of his own, so that the Adelaide and Lick parties might detect changes in the solar corona by comparing observations made at the two sites.[67] The Sydney Observatory went to Goondiwindi, near the southern border of Queensland, to look into the Einstein problem. Campbell helped this group too. He had Trumpler ship equipment to Sydney after completing the Tahiti phase of the project. "It is our desire [he wrote Trumpler] to cooperate and assist the Australian astronomers in their eclipse plans, in so far as this is practically possible."[68] The Canadian party under Chant located with Campbell at Wallal. Curtis supervised the construction of Chant's lens by the Brashear Company, and Campbell arranged to have Trumpler take the comparison plates for the Toronto expedition and bring them on with him to Australia. "I suppose you have your hands full advising some of the weak sisters who are going to observe the eclipse," Curtis remarked wryly to Campbell during the long months of consultation and preparation. "I have sent small volumes to one or two of them myself."[69]

By preparing his expedition far in advance, Campbell was able to perform experiments on Mount Hamilton to test important features of his eclipse program. For instance, to determine the maximum exposure possible for photographing stars illuminated by solar corona without fogging the plates, Campbell used the Crossley reflector to photograph star fields near the full Moon with varying exposure times. He ascertained that the two-minute exposure he had planned for his program for the 15-foot camera lenses would not fog the plates. Campbell also experimented with developers and lengths of time in the developing process to ensure optimum results in the darkroom stage of the work, which would be carried out in Australia.[70] None of his competitors had prepared as thoroughly as he had. None felt, as he did, anxiety to brighten a reputation that in his opinion had been tarnished by his "mistake of having reported, though guardedly, on Curtis's results at the meeting of the Royal Astronomical Society [in 1919]." Curtis sensed the drama as sailing time approached and admitted that it was "one of the big disappointments of my life that I am not 'in' on this particular eclipse."[71]

Nature rewarded her inquisitor's careful preparation by storming over his chief competition and smiling sunnily on him. A storm greeted the British at their landing site, preventing their steamer from unloading equipment for ten days. Bad weather plagued all the expeditions on Christmas Island. "The sky . . . is *never* entirely free from cloud, either day or night [Spencer Jones wrote to Dyson] so that it is of no use to try and disguise the fact that our chances on the day are a pure gamble." The British could not do their preliminary photometric work, and at eclipse time they lost the gamble—both the British and German-Dutch parties were clouded out. "The first few seconds of the eclipse the sky was still clear at Christmas Island, so that the second contact could be observed, but 6 or 7 seconds after the beginning of totality the sky became overcast and no successful photographs were obtained."[72] Campbell had perfect weather. (Fig. 8.3.)

Clear skies were not enough for success for Evershed, who encountered instrumental difficulties. They had started back in India when tests on

Figure 8.3. Campbell's Einstein cameras set up at Wallal to observe the 1922 Australian eclipse. (Courtesy Mary Lea Shane Archives of the Lick Observatory, University Library, University of California–Santa Cruz.)

Dyson's coelostat revealed that the driving screw was worn, causing a periodical error in the motion of stars observed with the instrument. A new screw and other parts were completed on the day that the instruments had to be shipped. Evershed started setting up at Wallal less than twenty days before the eclipse, leaving no time for rehearsals. Tests made about a week before the eclipse with the coelostat mirror showed "marked astigmatism" when it was set at the angle it would be at eclipse time. The only remedy was to drastically cut down the aperture (from 12 inches to 6 or 8 inches).[73] The driving screw began to act up again so that the star images would remain stationary for about 20 seconds and then begin to wander. Evershed could not cure the problem so he cut the exposure time to less than 20 seconds. He hoped that "might with luck give good images." During the eclipse he took five exposures. When developed, "all were found to have failed for one reason or another." His written report was blistering: "Failure under the ideal conditions of a perfectly clear sky, with excellent definition, and a long duration of totality, is deplorable, especially when public funds have been risked." The technology existed to make excellent coelostats for use in the Einstein problem, Evershed insisted, if only it would be used. He remarked angrily, "If British manufacturers could be induced to abandon the old methods and apply ball bearings to all moving parts in astronomical instruments, as should have been done thirty years ago, an enormous gain would result in the uniformity of movement so essential in this research." He reported "with envy" that the American installation was "fitted with ball or roller bearings, and with a simple and most effective method of driving without the use of any gearing whatever."[74]

Of the seven expeditions that were to try the Einstein problem, four obtained useful observations: the Lick and Canadian observers at Wallal, the Sydney astronomers at Goodiwindi, and the Adelaide party. Only three obtained useful results. By February 1923 the Sydney astronomer in charge of the Einstein test, William Ernest Cooke, had to announce that he could conclude nothing from the eight plates he had taken under excellent conditions. His equipment, like Curtis's in 1918, was not adequate. Cooke expressed the opinion that "the first satisfactory results would be those of Dr. Campbell, of the Lick Observatory." He predicted that the public would have to remain in a state of suspense for probably several years "because the most minute calculations and measurements would require to be carried out before any announcement on the subject could be accepted as being correct."[75]

Astronomers had to wait almost a year before they heard anything from any of the other expeditions. Knowing of the immense interest, Campbell had hoped to get out a preliminary statement of results before leaving Australia. He had planned for Trumpler to go to Perth Observatory after the Tahiti phase and spend about five weeks measuring the brighter stars on the Tahiti plates. Campbell and Moore were to reach Perth in August, leaving a week for pre-eclipse measurements and calculations before the voyage to Wallal. Trumpler had left Tahiti on schedule, but shipping delays forced him to start setting up his apparatus at Perth weeks later than intended. Then the travel plans to Wallal were altered at the last minute. Other expeditions decided to take advantage of the services of the Australian navy, so the rendezvous point was changed from Fremantle to Broome. This meant that Campbell's party would now have to take the commercial steamer from Fremantle to Broome, and that the Lick measuring apparatus had to be sent off earlier than expected. Trumpler had to take down the equipment for shipping before he could begin his measurements. The Tahiti stars remained untouched before the eclipse.[76]

The new travel arrangements meant a delay of more than a week at Broome on the return journey to wait for the commercial steamer to Fremantle. Unaware that Trumpler had not yet started on the Tahiti measures, Campbell wrote him from Sydney early in August, suggesting that they measure the Einstein eclipse plates at Broome while waiting for the steamer.

> I should prefer that you measure on the Tahiti plates at Perth, or extract from your complete measures of the Tahiti plates, the data for from twelve to twenty selected stars, and set up the equations of condition from them, so that the corresponding stars on the eclipse plates may be measured in five or six days at Broome, and these data be entered in the equations for solution at Broome or on our southbound steamer. I advise strongly against your remaining behind at Broome. If further measures seem to be desirable before making announcement, Perth is, in my opinion, the place for you to make them. In other words, the Broome program of measurement should not include too many stars.[77]

Unfortunately, Trumpler had already decided to return home via the northern route to Switzerland, where after a visit with family he would return to the United States.[78] The only time for measurement would be in Broome.

Trumpler and Campbell devoted eighteen hours a day to measuring one of the Einstein plates and the corresponding Tahiti comparison plate; but there was only time for measures in one direction, and they could not carry out the usual procedure of repeating the process in the reverse direction. Nonetheless, they obtained some numerical results from these rushed and incomplete measures, indicating a definite light deflection, larger than the Newtonian value, but smaller than the Einstein deflection. Trumpler fully reduced the measurements and left a hurried note for Campbell with the final figures.[79] He used the check region to determine the second-order terms, obtaining the first-order terms from nineteen stars in the eclipse field located 2 to 3 degrees from the center of the plate. Trumpler excluded one "faint discordant star" and found the deflection at the Sun's limb by a least squares solution using the radial displacements of all stars, giving the faint ones half weight. He reduced Campbell's and his own measures separately and secured the limb deflections shown in table 8.2. "Check region reduced exactly same way shows no appreciable deflection," Trumpler recorded; "calculation gives minus thirteen [−0.″13]." He concluded: "Some light deflection beyond doubt, but amount smaller than predicted."[80] Campbell, however, chose not to announce these preliminary results. For months afterwards he felt intense frustration that his detailed plans for extracting an Einstein result he could announce from Australia had "failed miserably."[81]

As he had feared, Campbell found himself under immense pressure from colleagues and the press. A telegram from Science Service awaited him at Lick asking for "first announcement of Einstein Eclipse results."

TABLE 8.2.
Trumpler's Preliminary Unpublished
Results (Wallal, 1922)

Trumpler	1.″38 (79 stars)
Campbell	1.″17 (72 stars)
Mean	1.″28
Probable error	0.″18

Campbell cabled that the Einstein eclipse negatives were on the steamer, due to reach Mount Hamilton about 10 December. "Probably 2 or 3 months measurement with powerful microscope necessary," he added, "and considerable computation before results available."[82] Campbell sent Trumpler an urgent telegram: "Intense public and academic pressure for Einstein results. Please make Swiss visit minimum possible. Write estimate when return." Trumpler had planned a fairly long stay visiting family and friends whom he had not seen for years. He replied that after careful consideration, the earliest he could leave was on the 27th of January, reaching home by the 6th or 7th of February. "I am sorry that we did not measure the four plates obtained with fifteen-foot cameras before leaving Australia, as it had originally been planned," he added: "My visit in Switzerland as well as the trip would have been more pleasant. You may however be assured that I am doing my best to follow your wishes with respect to my return trip."[83]

The shipment of equipment and plates reached Mount Hamilton on 16 December. A week later one of the Einstein plates sat on the measuring microscope ready for adjustment. "Of course," Campbell wrote Trumpler, "I wish you were here to proceed with the adjustments and measures. We apparently need your more extended notes before we can undertake the adjustments without considerable loss of time."[84]

The delay caused many people to suspend important plans. Samuel Alfred Mitchell, writing a book on eclipses, intended to include a large section on the Einstein theory. "When I saw Dyson last," he told Campbell, "he said that he would not be in the least surprised if the 1922 photographs did not confirm the Einstein effect. He thought that possibly they in England had stressed Einstein a little too much." Mitchell wanted to include an account of the Lick trip and results in his book. As for Dyson, he had an entire eclipse expedition weighing in the balance. The next eclipse, which would be visible from Mexico, Baja California, and southern California, was to take place in eight months, which did not leave much time to make arrangements should Campbell's results not agree with relativity. "The results of the 1919 eclipse were in accordance with Einstein, but it may be that your results of 1922 will not confirm this. If so, will you be kind enough to let me know as early as you can. My reason is, of course, that if there is disagreement we must regard the point as unsettled and every endeavour must be made to test the matter again next September. In that case it will be desirable for us to send an expedition, but it will hardly be necessary if you confirm the 1919 results in which case the question may be regarded as settled." Campbell told Dyson that it would be at least six weeks before he could cable any results. "We have some indications of the outcome," he remarked, "but any statement whatsoever at the present time would be scientifically unjustified."[85]

The press also clamored for results. The editor of the *Michigan Chimes* hunted Campbell for a brief article he had promised about the Australian expedition. In response to pressure from a San Francisco newsreel firm, Campbell declared himself averse to further publicity. "Astronomers and other scientific men do not want to appear before the public in relation to 'what they are going to do,' and before they do it. I am sure you will see this principle which throughout the high-class scientific world, is followed as closely as possible."[86] Henry Norris Russell asked Campbell to give a public evening lecture on his results during the April meeting of the American Philosophical Society in Philadelphia. "The reporters are at me already for comments on your observations," he wrote, but assured his publicity-shy colleague that "I tell them just one thing—that you will not announce any results until you are quite sure of them and that when you do they will be entirely decisive."[87] Campbell did not answer Russell's letter.

T.J.J. See wrote Campbell with his own helpful suggestion to look for refraction effects on his eclipse plates:

> As you are now measuring the plates taken during the recent eclipse, it occurs to me to ask you if it would not be well to test the amount of the refraction of the stars near the Sun in various directions from the Sun's centre, as seen upon the plates? The magnetic field of the Sun will not be the same in all directions, but stronger near the poles, according to the laws I have given; and as the cause of the refraction will be a matter of importance, when your work is done, the checking up of any difference of refraction in the different directions, if sensible, would be valuable data, and add to the conclusiveness of your eclipse results.

See had just published a new theory of the ether. Campbell knew him well enough to know he was looking for ways to connect his work to the high-profile Lick project. "We shall hold in mind your suggestion," he wrote diplomatically, "and, if the measures appear to have sufficiently high accuracy, endeavor to draw some conclusion in that connection." Referring to See's theory of the ether, Campbell added: "It will be interesting, within a few years, to see whether the further progress of physical science has confirmed your theory."[88]

Added to the fuss over the eclipse expedition came an invitation to accept the presidency of the University of California. Campbell accepted on condition he retain the directorship of Lick. The university agreed, and Campbell planned to assume his new duties on 1 July. "In the meantime the indirect consequences are bearing heavily upon my time," he told Dyson, "which I am trying to save for devotion to the astronomical problems which surround me."[89]

Trumpler reached Mount Hamilton in early February.[90] He gave an extensive interview to a San Jose newspaper, relating details of the Tahiti and Australian expeditions. He explained why interest in the pending results abounded, but predicted that it would be "at least two months before we can give out anything on it."[91] Then he and Campbell plunged into the work of measuring the plates and reducing the measures. (Fig. 8.4.)

While Campbell measured, two prestigious organizations vied to get him to present his results at their annual meetings in the coming spring. Russell pursued his earlier attempt on behalf of the president of the American Philosophical Society to get Campbell for a public evening lecture in Philadelphia on the eclipse results. "I share, of course, in the general eagerness of all the world to know how the plates come out," he wrote, "but I am glad to see that as I expected, no hasty or provisional announcements are being made from the Lick Observatory." Charles Greeley Abbot, home secretary of the National Academy of Sciences, wanted an evening lecture during the academy meeting in Washington. He hoped that Campbell would "decide for us rather than for the Philadelphia people. . . . It makes no difference whether your results are positive or negative as regards their application to the relativity problem. We are not advocates but seekers of truth in such matters. Even if you should give the lecture at Philadelphia, which I very much hope will not be the case, we shall at least depend upon hearing the substance of it at the regular program here." Campbell declined Russell's invitation and agreed to lecture at the academy.[92]

Abbot relayed the news to Slosson at Science Service, who immediately asked Campbell for first news of the results. He also asked Campbell to write a small book on relativity. "Since you have in your hands the latest and most authoritative evidence on the subject," he wrote, "you would be best qualified to write such a book." Campbell refused, reacting testily to the implication that he would sit on results and wait to announce them in Washington: "In some way the impression seems to have reached Washington that our Einstein results are to be withheld until and made public in my lecture of Monday evening, April 23rd. This has not been my intention. Dr. Trumpler and I are hoping that the results will be ready

MEASURING THE STARS

San Francisco Examiner--Wed. Jan. 30, 1924

Prof. Trumpler of Lick Observatory at work with instrument which corroborates the Einstein theory of relativity.

Figure 8.4. Robert Trumpler measuring Einstein plates at the Lick Observatory, as depicted in the *San Francisco Chronicle*, 30 January 1924. (Courtesy Mary Lea Shane Archives of the Lick Observatory, University Library, University of California–Santa Cruz.)

for announcement before that date, and that they can be made from the Lick Observatory. Many scientific interests are demanding the knowledge on the first day that it is available. Several least squares solutions are now in process, and just how the results are coming out I cannot today say."[93]

Meanwhile, Chant had been busy on his Wallal plates. On 6 April the Associated Press quoted "Prof. C. A. Chant of the University of Toronto as commenting upon as 'distinctly favorable' to the Einstein theory observations made at Wallal." On 11 April Herbert Hall Turner at Oxford, having written Campbell a note on other business, penned a postscript: "We have had no hint of results of your measures up to now, tho' the Canadians have voted for Einstein." Campbell was set to leave for the East Coast on the same day, but he and Trumpler had not completed

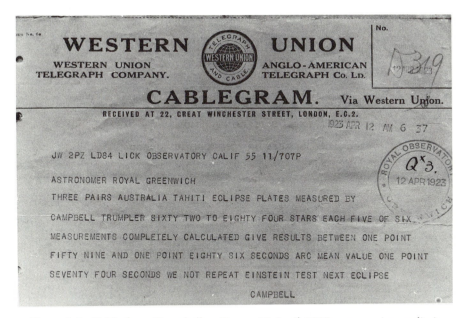

Figure 8.5. Cable from Campbell to Dyson 12 April 1923, announcing prelimi-nary results in favor of general relativity. (Courtesy Royal Greenwich Observatory archives, University of Cambridge.)

measuring all four sets of eclipse and comparison plates. Campbell de-cided they had enough measures to go public. News releases went from Lick with the preliminary 15-foot camera results, and the next day Camp-bell cabled Einstein in Berlin, confirming his prediction.[94]

Campbell sent a similar message to Dyson, adding "we not repeat Ein-stein test next eclipse" (fig. 8.5). The news flashed across the United States in a press release from Lick. The *New York Times* jumped on the story: "The agreement with Einstein's prediction from the theory of relativity . . . is as close as the most ardent proponent of that theory could hope for. In fact the agreement of our observed value with the predicted value is so satisfactory that the Lick Observatory does not plan to repeat the Einstein test at the total solar eclipse due to occur in extreme Southwest-ern California and in Mexico on September 10, 1923."[95]

At the Academy of Sciences meeting in Washington, Campbell gave suitable details, as discussed below.[96] There had been four plates for the 15-foot camera observations, two for each camera. Trumpler had com-pletely measured three plates, and Campbell two. Though the star images on all the plates were round, symmetrical, and well defined for the brighter stars, they were fuzzy and diffuse for the fainter stars, especially near the edge of the plate. In order to be able to use as many stars as possible, Campbell and Trumpler devised a system to account for the variability in image quality. Each assigned weights independently for their own measures using a method similar to one used at Allegheny Observatory for parallax photographs. Each measured eclipse and comparison plates relative to the intermediate plate four times: direct and reverse for differences in right ascension, and direct and reverse for differences in declination. They used as many stars as possible in each eclipse field. For comparison, they measured thirty-seven stars of the check field "well distributed over the plates and of suitable brightness." As a precaution against changes in the setup during measurement, two to four of the eclipse stars nearest to the Sun, "which are of the greatest importance to the present problem," were measured at the beginning, in the middle, and at the end of each series.

Campbell and Trumpler determined the mean eclipse-comparison differences for the two directions (right ascension and declination) by averaging the direct and reverse measures, corrected for proper motion and parallax of the stars. They used the check field of stars (90° from the eclipse field) to control for differential refraction, aberration, and inclination of the plate to the optical axis. They included terms of the first and second order using equations of the form

$$\Delta x \text{ (eclipse—comparison in R.A.)} = a + bx + cy + dx^2 + exy + fy^2$$

$$\Delta y \text{ (eclipse—comparison in Dec.)} = g + hx + iy + jx^2 + kxy + ly^2.$$

(3)

The coefficients came from least squares solutions using the check field stars with certain simplifying assumptions that reduced the number of constants to be determined for each plate. They applied the constants to correct the measured displacements of the eclipse field stars.

They then redetermined the linear plate constants (zero point, scale, and orientation) using from twenty-eight to thirty-eight eclipse stars (depending upon the plate). These stars were all more distant from the center than 2 degrees, where any deflection according to Einstein's theory would be small. This innovation assured that the plate constants would be reasonably independent of any light-bending law that existed in the neighborhood of the eclipsed Sun. Only the scale value thus determined would be slightly in error if any light deflection existed, and would be revised later when establishing the actual deflection law. This technique was possible only because Campbell had chosen to take long exposures to get a large number of stars out to fairly large angular distances from the Sun. He and Trumpler determined the linear plate constants by least squares solution of the standard formula (their equation 2; see p. 135) using the so-called reference stars 2 degrees or more away from the center.

TABLE 8.3.
Light Deflection at Sun's Limb (preliminary Lick published results)

Plate	Trumpler	Campbell	Mean
CD22-CD15	1″84 ± 0″28	1″70 ± 0″13	1″77
CD23-CD17	1″59 ± 0″22		
AB18-AB12	1″86 ± 0″20	1″71 ± 0″22	1″78
Mean	1″76 ± 0″13	1″71 ± 0″18	
Mean of five sets of measures			1″74
Deflection predicted by Einstein			1″75

Campbell and Trumpler applied the corrections to the measured displacements, yielding the so-called residuals in x and y. The radial component of these residuals represented the displacement of the stars owing to any light deflection. Assuming a linear law of radial displacement of the form

$$\Delta r = ad + b(1/d), \qquad\qquad (4)$$

where d is the angular distance of the star from the Sun's center, b is the deflection at the Sun's limb, and a is the scale value correction, Campbell and Trumpler found a and b for each set of measures by a least-squares solution using all the eclipse stars. The deflections at the Sun's limb (b) came out as shown in table 8.3. The mean probable error, determined for a star of weight 1 (good), was ± 0″18 for Trumpler and ± 0″20 for Campbell. These values represented probable errors (p.e.) of the difference of two measured star images. To get p.e. of one star image, they assumed the errors of the intermediate plate to be eliminated, and divided the mean errors by $\sqrt{2}$, getting ± 0″125 for Trumpler and ± 0″140 for Campbell. The report concluded that the accuracy of the data "compares well with other photographic measures; the Paris Zone of the Astrographic Catalogue for instance has a probable error of ± 0″13 for the mean of two images."[97] The final report for the 15-foot camera observations, completed in May and issued in July as a *Lick Bulletin*, gave the results for the four plates (see table 8.4).[98] As can be seen by comparing these data with the table Campbell presented in Washington (table 8.3), the mean result for the more complete measures was now 1″72 instead of 1″74 for the limb deflection, still in excellent agreement with Einstein's prediction.

TABLE 8.4.
Light Deflection at the Sun's Limb (final Lick published results)

Plates	Campbell	No. of Stars	Trumpler	No. of Stars	Plate Mean
CD22–CD15	*1″72 ± 0″32	62	*1″88 ± 0″27	69	1″80
CD23–CD17	1″35 ± 0″22	77	*1″62 ± 0″22	81	1.48
AD18–AB12	*1″78 ± 0″22	80	*1″91 ± 0″19	84	1.85
AB17–AB10/9			1″76 ± 0″22	85	1.76 (weight 0.9)
Mean for each observer	1″60 ± 0″14		1″78 ± 0″11		
Mean from four plates					1″72 ± 0″11
Einstein's predicted value					1″745

* These five values, very slightly modified, were the only ones available when the preliminary announcement of our results for the Einstein eclipse problem was made through the press associations and otherwise, on April 11, 1923. At that date we had also determined from measures of the check-region star images, that any corrections suggested or demanded by existing small plate errors could not operate to diminish any of the five values of the Einstein coefficient.

Chapter Nine

EMERGENCE OF THE CRITICS

REACTIONS TO THE LICK RESULTS

The Lick announcement marked a turning point in attitudes toward relativity. Opinions held tentatively after the British eclipse announcement became entrenched after Campbell's corroboration. When Dyson received Campbell's cable, he replied: "I don't think there is 'any possible probable shadow of doubt' about the correctness of Einstein's prediction of the deflection of light, whatever difficulties may be found with the rest of his theory. It is hardly likely that anyone will be coming from this side for the eclipse in California."[1] Harold Spencer Jones, disappointed at the 1922 eclipse, asserted that the nature of the debate over the correctness of Einstein's theory had changed. "The opponents of the theory must now take up the position that, although it has successfully accounted for the motion of the perihelion of Mercury, and although it has *predicted* the correct amount of the deflection of rays of light in passing through the field of gravitation of the Sun, it is not correct. It is no longer open to them to say that the prediction of the amount of the deflection has not been fully confirmed."[2] Turner commented on the implications of the event in his regular column in the *Observatory*:

> It is of no disparagement to the Canadian observers, who had already announced a similar result, to say that we were all hanging on the utterance which should come from the Lick Observatory; and now that it has come we feel more than ever how much was at stake. The English observers were resolute to go on even if the American verdict had been against them; there are no Courts of Appeal or Houses of Lords in Astronomy for getting verdicts reversed—the only recourse is to hammer on as before with renewed vigilance for possible flaws in the evidence but for this they were quite prepared. It is, however, a considerable relief to find that the necessity for this further campaign is now removed, and if any English observers are able to visit America this year it will no doubt be merely as a return to their old love, the Corona.[3]

In the United States reactions were mixed. Russell was delighted. "Just what I expected! *Now* where are the scoffers? I don't believe it will ever be necessary to do the thing again."[4] Charles St. John agreed. "Of course there is much talk of the eclipse," he wrote Hale. "Campbell's results look very definite as to the reality of the bending to the proper amount." St. John was just beginning to find evidence for a relativity shift in the solar spectrum though it would be some months before he presented any details publicly.[5] Mitchell changed from "betting on your [Campbell] confirming the half-deflection and not the full Einstein amount"[6] to asserting in his now up-to-date book on solar eclipses: "The consequences of relativity have been so thoroughly substantiated by observations that he who has a scientific reputation at stake must indeed be very rash to state that physicists must all be mistaken and that the whole theory of relativity is 'tommyrot.' " Mitchell's conversion was primarily due to the "magnificent results obtained by Campbell and Trumpler. . . . The amount of the deflection agrees remarkably closely with that predicted by the Einstein theory. The careful methods of the Lick astronomers, both in the taking of the plates and the measurement of them, seem to indicate that all possible sources of error have either been eliminated or been allowed for. There remains the observed deflection of 1.″72 at the Sun's limb to be explained."[7] Yet he remained cautious. He noted the still nonverified shift of the solar spectrum lines and a number of other developments that critics of relativity were reporting about that time.[8] Mitchell concluded that relativity was too important to leave to just two eclipses and he expressed regret that Lick was not planning to repeat the test at the 1923 eclipse. "Fortunately, other astronomers will continue the attack, for no doubt it will be many years in the future before astronomers and physicists are agreed on the exact status of the theory of relativity."[9]

Some of Campbell's colleagues had stronger reservations. Perrine had congratulated Campbell on his success in Australia months before any results were available. "I am very curious to know how your measures of the relativity plates will turn out. I will have more confidence in them than anything which has gone before. I have always been, and am yet, skeptical of any such effect, although I went into it at the Brazilian (rainy) eclipse of 1912, at the request of Freundlich, with two Vulcan cameras." Yet when Campbell cabled his preliminary results, Perrine found that "notwithstanding the great weight of the observations and the excellent agreement with theory" he was still "somewhat skeptical. . . . Not of the fact of the bending but of the explanation." "The whole relativity business has seemed to me unreal and so purely philosophical that to accept it is to upset our previously carefully constructed and very material systems. Indeed it seems to me like a 'near' return to the era of 'inductive reasoning' and an undermining of the material foundations of our science particularly. If it is true that our foundations are faulty the sooner we find it out the better. I am open minded but conservative in this matter."[10]

Though Curtis had told Campbell that his quadruplet lenses would "settle it," he found that he could not accept Campbell's and Chant's confirmation of the British measurements. "There may be a deflection, but I do not feel that I shall be ready to swallow the Einstein theory for a long time to come, if ever. I'm a heretic."[11] After the *Lick Bulletin* came out in July, another former Lick astronomer, George F. Paddock, wrote to Curtis expressing doubts whether Campbell's and Trumpler's published results confirmed Einstein's light-bending law. The Lick astronomers had plotted all their data on a graph of deflection versus angular distance from the Sun and superimposed the theoretical curve from Einstein's theory. The scatter in the observations was rather large, and Paddock asked: "Can you really think that in Fig. 2 the observed values substantiate the theoretical (dotted) curve rising to 1″.7?" Curtis replied: "No, I can't say that I regard Fig. 2 in Campbell's paper as giving any too strong a support to the theory of a deflection of 1″.7 at the Sun's limb. It does seem to me, however, that his results pretty well establish the existence of a deflection, due to some cause or other, and larger than the predicted deflection of 0″.87 predicted on the Newtonian Theory. Just what causes it, no one can tell." He told Paddock that Charles Lane Poor, "another one of the irreconcileables [*sic*]," thought that it was "some sort of refraction effect" and that he himself had tried to test this suggestion out at the recent Mexican eclipse using equipment that Poor had supplied.[12] The basis of Curtis's heresy was the belief that Einstein's fictions had trespassed the limits allowed in science:

> I have never been able to accept Einstein's theory. This in spite of the fact that many eminent mathematicians regard it as the greatest advance since Newton's time. I regard it as a beautifully worked out alternative "reference frame," apparently adequate, but by no means essential and by no means necessarily the correct system of reference. I regard it as like non-Euclidean geometry; we can form not one but many systems of geometry based on curved or hyperbolic space; we can explain every geometrical theorem by such geometries as well as we can by the Euclidean; they are alternatives simply. We do not force ourselves to accept non-Euclidean geometry simply because it seems to "fit." Perhaps I am wrong, but it does not seem to me at present that I shall ever be willing to accept Einstein's theory, beautiful but bizarre,—clever but not a true representation of the physical universe.

As for Campbell's results, Curtis had this to say: "I am firmly of the opinion . . . that we shall eventually be able to explain this and the motion of the perihelion of Mercury by ordinary Newtonian mechanics. I find it impossible to believe that gravitation is not a force, but a property of space, that space and time are 'curved,' that the universe which appears to us as three-dimensional is really a 'four-dimensional manifold in space of six-fold curvature,' and all the rest of it."[13]

The Lick's vindication of Einstein raised the stakes for the critics. The British verification of the light-bending prediction just after the end of the war had attracted unprecedented public attention, launching Einstein and his theory to world fame. Yet the scientific community had demanded that the results be repeated before making a final decision. The Lick confirmation three and a half years later influenced scientific opinion as much as the earlier British results had affected public opinion. Now relativity critics were under the gun to find ways to extricate themselves from the empirical juggernaut corroborating Einstein's predictions. As the scientific debate heated up, nonscientific issues began to enter the discussion. Several arose from the war and echoed controversies about Einstein and his theory that were raging in Europe. Nationalist concerns at home involving American education and research also fueled some of the rhetoric directed against relativity theory. Whichever side of the fence scientists were on, there was a lot at stake. The ubiquity of news reporters polarized their positions. There was much posturing. In many ways, the appearance of Einstein's theory crystallized the national character of the scientific enterprise. It focused disciplinary leaders' attention upon the strengths and weaknesses of their professional community.

T.J.J. See versus the Lick Observatory

On the same day that the Lick released Campbell's and Trumpler's preliminary results confirming Einstein, T.J.J. See issued a detailed statement condemning the Lick action. In 1922 See had published his "New Theory of the Aether" in *Astronomische Nachrichten* and in monographs printed in France, the United States, and England.[14] Believing that he might be able to utilize the eclipse results to bring attention to his own work, he attacked Campbell's April 1923 press release favoring Einstein's theory and accused Einstein of trickery and plagiarism. The press loved it. A Philadelphia news headline blared: "Government Scientist Exposes Einstein Trick" (fig. 9.1a). Campbell's announcement was only briefly mentioned. Most of the column discussed the fact that it had been "contested vigorously" by Captain T.J.J. See, government astronomer at Mare Island Navy Yard. The *San Francisco Chronicle* ran the headline, "U. S. Scientist Attacks Test on Einstein Theory as 'Piece of Humbuggery' " (fig. 9.1b).[15] Photos of the two astronomers appeared side by side, with captions taken from their respective press releases. Most of the article was devoted to See's views.

See's assault was inspired in part by the anti-Einstein propaganda of Philipp Lenard and Ernst Gehrcke, the only two prominent physicists who had joined the Study Group of German Natural Philosophers that had

GOVERNMENT SCIENTIST EXPOSES EINSTEIN TRICK

Declares German Astronomer Was Detected in Plagiarizing Von Soldner in 1911.

Philadelphia 1923

NEWTON THEORY HELD TRUE

Apr 12

Says Bending of Stars' Rays Do Not Bear Out Claims Which He Terms Crazy Vagaries.

Vallejo, Calif., April 12 [By Associated Press].—A statement saying that he "vigorously contested" the announcement made last night by Dr. W. W. Campbell, director of Lick Observatory, that development of photographs taken during the last total eclipse of the sun confirmed the Einstein theory of relativity, was made public here today by Captain T. J. J. See, Government astronomer at Mare Island Navy Yard.

After reciting the Campbell statement, Professor See said:

"The celebrated English physicist, Henry Cavendish (1731-1810) calculated the effect of Newton's theory that the corpuscles of light are bent toward the sun in passing near it; and in 1801 Dr. J. Von Soldner, a German physicist of eminence in his day, actually derived the formula recently used by Einstein. This was 122 years ago. Einstein never once mentions Soldner in his writings. This is bad enough, but the worst is yet to come.

"It has been shown by Professor Dr. E. Gehrcke, director of the Imperial Physical and Technical Institute of Berlin (a position first filled by Helmholtz) and by Professor P. Leonard, of Heidelberg, winner of the Nobel prize in physics, that Soldner omitted a certain factor in his formula of 1801, which error Einstein also copied when he appropriated the Einstein Soldner formula in the Einstein paper of 1911. In a subsequent paper to the Berlin Academy of Sciences, 1915, Einstein camouflaged this fraud as best he could, yet could not prevent its discovery and exposure by Professor Leonard, of Heidelberg; Gehrcke, of Berlin, and Westin, of Stockholm. Professor Westin charges Einstein with downright plagiarism, saying: 'From these facts the conclusion seems inevitable that Einstein cannot be regarded as a scientist of real note. He is not an honest investigator.' Thus Westin protested to the directorate of the Nobel Foundation against the reward of Einstein."

"In considering the Newton-Von Soldner refraction of starlight from the eclipse in Australia, the value of the eclipse observations is recognized, but the refraction of the starlight redounds to the credit of Newton-Soldner, not of Einstein.

"It only remains to be pointed out that the Einstein theory of relativity is not confirmed and cannot be confirmed. A fundamental postulate of Einsteinism is that the ether does not exist and gravity is not a force but a property of space. These crazy vagaries scarcely require mention, beyond the remark that such discussion is a disgrace to our age. Is it any wonder that the Paris Academy of Sciences, October 24, 1921, came out with conspicuous proclamations by Professors Picard and Painleve against Einsteinism and in favor of Newtonian mechanics?

"Everybody from Huyghens, Newton, Herschel, Maxwell, Helmholtz, Tisserand, Lord Kelvin, Poincare, etc., to our own Michelson, knows very well that ether exists and acts with forces equivalent to the breaking strength of millions of cables of the strongest steel, for holding planets in their orbits."

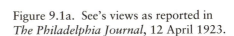

Figure 9.1a. See's views as reported in *The Philadelphia Journal*, 12 April 1923.

Figure 9.1b. The battle between Campbell and See as fought out in the *San Francisco Chronicle*, 12 April 1923. (Courtesy Mary Lea Shane Archives of the Lick Observatory, University Library, University of California–Santa Cruz.)

almost driven Einstein out of Germany in 1920.[16] They capitalized upon a numerical coincidence: Einstein's general relativistic prediction of a 1.″74 light bending at the limb of the Sun was exactly twice that of his 1911 prediction. They claimed that in his 1911 paper Einstein had copied Johann von Soldner's prediction of 1801, which had coincidentally contained a mathematical error of a factor of 2, and then changed the numbers in his 1915 paper to cover up the mistake. See repeated at face value all that the Germans said, implying that because of their high reputations as physicists, people should believe them.

> It has been shown by Professor Dr. E. Gehrcke, director of the Imperial Physical and Technical Institute of Berlin (a position first filled by Hemholtz) and by Professor P. Leonard [sic] of Heidelberg, winner of the Nobel prize in physics, that Soldner omitted a certain factor of his formula of 1801, which error Einstein also copied when he appropriated the Einstein-Soldner formula in the Einstein paper of 1911. In a subsequent paper to the Berlin Academy of Sciences, 1915, Einstein camouflaged this fraud as best he could, yet could not prevent its discovery and exposure by Professor Leonard of Heidelberg, Gehrcke of Berlin, and Westin of Stockholm.[17]

In blindly following Gehrcke and Lenard, See was either a fool or a cheat. The same may be said for his allegation that in October 1921, the Paris Academy of Sciences "came out with conspicuous proclamations by professors [Emil] Picard and [Paul] Painlevé against Einsteinism, and in favor of Newtonian mechanics."[18] Again, there was a larger context. Picard had been instrumental in refusing to admit astronomers from the former central powers or neutral countries into the newly created International Astronomical Union. Intense anti-German feelings created by the experience of 1914–1918 colored his judgments on scientific matters.[19] It was years before the Paris Academy elected Einstein as a foreign associate. Even in England, astronomers who were leery of honoring a German, especially one with such a controversial theory, blocked a move to award the 1920 Gold Medal of the Royal Astronomical Society to Einstein.[20]

While Campbell was still in the East, See stormed again at news from Europe that Einstein was about to publish a "new discovery" concerning "the connection between the earth's power of attraction and terrestrial magnetism."[21] Einstein had recently delivered a preliminary discussion of electromagnetism and gravitation to the Berlin academy and the press made a big fuss about it.[22] See had put forward his ether theory as the final connection between the Earth's gravitation and magnetism. In a letter to the *London Times* later reprinted by U.S. newspapers, See cried out that he was being robbed. The press approached Lick Observatory for comment on See's claims. In Campbell's absence, Aitken told reporters that

"Professor See is at liberty to have his own opinions. . . . We don't wish to enter into any controversy."[23]

See would not allow the Lick to stay silent. During April 1923, the *San Francisco Journal* printed a series of attacks by See on Einstein, relativity, and the Lick Observatory. After the series had completed its run in the newspaper, See presented an eight-page pasteup copy to the library of Lick Observatory, and one to the Lowell Observatory in Flagstaff, Arizona (fig. 9.2a,b).[24] Campbell was doubly concerned as director of the Lick and soon to be president at the university. See had written: "Let us Californians watch the publications of Lick Observatory, which is a public institution, supported by the state of California, to see if they will bear witness to the truth of history, or attempt to cover up Einstein's appropriation of Von Soldner's work of 1801."[25]

Campbell chose to point out See's errors about the Soldner formula, and Trumpler was clearly the man for the job. Campbell asked him to "write down for me the supposed relations of Einstein and Soldner, from the point of view which you recently expressed to me."[26] Campbell arranged with Aitken to publish an article by Trumpler in the *Publications* of the Astronomical Society of the Pacific, and to have it reprinted in *Science*.[27] "A series of articles recently published by Professor T.J.J. See, U.S. Navy," Trumpler began, "gives a quite incorrect impression of the relation of J. Soldner's and of Einstein's work in connection with the deflection of light in the Sun's gravitational field." Trumpler detailed the methods by which Soldner and Einstein (in 1911) had arrived at their results. He showed that the approaches and goals were completely different. Referring to Einstein's value of 1916 Trumpler remarked: "The increase of this value over that in Einstein's 1911 paper is not due to a mistake in calculation in the earlier paper but is an effect of the difference between Einstein's and Newton's law of gravitation." Trumpler concluded that his comparison sufficiently showed the independence of Einstein's work "even if he knew about Soldner's paper, which is not likely, as Soldner's results had fallen into oblivion following the rejection of the corpuscular theory of light on which it is based." He explained See's misconception as arising from Lenard's version of the story. "Professor See, accusing Einstein of plagiarism, clearly has not read Soldner's original paper and has been misled by a fragmentary reprint of it published in 1921 together with comments by a German physicist, P. Lenard. In these comments Lenard transforms Soldner's formula into a notation and form similar to those employed by Einstein. Professor See mistakes Lenard's transformed formula for Soldner's and bases his unfounded accusations upon its similarity to Einstein's result."[28]

See's public attack on Einstein and his theory forced Campbell to engage in the general debate over relativity. His preferred position had al-

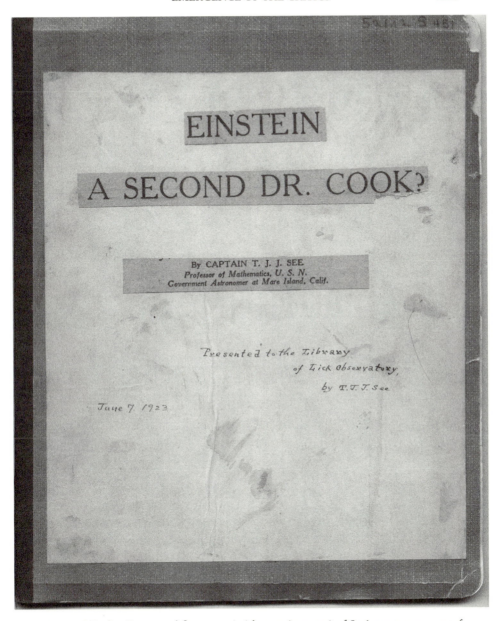

EINSTEIN

A SECOND DR. COOK?

By CAPTAIN T. J. J. SEE
Professor of Mathematics, U. S. N.
Government Astronomer at Mare Island, Calif.

Presented to the Library
of Lick Observatory,
by T.J.J.See

June 7, 1923

Figure 9.2a,b. Cover and first page (with margin notes) of See's paste-up copy of his *San Francisco Journal* series of articles as presented to the Lick Observatory. (Courtesy Mary Lea Shane Archives of the Lick Observatory, University Library, University of California–Santa Cruz.)

EINSTEIN A TRICKSTER
Objections to Relativity Theory
By Capt. T. J. J. See

Professor of Mathematics, U. S. N.

Government Astronomer at Mare Island, Cal.

This is good!

THE magazine and newspaper press for the last eight years has been so filled with systematic propaganda, undoubtedly organized and directed by Einstein and his agents, that the public has become familiar with the name of Einstein and with the phrase "Theory of Relativity." Not one lay person in a thousand has any idea what this all means; and as the people do not understand it, the phrases are passed on in joke, or assumed to represent something important in the higher lines of physical science. It is well known that about six years ago Einstein tried to cast a halo of glory about his head by allowing the report to go forth that not over twelve mathematicians in the word could understand his benighted theory of relativity. Of course this is preposterous, and nobody knows it better than Einstein himself.

We shall show in a simple way that the theory of relativity is unsound, and therefore rejected by the competent mathematicians and natural philosophers of our day, just as it would be by such great historic authorities as Kepler, Galileo, Huyghens, Newton, Euler, Lagrange, Laplace, Sir W. Herschel, Poisson, Bessel, Gauss, Hansen, Sir John Herschel, Maxwell, Airy, Adams, Leverrier, Tisserand, Poincare, Lord Kelvin, Newcomb, Hill, etc. Not one of these great men would lend the slightest sanction to the theory of relativity if they were living today; and hence it is the duty of any competent investigator to denounce the fraudulent trains of thought which Einstein and his deluded followers have spread about with no other effect than to confuse the public mind.

In short, I have at length become convinced that Einstein is a fakir, with considerable skill in deceiving the press and public, so as to ding-dong into the unthinking the idea that he is a great mathematician and philosopher, who is improving on Newton. Let us first notice the errors of Einstein, and the cunning way in which he gets away from them, owing to the layman's inability to pin him down.

Napoleon used to say that he put a stop to humbuggery in the arguments presented to him by confronting pretenders with the statement that two and two made four, not five. We propose to handle Einstein in this direct fashion, which will not allow him to wiggle out of his misleading teaching.

1. In 1919 it was oracularly heralded abroad by Einstein that there is no aether. This pernicious proposition was echoed in Holland, and repeated in England, by certain mediocre physicists in the Royal Society, more especially by Eddington and Jeans, who have since done so much to spread errors over the world. In fact, a joint meeting of the Royal Society with the Royal Astronomical Society was called and a formal debate held, November, 1919, on the proposition for the abolition of the aether.

?!!

Opposed by Lodge

In vain did Sir Oliver Lodge and other experienced physicists protest against this folly, but the misguided zealots shouted "On with the dance!" —so determined were they on innovation and error, rather than truth! Lodge pointed out that if we accept Einstein's theory, "the death knell of the aether will seem to have been sounded, strangely efficient properties will be attributed to emptiness, and theories of light and of gravitation will have come into being unintelligible on ordinary dynamical principles." In other words, Einstein's theory was dynamically impossible, as Newton himself points out in the passage cited below.

In an interview at Chicago, December 19, 1919, Professor Michelson, the chief authority on light, openly rejected Einstein's theory just because it proposed to do away with the aether. "Einstein thinks there is no such thing as aether," remarked Michelson. "He does not attempt to account for the transmission of light, but holds that the aether should be thrown overboard."

These citations are evidence enough that Einstein committed the unpardonable philosophic sin of proposing to do away with the aether, and we see that this stupid proposal was at once rejected by Lodge, Larmor, Wiechert, Michelson, See and other investigators.

To judge how absurd it was to suppose all space to be filled with mere emptiness, without the aetherial medium, we cite the remarks of Sir Isaac Newton, Letter to Bentley, February 25, 1692-3: "That gravity should be innate, inherent and essential to matter, so that one body may act upon another at a distance through vacuum, without the mediation of anything else, by and through which their action and force may be conveyed from one to another, is to me so great an ab-

ways been to provide the best observations and leave the theoretical wrangling to others. Now he was being cast in the role of a defender of the theory itself, and drawn into theoretical issues such as competing ways of deriving the light-bending effect. Fortunately in Trumpler, Campbell had not only a skilled observer, but also a competent relativist and controversialist. After the See incident Trumpler studied the details of the Soldner episode and began to monitor the literature for new developments regarding relativity. From 1922 to 1924 he compiled an annotated bibliography on the theory, dividing publications into five categories: (1) theoretical developments; (2) perihelion of Mercury; (3) light deflection; (4) redshift of spectral lines, and (5) other tests, including ether-drift experiments and cosmological considerations.[29] Trumpler would do valuable service in later controversies.

The press aggravated the situation, gleefully reporting the comments of each new critic. In July one journal reported: "Father Ricard, the eminent astronomical scientist of the University of Santa Clara . . . has branded Einstein's theory of relativity as an insult to common sense. Father Ricard also declared that the starlight photographs taken by Professor Campbell of the University of California during the Australian expedition were very rickety and not conclusive."[30] Jerome Sextus Ricard was an elderly trustee of the University of Santa Clara, who had come over from his native France in the 1870s to teach mathematics and moral philosophy. His claim to the status of astronomical expert was research on long-range forecasting by sunspots and an editorship of a magazine called *Sunspot*.[31]

Toward the end of 1923 another brush with See convinced Campbell and his colleagues that dissimulating a neutral stand in the public debates would be useless. See was to give a public address in San Francisco. He wrote to Robert Aitken (who was then associate director of Lick) that he wanted a statement of his position on relativity. He claimed that Aitken "took refuge under the authority of Eddington" when asked to defend relativity and that "leading men in San Francisco" agreed with him that such conduct was "the surrender of Americanism to low and discredited foreigners." He threatened public denunciation at his lecture, should a statement not be forthcoming from Lick that such allegations were untrue.[32] Aitken did not rise to take the bait. He returned See's relativity papers, noting that they were "apparently expositions of your personal ideas and convictions, with some references to, and quotations from papers by other men." He did not see that they called for any comment on his part. "Nor do I see that your letters require any special answer," he added, "though I might, perhaps, remind you that hear-say evidence is not accepted in a court of law and add that I question its weight with scientific men of standing. If you care to base public statements or arguments upon such evidence the responsibility must rest with you."

See's remarks resonated with antagonism in certain circles toward the influence of European science and culture on American institutions. The postwar period saw a great influx of Europeans into academic and research positions, and some Americans did not like it. When Campbell informed Curtis in the fall of 1921 that the sudden resignation of one of the junior astronomers had left a vacancy at Lick, Curtis urged Campbell to hire an American. "Yes,—I know that science is international and all that, and also that every time we have disagreed in the past twenty years it has turned out that you were the right of it. But it seems that the personnel list of Lick and Mount Wilson are getting to read too much like a page of a Swedish directory. There are plenty of good youngsters who would give their eyes for a job at Lick or Mount Wilson." Curtis claimed that if he ever got enough money to hire another staff member, "they are going to be American or Canadian citizens, *born in* the U.S. or Canada, or England." By all means "a foreigner" should be brought in once in a while to prevent "inbreeding," he allowed, "but the actual bread-and-butter jobs at the only two places on earth where it's really worth while to be an astronomer ought to go by preference to our own sort. If we can't wallop our brethren across the water with native born talent, let's acknowledge our inferiority frankly."[33]

Some of Curtis's spontaneous remarks echoed See's harsher words. Campbell had been obliged to answer Curtis that although he sympathized strongly with his comments, it was hard to find an American to hire. The problem continued well into the 1920s. Five years later Hale told Campbell that he had been asked to suggest a professor of physics for a research post at M.I.T. and had mentioned several leading British men, all of whom had declined. "You of course know the difficulty of finding an American of the necessary caliber."[34]

By the end of 1923, Campbell was ready to endorse Einstein's theory officially in face of the commotion See had created. Newspapers received the statement from "the dean of western astronomers" that "criticisms of Professor Einstein's laws are based on prejudice with which I have no sympathy."[35] The *San Francisco Journal*, which had published See's complete manifesto months earlier, refused to look on the matter as settled. Instead, the editors announced "Einstein Again Supported" and boasted that California could claim the distinction of furnishing both his attacker and defender: "It is a tribute to science in California that we should furnish to the world the scientific ammunition for both sides of the controversy over this epoch-making discovery" (fig. 9.3).[36]

Figure 9.3. Page from a newsclippings file at the Lick Observatory. Left-hand article from the *San Francisco Journal* reports Campbell's official endorsement of Einstein's relativity theory on 9 December 1923. (Courtesy Mary Lea Shane Archives of the Lick Observatory, University Library of California-Santa Cruz.)

An Antirelativity Coalition in the East

While See was battling Lick astronomers in the West, another relativity critic launched his own campaign in the East. Charles Lane Poor at Columbia University in New York began attempts to disprove the British eclipse observations immediately after they were announced. During the 1920s he mounted a concerted effort to convince astronomers to try different tests to disprove the theory and find alternative explanations of the predicted effects. Trained in celestial mechanics, Poor had spent years trying to account for Mercury's perihelion motion. One of his approaches was to look for oblateness in the shape of the Sun. Poor reasoned that centrifugal forces at the solar equator due to the Sun's rotation would expand the surface slightly, distorting its spherical shape into an ellipsoid. If the effect was large enough, Newtonian mechanics could explain the precessing of the planets' orbits. Poor was dissatisfied with all previous attempts to measure the figure of the Sun and spent several years trying to convince an observatory to take photographs according to his specifications. He succeeded in making provisional arrangements with Yerkes Observatory. He provided a specially made lens and shutter, and agreed to pay a monthly sum toward the cost of the project. He never got more than a "very few test plates." When Samuel Alfred Mitchell, who had been Poor's student at Johns Hopkins University during the late 1890s, assumed the directorship of Leander McCormick Observatory in Virginia, he agreed to take on Poor's project. The instrument was transferred from Yerkes, and Poor gave Mitchell "several sums of money for building a new shutter, plate holders, etc." Yet he never saw a photograph.[37] After the British announced that starlight had been bent by the Sun's gravitational field, Poor tried to come up with a refraction explanation. He also renewed his efforts to account for the perihelion motion of planets using Newtonian theory.[38]

In 1922 Poor published a book criticizing relativity, claiming that "no substantial experimental proofs have yet been submitted by any of its adherents." The first chapter explained the theory of relativity as Poor understood it, which he clearly did not. With implied competence, he asserted that the general relativistic study of gravitation "involves the most intricate mathematics, and the mathematical processes and methods, used by Einstein, cannot be explained in untechnical language." He claimed that because "the velocity of a body enters into every formula and into every measure of its position in space" the mathematical expression for the law of gravitation "is not the same as that formulated by Sir Isaac Newton." Referring to the Lorentz transformation equations from special relativity, Poor noted that they contain terms involving the ratio of the velocity of the body to that of light. "When the velocity of a body is extremely small as compared to the velocity of light, then this ratio becomes small and these terms become negligible in comparison with the other terms of the expression. In this case the formulas of Einstein degenerate into those of Newton, and, thus, for small velocities the two laws give identically the same results."[39] Poor did not go any further, because he could not. "I have no idea as to the method used by Einstein et al.," he admitted privately months after his book was published, "as I have never checked their work."[40] Even when he turned to Eddington's technical treatise on relativity, Poor received no further illumination. "I cannot follow his mathematical gymnastics."[41]

In his 1922 book Poor reviewed the empirical evidence for relativity. He rejected any appeal to the redshift for support of relativity, referring to St. John's published work refuting the relativity explanation. Poor asserted that St. John had used "equipment far surpassing anything to be found elsewhere" while others used "ordinary, average equipment of a small laboratory or observatory."[42] Poor admitted that the perihelion motion and the light bending appeared to be the "only tangible evidence in favor of the theory," but cautioned that the evidence must be carefully examined.[43] To discount the perihelion result, Poor resurrected his ideas about the shape of the Sun. He claimed that an equatorial oblateness could account for $3\overset{''}{.}5$ of Mercury's precession, reducing the unexplained perihelion motion of Mercury to "not over $36\overset{''}{.}3$, or 16% from Einstein's prediction." "Such a difference is very nearly fatal to the Relativity Theory," he concluded, "for that theory contains no arbitrary constant by which Einstein can, in the future, readjust his figures to fit the real and not the imaginary facts."[44] Poor sarcastically asserted that although there were other anomalies in planetary motions about which Einstein's theory had nothing to say, "if the methods of the author of relativity are to be admitted, there is no necessity of explaining the perihelial motion of Mercury. If it is troublesome to our theories, it can be discarded along with all the other discordances. Why even bother about Mercury itself! Copernicus is said never to have seen the planet; and the solar system would really be much simpler without it!"[45]

Poor also attacked the British eclipse results on the light bending. He appealed to Russell's explanation that distortion of the coelostat mirror due to heating had caused the nonradiality of the Sobral astrographic deflections and concluded "the entire set of plates is worthless for proving the existence or non-existence of the 'Einstein effect.'"[46] He considered the refraction explanation in great detail, ignoring all arguments that had been discussed for months in English journals in response to Newall's attempts to find one. "This possibility of accounting, in a perfectly normal way, for the observed light deflections has been dismissed by the relativist in a few words as a matter scarcely worth mentioning," he wrote.[47] Poor's brief concluding chapter emphasized that Einstein's hypothesis and formulas were "neither *necessary* nor *sufficient* to explain the observed phenomena," and that all conflicting evidence was blithely dismissed by the relativists.[48] "But for the true relativitist the pathway through all the difficulties of conflicting evidence is smooth and clear; for does not everything depend upon the observer? Nothing is absolute, everything is relative; the statue is golden for one observer and silver to the other."[49]

In December 1922 Poor sent a copy of his book to Campbell, introducing it as his attempt "to make a complete investigation of the astronomical evidence cited as proof of Relativity Theory, and to place the pros and cons fairly before the readers." Shortly after, having heard of Campbell's recent success in obtaining plates in Australia, he wrote again, asking Campbell if he would try out an idea on his "wonderful photographs." His plan was to measure the diameter of the Moon to see if it changed in the same way that the star positions had changed. "Would it be practicable to measure on your photographs a number of points on the edge of the Moon, measure them in identically the same way as the star images? If practicable, such measurements could be reduced in a manner entirely similar to that used for the stars, applying the same corrections for orientation, scale value, differential refraction, etc. From these measures could then be obtained the values for diameters of the Moon, and such observed diameters could be compared with that calculated from the lunar tables." If the diameter increased in direction and amount similar to the way the stars were deflected, then one would have to look for the cause between the Moon and the Earth rather than to an effect of the Sun's gravitational field.[50]

Campbell did not reply until April, a week before his preliminary announcement of results. He thought that Poor's suggestion was impracticable for the Lick plates: "Our Einstein exposures in Australia varied from one minute to two minutes in length, and the driving clocks were adjusted to the stellar rate. Consequently, the images of the inner corona are not only horribly over-exposed, but they eat into the edge of the moon's images, and, further, the drifts of the moon during exposures were very large. . . . Perhaps a program of observation formulated with reference to your proposal could be made useful, but I am not sure." Campbell told Poor that he hoped to make a preliminary announcement of his and Trumpler's Einstein results "within two or three weeks." For the benefit of his antirelativity correspondent, he added: "Our philosophy throughout the entire campaign has been that of Alexander Pope—'Whatever is, is right.' "[51]

Poor answered Campbell's letter after the preliminary results had been announced. He asked if he could have a copy of any statement of methods used and tables of results that may have been prepared. Campbell replied that he and Trumpler were just "putting the finishing touches" on an article for publication.[52]

Poor approached other astronomers about his Moon test for refraction effects.[53] When he wrote Curtis, offering to pay the costs, he struck gold. Curtis was planning to redo the Einstein test at the coming eclipse in September with John A. Miller, director of Sproul Observatory at

Swarthmore College. He told Poor that if his letter had come a month or two earlier, "I should have been glad to make use of your offer, as I tried unsuccessfully to get the funds for an expedition of my own." He suggested that Poor contact Miller directly, offering "to aid in any way, at the eclipse, or by assisting in the design of apparatus beforehand." Curtis went even further:

> I may say, first,—that I put in a year of heart-breaking work measuring the Goldendale plates, which were not good enough to give a decisive result. I have heard Dr. Campbell's recent paper, and talked with him on the methods used, etc., quite fully. My conclusion is that there is absolutely no doubt, in my own mind, that a deflection exists essentially as shown on his plates, of about 1.″75 at the Sun's limb. This does not, however, make me a believer in the theory of relativity. I am still an irredeemable heretic, and it does not seem now that I shall ever swallow that theory unless chloroformed first.

Curtis related that "after a short period of hesitation after hearing Campbell's paper" he and Miller had decided to "go right ahead" with their plans to repeat the Einstein test. "No expense or care is being spared to make this the best Einstein outfit ever used." "I wish some one with more time and ability in that line than I have would check over the amount of deflection due to the ordinary gravitational theory. Einstein made a mistake of the factor 2 in his work at first. Maybe someone else has done it."[54]

Poor replied immediately. "The day the British announced the results of the 1919 eclipse I called Einstein 'The Bolshevist of Science,' and I have been pounding relativity ever since; so I am very glad to have your strong statement of unbelief." He admitted that after Campbell's announcement "I was rather knocked off my feet." Now he asserted that Campbell "had not used all possible checks."

> If Campbell's individual measures are anything like those of Chant then we need not worry very much, for the Canadian results, as published, are very discordant and can mean almost anything. . . . I am not an observer and the various suggestions I have made have been totally disregarded, and my repeated remarks, in lectures and in print, about the abnormal conditions under which the eclipse plates are taken have been ignored by all relativists. . . . I hope that my suggestions may prove of value and that we may be able to cooperate in a vigourous and sustained attack upon the most dangerous doctrine of modern times.[55]

Poor had several tests in mind. His first was a "full and complete check" for abnormal refractions utilizing his idea of measuring the Moon's diameter during an eclipse. Second, he wanted to test the possibility of film distortions "raised by the experts of the Eastman Kodak Company."[56] However, Frederick Slocum had definitively refuted that possibility. He calculated the maximum amount that star images might shift due to an emulsion effect. It would be too small to account for the observed deflections of stellar images on the eclipse plates. Curtis told Poor that astronomers experienced with using photographic plates for precise measurements could rule out the suggestion because possible errors from this source are very small and "such distortions are purely random effects. . . . The many thousands of parallax plates which have now been measured and reduced are sufficient proof of this." Curtis was interested in how a plate emulsion would be affected by a large intense image at the center of the plate. Would the general pattern of a star field surrounding it be affected? "One thing I would like to see tried, and if it has not been done yet I think we must try, is to print a husky artificial corona, and considerable sky blackening, in the middle of an artificial star field, and compare with the same artificial star field without such central film change."[57] Poor also wanted to conduct spectroscopic observations of the inner corona to determine whether the matter in the corona rotates with the speed of the Sun or with varying speeds due to gravitational forces. "If the corona be a swarm of solar moons, so to speak, revolving about the Sun under gravitational forces, then most of the stock arguments for an extremely low density fail." In this case, the corona could have a larger effect on passing light rays. Poor told Miller that he had suggested it to Hale before the 1922 eclipse. However, ideas of coronal structure had long since passed the stage of particulate models, and Poor's idea was likely regarded as outdated.[58]

Curtis and Miller considered only the Moon test, but there were problems. Poor noted that it might be difficult to get both star images and the Moon on the same plate with one exposure. He suggested an interlocking series of plates, some with long exposures for the stars, some with short ones for the Moon. Curtis and Miller were adopting Shane's procedure of using an auxiliary star field 10 degrees away from the eclipse field. They wanted to eliminate any possible errors due to Campbell and Trumpler's method of doing this check the nights before and after the eclipse. Curtis thought the check stars would provide the most precise measures of any abnormal refraction effects due to atmospheric cooling in the Moon's shadow cone during the eclipse. Poor didn't think that the check field would be useful for his test. His hypothesized refraction would be symmetrical with the axis of the eclipse shadow and would decrease very rapidly with increasing distance from the axis. "They [star displacements due to refraction] would, I fear, be very small, if not inappreciable, at 10° from the axis. There is thus a chance that a star-field, 10° from the Sun, might not show any abnormal effects, or such small effects as to be confused with accidental errors, and yet, at the distance of the MOON's limb the abnormal refraction effects might be measurable."[59] Curtis worried that it would be impossible, even on short exposures, to prevent the bright chromosphere from interfering with other parts of the negative, including the Moon's limb; and on long exposures the effect would be considerable "whether the plates are backed, or double or triple coated plates used."[60] Furthermore, the program was already too full.

> It is . . . impossible to use any time at this eclipse for such short exposures with the Einstein cameras proper; it is taking all kinds of planning to get two exposures with each lens and allow sufficient [time] for changing plates, moving to the auxiliary field, and taking a short exposure on it. To get such short and long exposures at the coming eclipse would practically necessitate the duplication of the entire Einstein outfit and polar axis, clock, etc. . . . Perhaps I am unduly pessimistic and of course everything possible should be tried which shows any promise whatever of helping us out of our dilemma, but the measurement of moon diameters appears to offer many difficulties, both in its practical application and in our right to interpret rigorously the results of our measures.[61]

Poor persevered, raising the money for a separate piece of equipment from a friend. Miller was able to order a camera to be made by the James McDowell Company that was "almost an exact duplicate" of the Einstein cameras constructed for the Sproul expedition. By the time all expenses were met, Poor himself had contributed some of his own money as had two of his acquaintances, H.G.S. Noble, after whom the camera was named, and E. Vaile Stebbins. At the eclipse, Curtis and Dinsmore Alter of the University of Kansas operated the Noble camera for Poor's tests.[62]

THE ETHER ATTEMPTS A COMEBACK

The widespread interest in relativity prompted a number of diehard ether enthusiasts to resurrect interest in ether drift experiments as another line of attack to test relativity. In 1905, the year Einstein published his first paper on relativity, Dayton Clarence Miller, a physicist at Cleveland, had teamed up with E. W. Morley to confirm the original 1887 Michelson-Morley experiment. Michelson's and Morley's famous result had been that the Earth's motion through the "ether" is undetectable. Miller and Morley obtained the same result in their basement laboratory on the campus of the Case School of Applied Science. They then tried the experiment on an elevated plot of land. Some physicists had suggested that the

Earth drags the ether partially. At the surface, the drag would be 100 percent, hence no detectable motion through the ether. At higher elevations, the drag would be partial. Motion relative to ether might be measurable. Miller and Morley reported a "definite positive effect" at the higher elevation, but acknowledged that a temperature effect might have caused the result. After Morley's retirement in 1906, Miller was left alone to carry on the work, but "numerous causes prevented the resumption of observations."[63]

Widespread interest in Einstein and relativity after the British eclipse results appeared in 1919 gave new life to Miller's project. Hale was aggressively building the scientific capabilities of Mount Wilson, which was increasingly becoming a Mecca for scientists keen on the advanced equipment and excellent climate. Having established Mount Wilson as the final arbiter on the gravitational redshift, Hale saw an opportunity to weigh in with another relativity test. He invited D. C. Miller to come to his mountain observatory, whose nearly 6,000-foot elevation made it an ideal location. Miller repeated his experiment with essentially the same apparatus that he and Morley had used years earlier. He obtained a positive result, but there was also an unexplained periodic effect of half the frequency.[64] Einstein was visiting the United States for the first time, delivering lectures in Princeton around the time Miller's results became known. When reporters asked him to comment, he delivered a now famous line that is chiseled in German above the stone fireplace in the former mathematics building at Princeton University: "Subtle is the Lord, but He is not malicious." Einstein was commenting that Nature hides her secrets through subtlety, not slyness. In other words, he did not think Miller's experiments would prove to be correct, as the situation would be too complicated to explain. Nonetheless, Einstein visited Miller in Cleveland, where they had an amicable discussion.[65]

Hale was always on the lookout for theoreticians to support his observational researches. He was therefore pleased to receive the following from Ernest Merritt, head of the department of physics at Cornell University:

> I recently heard a rumor to the effect that someone, possibly D. C. Miller, was to repeat the Michelson and Morley experiment at Mount Wilson. Upon the assumption that the rumor is true I am writing to suggest that there is an unusually good opportunity of getting available help on the theoretical side in the presence at Rochester of Dr. L[udwik]. Silberstein who has recently joined the staff of the Eastman Kodak Research Laboratory. Dr. Silberstein is giving a course of lectures on relativity before our graduate students and I have come to feel pretty well acquainted with him. I have been pleased in

conversations with him to find that, in spite of his remarkable ability in the mathematical line, he seems to approach the subject of relativity in a manner that is very much that of the experimental physicist.[66]

Silberstein had recently emigrated from England, where he had been active in the British astronomical community. He had initially been favorably inclined toward Einstein's 1905 theory of relativity. When the generalized theory appeared, he had reacted negatively, especially to the geometrical interpretation of gravity. He resented the uninformed public attention it received as a result of the British verification of the light-bending prediction. After the announcement of the British eclipse results, Silberstein tried to explain them using a model of condensed ether around the Sun. He argued that scientists should not be carried away by the popular interest in relativity.

> Let us imagine for the moment that Einstein had never published his debatable, though undoubtedly beautiful, new theory—not even that of 1905. Then it is almost certain that the Eclipse result would readily be acclaimed as an evidence of the condensation of the aether near the Sun, as required by the theory of Stokes-Planck, and would encourage physicists to work out in detail the optical and associated consequences of such a condensation. But even though Einstein's theory has been published, and is being made popular in a most sensational way, we cannot help clinging to the said idea.[67]

Soon after he arrived in America, Silberstein was in demand as a lecturer on relativity. In January 1921, he gave a three-week course of fifteen lectures on "Einstein's Relativity and Gravitation Theory" at the University of Toronto. During the summer of that year he delivered a course of forty-seven lectures on "Relativity, Gravitation and Electromagnetism" at the University of Chicago. He expanded his lectures on the general theory into a book, where his critical stance toward relativity was explicit: "Some of my readers will miss, perhaps, in this volume the enthusiastic tone which usually permeates the books and pamphlets that have been written on the subject (with a notable exception of Einstein's own writings). Yet the author is the last man to be blind to the admirable boldness and the severe architectonic beauty of Einstein's theory. But it has seemed that beauties of such a kind are rather enhanced than obscured by the adoption of a sober tone and an apparently cold form of presentation.[68] In his book, Silberstein continued to hold the gravitational redshift as contrary evidence against the theory. While giving his Chicago lectures, he created a stir by claiming to have discovered a fatal flaw that "has completely knocked out 'Relativity.' " He subsequently found an error in his calculations.[69]

Michelson attended one of Silberstein's Toronto lectures, and soon the two men were discussing ether-drift experiments. Silberstein proposed a method to test for a partial drag of the ether due to the Earth's rotation. It would also serve as a test of Einstein's general theory of relativity. Michelson was reluctant to try the experiment at first, but agreed to supervise a project in view of the current interest in Einstein's theory.[70] Silberstein announced the plans in a lecture he gave at Princeton in May 1921, emphasizing that the experiment "may prove Einstein's theory of relativity to be all wrong." The *New York Times* duly noted Silberstein's remarks. Michelson wrote to Hale: "I have also a big relativity experiment in view—for which the Eastman Kodak people are providing funds through Dr. Silberstein."[71]

Hale was particularly interested in the news because of his project with D. C. Miller. When Michelson went to Pasadena in the summer of 1921 to make preliminary tests on Mount Wilson, Miller consulted with Hale and Michelson. He designed a new interferometer using a concrete base instead of steel to eliminate any magnetic effects that might have caused the anomalous periodic displacements. Observations with the new equipment showed a positive effect "substantially the same as in April," including the mysterious periodic result. At Hale's suggestion, Miller wrote to Lorentz in the hope that he "might perhaps offer suggestions that would save time in discovering the source of the effect which is periodic in one revolution of the apparatus."[72]

Meanwhile, Michelson's tests were going well. By the end of July he could write to Silberstein that "contrary to my previous skepticism . . . the chances are favourable."[73] Hale described the scheme to Miller: "Michelson is preparing to try a different form of ether-drift experiment, in which the interferometer mirrors are to be at the angles of a triangle more than a thousand feet on a side. The first tests here in the valley gave beautifully sharp and steady fringes when the total path was about a thousand feet. As this was in the middle of the day, in full sunlight, I feel confident that the fringes will be sharp over a much longer path at certain hours of the day on the mountain."[74] Hale was initially willing to consider funding the project, but Michelson told him "Dr. Silberstein has begun a sort of subscription fund for the work tho I doubt if he will be able to accomplish much."[75] In the end, the project did go to Chicago. The university contributed $17,000, the city furnished free pipe for the optical paths, and the Chicago telephone company donated a telephone system. Silberstein was able to raise $500 more.[76]

Miller returned his apparatus to Cleveland in 1922, where he began further observations to find the causes of the extra periodic displacement of the interferometer fringes. In April 1922 he presented results to the

National Academy of Sciences. Although no ether-drift was found in the basement at Case in 1905, one was found 6,000 feet above sea level on Mount Wilson. His interpretation was generally conservative, and he mentioned the anomalous fringe displacement that must be explained before reaching any conclusion.[77]

By 1922, then, plans were in the works to conduct two ether drift experiments, both with the intent to disprove Einstein.

Chapter Ten

THE DEBATE INTENSIFIES

ANOTHER CHANCE TO TEST EINSTEIN

After his Goldendale debacle, Campbell had originally intended to attack the Einstein problem close to home again at the eclipse that would be visible in southern California and Mexico on 10 September 1923. The British announcement in 1919 raised the stakes and he seized the earlier opportunity in Australia. While preparing for the 1922 expedition to Wallal, Campbell had surveyed regions in California and Mexico in his capacity as chairman of the American Astronomical Society's eclipse committee. He decided that the American islands lying to the south of Santa Barbara and west of San Diego would be good locations. After Campbell and Trumpler's announcement in the spring of 1923 that their Australian results had verified British conclusions from the 1919 eclipse, European astronomers chose not to go to the 1923 eclipse. Only American and Mexican parties went out. Herbert Couper Wilson, director of Goodsell Observatory at Carlton College, decided to try his luck on the Einstein problem. His specialty was celestial photography and the measurement of photographs in search of asteroids, skills well suited to the task. He decided to locate on Catalina Island on Campbell's recommendation, as did several other expeditions. John A. Miller's party from Sproul Observatory, with Curtis and Poor's collaboration, had two separate tests planned—the regular light-bending program, and Poor's lunar test for abnormal refraction. Miller decided to locate in Mexico, despite Campbell's insistence that chances for success south of the border were not good.[1]

Early in 1922, Sir Joseph Larmor wrote Hale at Mount Wilson that "I have been coming under the fascination of the Einstein theory of gravitation." As an unrepentant ether advocate he was not happy with its demise at the hands of Einstein. "From my point of view stars are still made of electrons and gravitation between electrons is *essential* to their existence in the aether so that one could not imagine a world devoid of it."[2] Since 1919 Larmor had been proposing a reconstruction of general relativity that would eliminate the need for a gravitational redshift. He was particularly annoyed with Einstein's "metaphysical" approach: "The question at issue seems to reduce to this: whether it is safe to mix up a strict mathematical analysis of the modified space-time with an extraneous semi-meta-physical and loosely expressed 'principle of equivalence.' "[3]

Larmor continued to ponder general relativity, and during the summer of 1922 he became convinced that the light deflection should be half the Einstein value. At the beginning of 1923 he wrote Hale that Einstein was wrong after all:

> I am convinced that if gravitation is to be introduced into the scheme of electrodynamic and optical (limited) relativity by introducing a varying space and time, it must be very different from Einstein's way. I worked it out by minimal Action last summer and sent a paper to the Royal Society in mid-October claiming that the optical effect must be halved. I fear the young guard . . . thought I had gone off my head: and on reflecting I had no means of assuring myself on that subject. However I withdrew it two months later and it is (I hope) in the *Philosophical Magazine* for January. In November J[ean Marie] Le Roux began an attack on the postulate that orbits could be 'shortest lines' . . . and after various exchanges he seems to be left in possession of the field. If they are not geodesics they are not invariant and Einstein becomes nonsense on his own principles.

Campbell and the other eclipse hunters were still measuring their Australian eclipse plates. Larmor made a strong case to Hale that results from these tests were crucial: "I think my theory of least Action stands self-consistent: but if they don't get the half value for deflection of light it must stand down. If light is deflected *at all* it indicates of course interaction between gravity and aether: so a purely negative result would be an emphatic but not likely addition to knowledge. Anyhow, confirmed by the destructive criticism of Le Roux, in the matter of my sanity, I believe that the Einstein metaphysics is gone for good."[4]

Hale was taking a medical leave for a year and he received Larmor's letter on board ship. It clearly made an impact. He copied the last paragraph verbatim into the back page of his completed 1922 diary before sending the letter on to Adams, then acting director at Mount Wilson. Hale and Adams had not put the Einstein problem on their observing program for the September 1923 eclipse because they knew that Campbell and others were repeating the test in Australia. Larmor's communication changed Hale's mind: "It is plain that the magnitude of the deflection by the Sun must be determined with the highest precision and by several observers, with independent material. Campbell's photographs at the last eclipse are said by the papers to be very good, but in view of Larmor's conclusions they will hardly settle the question. I therefore believe it should be put on our eclipse program, and that the most suitable instruments obtainable should be used for this work."[5] So Adams put the Einstein test on the Mount Wilson eclipse program.

Meanwhile, Larmor had trouble getting his ideas taken seriously in England. "My squib about halving the deviation . . . of light by gravitation duly appeared . . . but nobody has come to take any notice of it." He tried to interest the pure mathematicians at Cambridge but "they never believed it in any practical sense." He even attempted to "draw Lorentz on the subject in *Nature* but he would not take the bait. . . ." "I observe it goes on sporadically in the *Comptes Rendus* with feeble attempts to get round the criticisms. . . . Anyway if the stars prove *not* to be deflected it will exclude all tampering with space altogether. But I think the chances are strongly the other way."[6] Hale's colleagues back home were not so keen on Larmor, either. Adams consulted Caltech physicists Paul Epstein and Charles G. Darwin, as well as Arnold Sommerfeld who was visiting from Germany. "Epstein and Darwin do not think that Larmor is right," he wrote Hale, "and Sommerfeld was inclined to think the same thing. Originally Einstein believed that the displacement was the half amount but later came over to the larger value. . . . We shall certainly plan to do all we can on this effect at the eclipse." Adams expected to see Campbell in March and hoped he would have some results. "If they show the half displacement, this year's eclipse will be immensely important." Millikan also wrote Hale that "Epstein and Darwin are both interested, and I think a trifle amused" by Larmor's letter. "Neither of them is disposed to have very much confidence in the soundness of Larmor's conclusions, and I myself have had some reason to know that he is not infallible."[7] Larmor soon acknowledged to Hale: "I fear I have got into permanent trouble on the enclosed [*Phil. Mag.* article]. But they refused twice my request to read it publicly at the Royal Society and meantime the French were discussing cogent things in *C. R.* every week. . . . My only correspondence hitherto is [Friedrich] Kottler of Vienna . . . who was critical anyhow, and regards the whole scheme as premature and mystical."[8]

It was not until 17 April that Adams could communicate news of the Lick measurements to Hale: "The principle astronomical news of the day is the complete confirmation by Campbell of the relativity shift from his Australian eclipse plates. His final result is 1″74 against the predicted 1″75. The Toronto observers obtained similar though much less accurate results. Campbell considers his values so definite that he will not repeat the observations at the coming eclipse. The bearing of this upon our expedition is, of course, important, but everyone agrees that we ought to follow out our original plans." The apparatus was well advanced and was "more powerful than Campbell's." Adrian van Maanen, an expert measurer of stellar parallaxes, would operate the cameras. "It will form but a small portion of our eclipse program and there would be little gained by omitting it. There is always the possibility that something unexpected may turn up." The British took special note of the Mount Wilson participation and that Campbell was not going. "As his [Australian] results agree with those of the British expeditions of 1919, he will not again deal with this subject."[9]

It was a fitting conclusion to Campbell's long-standing personal involvement with the Einstein problem to decide not to include it on the 1923 eclipse program. Yet reactions to his positive announcements exposed him and his colleagues to the larger issue of the theory's validity and its acceptance by the scientific community. The Lick began to find itself cast in the role of expert and arbiter in the ongoing relativity debate. Whereas Hale at Mount Wilson worked hard to build theoretical expertise at home and drew on advice from abroad, Campbell initially focused entirely on the observational problem. In Robert Trumpler, however, he was fortunate in finding not only a skilled observer but also someone competent with the theory. Trumpler's theoretical abilities proved to be invaluable in the skirmish with T.J.J. See over the Soldner episode in Germany. As Campbell moved into the president's office at the university, Trumpler increasingly took charge of the continuing work on the eclipse data from Australia. He was in charge of preparing the final publications of the 15-foot camera results, as well as reducing the data from the 5-foot cameras and publishing the results. The 15-foot cameras were designed to measure stellar displacements near the Sun in order to determine the limb displacement and compare it to Einstein's prediction. The 5-foot cameras captured a wider field around the Sun in order to study in detail how stellar displacements decrease further from the Sun. This refinement of the problem was entirely due to Campbell.

Trumpler also evolved into a "relativity expert" as he took on the task of dealing with the press, giving public lectures, and corresponding with scientific colleagues. He expressed his views on relativity in an exchange with a columnist asking the Lick to check the accuracy of an article he had written for a local paper.[10] The writer argued that the Lick results from Australia favored relativity but not necessarily the philosophical implications about space and time that apparently had to come with the theory. He equated the philosophical aspects of relativity to speculations about space and time made throughout history by philosophers such as Berkeley, Kant, Hegel, and Schopenhauer, and remarked wryly that despite "this prolonged fusillade of words poor old space and time bore up exceedingly well."[11] Trumpler assured the author that his statements were "quite correct and well adapted to the popular frame of mind." Yes, the star displacements had essentially proved Einstein's law of gravitation, but he went further:

It must however not be overlooked, that the geometrical part of Einstein's theory forms at present the only foundation on which this new law of gravitation can be based. This geometrical part of the theory, had led to the prediction of the light deflection near the Sun, recently confirmed and deserves for this great accomplishment more credit than being classed as a philosophical speculation. There is nothing more philosophical about Einstein's space and time measurement than about Newton's force of gravitation, a force acting at a distance without any medium of transmission; and at Newton's time the acceptance of such a force found just as much resistance in the preconceived ideas of the people as Einstein's developed by a number of physicists and mathematicians out of purely experimental results and they differ very widely from philosophical systems like those you mention. It is rather our old geometry of motion that is based on philosophy instead of on experiment.[12]

As the 1923 eclipse approached, Campbell became less and less involved with the eclipse preparations and the work of getting results from the 1922 eclipse. With extra duties to perform as president of the university, and the light-bending test taken care of at last, Campbell asked William H. Wright to lead the Lick expedition to observe the 1923 eclipse. Wright decided to ask Trumpler to take the Ross 5-foot cameras and repeat that part of the 1922 program. In face of the continuing skepticism about relativity despite Lick's confirmation of the limb deflection, the Lick astronomers decided to continue collecting data on the more complex problem. Trumpler elaborated:

> Although the observations of the 1922 eclipse sufficiently prove the existence of light deflections near the Sun and show remarkably good agreement with Einstein's prediction, it seems desirable also to establish accurately by observations the law according to which the light deflections depend on the star's angular distance from the Sun's center. This is of importance in order to decide whether effects of other than gravitational origin (refraction in a circumsolar medium, Courvoisier's yearly refraction, etc.) contribute to the observed star displacements. Einstein's theory requires that these displacements be inversely proportional to the star's angular distance from the Sun's center. It must be proven that this is the only law that will satisfy the observations with sufficient accuracy, before the last doubt as to the correct interpretation of the observations as light deflections due to gravitation can be dispelled. Observations at several eclipses with different distribution of the stars around the eclipsed Sun will be needed to determine the law of the star displacements with the required degree of certainty.[13]

The Lick party went to Ensenada in Lower California (now called Baja California), Mexico. The majority of the observers followed Campbell's advice and stayed on the American side of the eclipse path, including two of the other three Einstein testers. The Mount Wilson group located at Point Loma near San Diego and the Goodsell Observatory went to Catalina Island. Only Miller and Curtis set up in central Mexico against Campbell's advice. They had the last word. Clouds and fog along the coast of southern California and Lower California from Santa Barbara to below Ensenada thwarted the observers located there. None of those parties obtained any Einstein plates. The weather in Mexico caused Miller and Curtis some anxiety, but they succeeded in getting plates with both the Einstein and Noble cameras. The critics were to get another chance.[14]

MOUNT WILSON AND LICK VOTE FOR EINSTEIN

The September 1923 eclipse marked the beginning of a period in the ongoing relativity debate that increasingly cast Lick and Mount Wilson in the role of supporters of Einstein's theory. The ground was laid at a joint symposium on "Relativity and Eclipses" in Los Angeles and Pasadena, held immediately after the eclipse by the American Astronomical Society, the Astronomical Society of the Pacific, and Section D of the AAAS. Campbell, Mitchell, St. John, Trumpler, and J. A. Miller all gave talks. Trumpler and St. John spoke about relativity. St. John created a stir by officially pronouncing the existence of a gravitational redshift in the spectrum lines of the Sun. He and others had sought this effect for years. It was the one empirical prediction of relativity that had not yielded a conclusive positive result. St. John had resolutely stood firm against interpreting his extensive data as favorable to relativity, despite more positive reports from Evershed and others. New information regarding the extremely low density of the solar atmosphere now allowed him to account for the observed line displacements as a mixture of effects. The relativistic contribution was the predicted amount from theory. "The conclusion is that three major causes are at work in producing the regular differences between solar and terrestrial wave-lengths, and that it is possible to disentangle their effects; namely the slowing of the atomic clock in the Sun to the amount predicted by the general theory of relativity, radial velocities of moderate cosmic magnitudes and of probable directions, and differential scattering in the longer paths traversed by the light coming from the edge of the Sun." Trumpler weighed in with the final Lick 15-foot camera results for the 1922 eclipse. He displayed a slide showing the Sun and displaced stars: the calculated mean for the limb displacement was $1.''72$ as compared to the predicted $1.''75$. Trumpler concluded that observations made at the last two eclipses "by entirely different instruments and method" agreed in showing "that light is subject to gravitation and that Einstein's law of gravitation is more accurate than Newton's law."[15]

The British commented in *Observatory*: "It is of great interest to learn that Dr. St. John, who at first took the contrary attitude, has now, as a result of his careful and painstaking researches at Mount Wilson, changed his views." They were satisfied that with the new Mount Wilson and Lick results, all three of Einstein's predictions "have been satisfactorily confirmed."[16] Evershed, who had by now retired from Kodaikanal and returned to set up a private observatory in his native Surrey, remarked wryly to Herbert Hall Turner on the sudden conversion to relativity now that St. John had finally pronounced in its favor.

> I may confess to you that for the last two or three years I have felt a little depressed by the fact or fancy that our Kodaikanal work was not considered of much value compared with Mount Wilson. St. John's first paper, which I criticized at length in the *Observatory* in 1918, vol. 41, 371, seemed to me to be taken much too seriously. I was all the time conscious of the fact that at Kodai we had made a much more extended study of the question, and had measured a great many more plates than had St. John, besides developing a more accurate method of measuring than that employed at Mount Wilson. Yet it seemed that anything from Mount Wilson must necessarily be superior to Kodaikanal, and so St. John's first conclusion has until now been accepted with little qualification.

"I certainly think you have been hardly treated," replied Turner, who had been among those who held more stock in Mount Wilson than Kodaikanal, "& I shall do my best to make amends."[17]

The combined effect of the Mount Wilson confirmation of the redshift and the Lick results on the eclipse test would completely change the nature of discussions about the theory. Critics now had to contend with verdicts from two of the most prestigious astronomical observatories in the world working on two entirely different tests of relativity. The Lick result carried immense weight as it came from Campbell, who had been involved with the eclipse test for almost ten years. The redshift test was more controversial and complex, yet St. John's public reticence to vote prematurely lent enormous credibility to his announcement. Some historians have suggested that the British eclipse results verifying Einstein's prediction in 1919 swayed spectroscopists working on the gravitational redshift to reinterpret their data in favor of relativity. In the case of Evershed and St. John, this is clearly not the case. While there was intense pressure on them to resolve the issue, their understanding of the complexities of the problem led them to be very careful before making any pronouncements on the presence of the gravitational redshift. In fact, despite the powerful vindication of relativity from the prestigious Pacific observatories, the critics did not relent.[18]

The Antirelativity Campaign Gains Momentum

While the Lick and Mount Wilson astronomers were pronouncing in favor of relativity at the joint symposium, Charles Lane Poor was at another session attacking the theory. He based his attack on a reading of Eddington's 1918 "Report on the Relativity Theory of Gravitation." Poor did not like the fact that Eddington had used an approximation to get an equation for the velocity of light near the Sun that was independent of direction. Noting that using the equivalence principle had led to the same deflection as the corpuscular theory of light, Poor insisted that Eddington's relativistic calculation of 1."78 for the light bending was "invalid" because it was based on an approximation.[19]

Poor sent a copy of his paper to Curtis, who encouraged him to "keep up the good work." Curtis complained: "Everybody seems to be 'falling for' the theory, without knowing very much about it. I am certainly still one of the irreconcileables [sic], and it does not seem at present that I can ever believe in it." He confided to Poor that he had an investigation in mind which he was anxious to get at, adding that "there may be nothing in it, but if there is I shall certainly give the other side something to think of."[20] Curtis related the events of the recent eclipse, informing Poor that he and Alter had run "your machine." The program of direct photographs "was almost 100% perfect," which "gives us hope for the Einstein plates as well." They would have to wait until J. A. Miller studied the plates at Swarthmore. As for Poor's experiment, Curtis suggested that it would be advisable to take much shorter exposures of the Moon. "For the 'next time' (second thoughts are always best) it would be advisable to have made by the Eastman Co., a large focal plane shutter, so that the short exposures on the moon could be properly timed. 1/20 sec. would be too long rather than too short." The next eclipse would be taking place in January 1925, and would be visible in the eastern United States. Curtis told Poor that he was planning to take "a modest expedition" to New Haven and that Miller "is thinking of it also."[21]

Miller took all the plates from the 1923 eclipse to Swarthmore after the meetings in September, but college work and outside lecturing kept him from looking at them until November.[22] By December he had come to the depressing conclusion that although his corona photographs had been excellent, the Einstein plates were unusable. Only fourteen star images of the eclipse field appeared on the plates, and most of these were on one side of the Sun. The images were not good, and to make matters worse, the entire reference field was lost because of clouds covering that region of the sky.[23] When Curtis heard the news, he urged Miller to discard the plates: "I strongly advise that you do nothing whatever in the way of measurement of the Einstein plates. Everything about it would make such a course worse than foolish, as you say. Not enough stars is itself fatal, even if you had a dozen check fields. Unsymmetrical distribution is another. I put in one heart-breaking year trying to measure Einstein plates which were not good enough, and I would never do it again. Charge them up to the red ink side, cuss the light clouds, and forget them." Curtis told Miller to take consolation in his excellent corona plates and the fact that he didn't locate in California.[24]

Miller had slightly better luck with the plates that Curtis and Alter took with the Noble camera. They had designed the program to yield two plates, each with a reference field, the eclipse field, and two short exposures of the Moon. The night before the eclipse, they exposed a reference star field on the first plate. During the first sixty-three seconds of the eclipse, they exposed the eclipse field on the same plate. Then, shifting the plate in the camera, they took two very short exposures of the Moon. They repeated the process in reverse on a second plate, with the reference field taken during the night after the eclipse. The two plates would allow them to obtain not only the diameter of the Moon, but also deflections of the rays of light "using the methods applied to the Wallal plates by Campbell and Trumpler."[25] Unfortunately, clouds prevented any night fields from being taken before or after the eclipse, and thin clouds at the beginning and end of the eclipse yielded very few stars. Only four stars plus Venus appeared satisfactorily on the plate. Venus and one star had to be rejected during measurement. The few stars were enough to obtain a satisfactory scale value for the plates.[26] The short exposures of the Moon were "very good indeed," even though there was "a suggestion of corona around the circular image, and the spread of light on the plate may have decreased the diameters slightly."[27]

Miller performed two separate sets of measures of the Moon's diameters on each of the two images. Since he had no measuring engine large enough for the plates at Swarthmore, Frank Schlesinger at Yale made one of his machines available. Miller used the few stars on the plates to determine the scale of the plates. Combining the plate scale with the mean measures of diameters for each set, he obtained values for the angular diameter of the Moon (table 10.1).

Poor arranged for W. S. Eichelberger, director of the Nautical Almanac Department of the U.S. Naval Observatory in Washington, to calculate the mean diameter of the Moon from occultations. He obtained a value for the semidiameter of 992.″96, or 0.″14 less than the mean observed value during the eclipse. Poor's test could possibly have nixed Einstein. Miller concluded: "There is no doubt that the method proposed by Dr. Poor furnishes a practicable test of his theory. The measures indicate a slight radial expansion of the moon, though the quantity found is about equal to the probable error. If the second set of diameters only are con-

TABLE 10.1.
J. A. Miller's Preliminary Results for the Moon's
Angular Diameter

First set of measures	1984″.93, weight 1
Second set of measures	1986″.67, weight 3
Weighted mean diameter	1986″.2 ± 0″.30
Weighted mean semi-diameter	993″.1 ± 0″.15

sidered, the difference, observed diameter *minus* the computed diameter, is 0″.37. It is our belief that this is nearer the truth than the quantity given above."[28] Miller cautioned that the results "can be considered preliminary only." He reiterated the problems: no night fields; scanty and badly distributed sample of stars from which to calculate plate constants (due to thin clouds); short exposures on the Moon were a little too long; and a "fringe of corona and some spreading of light." Nonetheless, he had confidence in his determination of the scale of the plate, and he was certain that "the diameters were not measured too large, and it is probable that they were measured too small." Given the enormous implications of this result, Miller announced his intention "to repeat this experiment on January 24, 1925, making the exposure of the moon practically instantaneous."[29]

When Trumpler saw the paper, he noted the comments regarding the spread of coronal light. "No account is taken of irradiation & spreading; and exposures are very short, it is not sure if irradiation is effective." However, he judged that "with proper care in determining & testing irradiation the method may serve as a test of atmospheric refraction."[30]

While Miller was measuring the 1923 eclipse plates, Poor launched an attack on the 1922 Lick eclipse observations in Australia. His first salvo was to try to refute the validity of the 15-foot camera results on the basis of nonradial displacements. Working with the data published in the *Lick Bulletin*, he divided the stars in the eclipse field into quadrants and averaged the tangential (nonradial) components of the observed displacements. He sent a diagram of his result to Curtis, claiming that he had found noticeable nonradial deflections that seemed to correlate with the structure of the corona. Curtis was enthusiastic and encouraged Poor to publish his critique of the Lick data "ere long." He referred Poor to a graph of deflection versus angular distance from the Sun that appeared in the Lick publication. "It seems to me that C[ampbell]'s results indicate some sort of a deflection, but if you take his diagram you will find it very easy to draw lines through his plotted points which may meet the Sun's limb almost anywhere from 0″.87 out, with very little increase in the probable error of his solution."[31]

Poor told Curtis that he was trying to coordinate a systematic campaign to investigate relativity. "I have in mind the organization of a group of scientists to investigate thoroughly all evidence of the so-called relativitists, also their mathematics. I also have in mind organizing financial assistance, both for experiments and publication. How does this idea strike you?"[32] Curtis decided to tell Poor about the investigation he had alluded to earlier that might "give the other side something to think of."[33] He insisted that the "matter should be regarded as confidential for the present and not published or quoted." "Our own share here will be on the question of the shift of the spectrum lines to the red. We are undertaking some new solar work, under the supervision of Dr. Burns, in which there will be no trouble at all to get the solar wave-lengths with an accuracy of one part in five million. If the shift that St. John thinks he has found is there, we'll get it. But, from some preliminary measures, we shall not be at all surprised to contradict him. Later results may change this."[34]

Curtis was referring to a joint project of the Allegheny Observatory and the Bureau of Standards in Washington that he had initiated in the fall of 1922. The International Astronomical Union had just established a new set of international standards for solar spectroscopists. Curtis saw an opportunity to develop a major spectroscopy project at his observatory utilizing the new standards. He planned to make detailed comparisons of standard solar lines with reference to the cadmium and neon standards from the laboratory. Curtis obtained a National Academy of Science grant to remount and adapt a coelostat and solar spectrographs for the project. He assigned Burns, who was by then his "right-hand man," to do the observations. Curtis proposed a collaboration in which the Bureau of Standards would assign William F. Meggers of the Spectroscopy Section to reduce the data. Bureau chief William F. Stratton had agreed to the plan.[35] Curtis's tentative remarks to Poor were based on several months' observations.

While Curtis and his colleagues worked the solar spectrum, Poor continued his attack on the Lick eclipse observations. He used a deceptive trick to create a strong visual impression that the Lick 15-foot camera results did not actually support Einstein's prediction. He traced a diagram from the *Lick Bulletin* paper showing the observed stellar displacements for ninety-two stars around the eclipsed Sun. It showed the Sun central, the outline of the brightest part of the corona, the outline of the limit of the faintest traces of coronal light, and the ninety-two star positions. Lines emanating from each point (star position) indicated the stellar displacements. But Poor added a few subtle changes. In the Lick publication, the measured displacements for all ninety-two stars appeared in tabular form. Each value had an assigned weight, based upon the quality of the star images and other factors bearing on the reliability of each individual result. In the Lick diagram, full lines depicted displacements of weights 2.0 to 3.9, dotted lines for weights 1.0 to 2.0, and only the stellar point, with no displacement, for smaller weights. Poor took the values for the lowest weight star displacements and put them in his diagram indicating all ninety-two displacements with solid lines. To the left of this "doctored" version of the observations, he displayed a similar diagram with the theoretical Einstein displacements for the same ninety-two stars. He did not mention anything about assigned weights or probable errors, stating that the right-hand figure "is a direct tracing of the star chart in Lick Observatory Bulletin, No. 346. In that chart, however, the displacements of 21 stars were omitted: these have been added to make the above diagram complete." In November 1923 Poor gave a public lecture entitled "The Errors of Einstein," where he presented his diagram to an audience of more than five hundred people at the American Museum of Natural History in New York. He also sent the diagram to Curtis, noting that "I used this a few weeks ago in a public lecture, and it has attracted considerable attention."[36]

Campbell objected strenuously to Poor's antics. Several years later Poor changed his diagram slightly in response to Campbell's objections. He still included the twenty-one star displacements with lowest weights, but indicated the displacements with dotted lines. He sent his revised manuscript to Campbell, explaining his changes and signing off cheerily: "Trusting that these changes will fully prevent any such misconstruction, as you appeared to place upon the diagrams and captions in the informal reproduction of the paper, I am very sincerely yours . . ." Campbell cabled back promptly: "Referring to last paragraph your letter I regret to say that the difficulty was not misconstruction but errors of statements in your earlier manuscript."[37]

Unlike Campbell, Curtis and Burns found the diagram "very interesting." Burns thought the displacements appeared random "like the arrangement of residuals to be found whenever one compares the measures on any two plates of the same star field." Curtis was more circumspect. "There is perhaps, however, a slight preponderance of arrows making less than 90 degrees with the radial direction, instead of there being just as many tangential as radial, as we should expect were there no deflection effect whatever." On his own project, Curtis reported to Poor: "We are very busy here doing some final 'clinching' on the matter of the shift of the solar lines to the red; hope to have things ready to announce at the April meetings."[38]

Curtis was referring to the annual gatherings in Philadelphia and Washington of the two prestigious societies that had vied for Campbell's public lecture on his eclipse results the year before. J. A. Miller had approached Curtis several weeks earlier for a paper of "popular interest" for the

American Philosophical Society's meeting in Philadelphia. Curtis had offered the project by Burns and Meggers: "As a by-product in the program of determination of precision solar wave-lengths which Dr. [Keivin] Burns is carrying on, we have some very interesting results which apparently negative one of the so-called proofs of the theory of relativity. While a couple of months' work must still be done to clinch things, there seems no doubt that Burns' very precise results (accurate to about one part in five million) show *absolutely no shift* of the solar lines to the red." Since he had received partial funding from the National Academy, he told Miller that "strictly it should first be given at Washington," but he told his colleague that he could not see anything wrong with "spilling it" almost simultaneously at the two places.[39]

Not long afterwards, Burns received conclusive results from Meggers. Contrary to what Curtis had told Miller, Meggers found a redshift; but it appeared to change with line intensity rather than being constant for all lines as called for by relativity: "I am convinced now that there is a difference between laboratory arc and solar values of wave-lengths. The puzzling thing is that this difference appears to be a function of line intensity. From my results the lines of solar intensity between 1 and 3 are shifted to the red by 0.003 A, lines between 4 and 6 displaced 0.007 A and the stronger lines (7 to 15) are lengthened by 0.016 A. Here we have the Einstein effect as well as half the Einstein effect and two times the Einstein effect. What does this mean?" Meggers ruled out instrumental effects and thought perhaps electron scattering might explain the result. He urged Burns "to call these results to the attention of astronomers and physicists as soon as possible, because it is good food for thought and speculation." While Curtis did not get his "absolutely no shift, " Meggers had introduced another complication that demanded clarification.[40]

Meanwhile, Poor found a new line of attack on the Lick eclipse observations. In the preliminary announcement that Campbell had made in April 1923, he reported the mean limb deflection as 1″.74. When he and Trumpler published the final measurements of the 15-foot camera plates in July 1923, the revised mean from all the plates was slightly smaller at 1″.72. In discussing possible sources of error, Campbell and Trumpler introduced another factor. They noted an apparent displacement of the check stars toward the Sun, due to some unknown instrumental effect. The Lick astronomers had reason to believe that the eclipse stars had not been similarly affected. For completeness, however, they calculated the limb deflection if the eclipse stars had been displaced toward the Sun by the same amount as the check field stars. The displacements *away* from the Sun due to light bending would then be larger than reported. They calculated under this scenario that the gravitational limb deflection would be 2″.05. In their discussion Campbell and Trumpler emphasized that they only tentatively suggested this result and would await the measurement of the 5-foot camera plates before making a final conclusion.[41] In Britain, Harold S. Jeffreys agreed that it was "doubtful whether corrections based on these [check field star] residuals should be applied to the observed displacements of the eclipse stars." He did note, however, that the corrected value of 2″05 for the deflection, "in excess of Einstein's value, is in close agreement with that obtained from the discussion of the 1919 eclipse results."[42]

Poor exposed this vulnerability of Campbell's data in a lengthy *New York Times* article. His goal was to discredit relativity, Einstein, the British, and the Lick as well as to highlight Miller's recent eclipse expedition as the final arbiter on the matter. He raised the familiar specter of relativity's incomprehensibility and its "transcendental conception of a fourth dimension and warped space." He made the worn assertion that Einstein's prediction of a gravitation bending of light contained "little that is new," mentioning Soldner's 1801 calculation, which he referred to as the "Newtonian" deflection. He repeated his claim that "recent investigations of the mathematics of relativity" had shown that Einstein had made "an erroneous calculation" in getting 1″75 for the Sun's limb deflection. "The relativists cling to the figure 1.75 seconds and claim that if such a deflection be observed it will completely prove the entire theory and the correctness of all their mathematical jugglery." Poor described in great detail how various refraction effects could explain any observed light bending. He raised doubts about the British 1919 eclipse results, claiming that their photographs showed deflections that "did not agree with the Einstein prediction either in direction or in amount." Poor laid similar charges against the Lick results from the 1922 eclipse, adding his new coup de grace. Referring to Campbell's preliminary announcement in April 1923, Poor stated:

> Less than three months later Professor Campbell retracted this statement to a certain extent, for in an official publication of the Lick Observatory he speaks of this April statement as a "preliminary announcement," and gives the true figure as 2.05 seconds of arc, or some 17 percent greater than the Einstein prediction. The "close" agreement of April has now become a roughly approximate agreement. . . . These observations of Campbell do not prove the truth of the Einstein theory: they indicate, it is true, that rays of light are bent during a total solar eclipse, but they do not give any indication as to the cause, or causes, of such bendings.

Poor concluded by drawing attention to the recent success of Miller's expedition to observe the 1923 eclipse in Mexico. "The expedition possessed an equipment far superior to that of any other expedition. Under the able direction of Professor John A. Miller, methods of observation were evolved and a program adopted to entrap the elusive deflections, to force them to record their origin and to determine whether the bending occurs in our own atmosphere or at the Sun." Poor announced that Miller's party had been successful, but it "will take many months of painstaking measurement and calculation to read the riddle of these plates, and not until then can any answer be given as to the cause or causes of these elusive light deflections."[43]

Poor's public attacks on the published Lick results began to have repercussions. In February 1924, Miller wrote to Campbell asking if he would present a paper on his Australian eclipse results at the American Philosophical Society meetings in Philadelphia. He felt compelled to explain: "I should like to add that I did not initiate this request but the truth is that there have been certain newspaper articles . . . in the east indicating that you were less certain about the results than you were a year ago and I find that it has made some impression, in fact one member of the committee said that he understood you had entirely changed your opinion as to the validity of your conclusions." Miller assured Campbell that he had tried to convince the committee that "I thought that certainly was not true" but he felt a statement from Lick "would be a very good thing and probably clear the atmosphere a little bit."[44]

Campbell received the letter in the President's Office of the university. He scrawled on top of the letter: "Send to Dr. Aitken [Associate Director of Lick] at once." A week later a telegram went from Lick to Miller saying that Trumpler would have a paper for the meeting. Miller told Aitken: "I hasten to assure you that we have not been disturbed and I think scientific people have paid little attention to those reports, but there have been rather persistent statements in the newspapers that the theory had not been so well confirmed as the Lick observers at first thought it was. Now, a paper of that sort before the annual meeting of the A.P.S. will put the quietus upon those reports and it seemed to me rather worthwhile to do." Aitken wrote Miller that as soon as he received his letter from Campbell he had "discussed it fully with Dr. Trumpler." Trumpler had completely measured two of the 5-foot Einstein plates and expected to measure a third one in time to use the results in his paper. "The first plate measures are completely reduced and strongly confirm the general results deduced from the 15-foot plate measures. I think it perfectly safe to say that we shall have no occasion to abandon the position taken as to the result of these earlier measures." Aitken added the following postscript: "You may be interested in this connection to know that Dr. See is to give a lecture before the California Academy of Sciences next Sunday, March 2, in which he promises to tell the world the whole truth about Einstein and his pernicious doctrines. I know that you, as well as I, will be greatly edified by his discourse."[45]

Curtis had meanwhile placed his solar spectral lines paper onto the American Philosophical Society program, and he also approached the home secretary of the National Academy of Sciences in March offering to present "a paper of some interest": "Dr. Burns' results on the wavelengths of solar spectrum lines are of the highest precision, and negative one of the so-called proofs of the theory of relativity. It is this evidence against the shift of the solar lines to the red predicted by Einstein that forms the subject of the paper. Dr. Burns is giving this paper before the American Philosophical Society, a few days in advance of the Washington meeting, but the paper may be of especial interest, should Dr. Campbell or Dr. St. John be planning for papers on related aspects of the theory at the N.A. meeting." The home secretary answered that he would enter the paper on the program. "You will be interested to know that Doctor St. John is on the program for a paper under the title 'Gravitational Influence on the Spectral Lines.' "[46] A few days after receiving this confirmation of a confrontation with Mount Wilson, Curtis received a letter from Campbell's secretary at the President's Office of the University of California: "President Campbell asks me to say that Mr. Allen H. Babcock spoke to him two days ago of your apparent plan to publish an article, based upon observations by you and Dr. Burns, to the effect that the Einstein line shift does not exist for the red end of the spectrum. He wants to express his concern that you do not join the society now composed of Thomas Jefferson Jackson See and Charles Lane Poore [sic]."[47]

Curtis was unperturbed. Writing to long-time friend and associate Paul Merrill at Mount Wilson about his and Burns' solar work, he commented: "It looks as though we were getting some 'by-products' of great interest; more work must be done, but we will inform an anxious and waiting public in a preliminary way at the Philadelphia and Washington meetings."[48]

In fact, things continued to go fairly well for the antirelativity coalition. A few weeks before the meetings, Poor gave a short paper at Columbia and at Princeton on "relativity mathematics," in which he claimed to have "unearthed the fundamental errors in the identification of the relativity formulas with the motions of the planets." He sent Curtis a copy, and told him: "The mathematicians of Princeton now admit that the work on this subject of both Einstein and Eddington contains serious flaws, and that there may be no such thing as the relativity motion of the perihelion of Mercury." He had even more exciting news. Poor had not forgotten his idea of organizing a group of scientists to undertake a systematic effort to undermine relativity and had been trying for months to get financial backing. He told Curtis that he had "at last succeeded." "On March 20th, the Council of the New York Academy of Sciences adopted resolutions agreeing to stand sponsor for such an investigation and to have the investigation carried out under my general direction. We are now negotiating for a large fund to cover the expenses of the investigation, including publication of results: we want at least $10,000 per year for several years. We have received decided encouragement, and, while we may not get the entire sum asked for, I feel confident that we will get enough to carry on the work for a year or two." Poor told Curtis that nonmembers of the New York Academy could be involved and that cooperating astronomers and physicists "may have the work carried on at the places equipped for each portion of the investigation." He asked Curtis for a short note of support, which he gladly gave, noting: "It would be difficult to select a field for assistance which would be more timely and appropriate just now." Curtis was "greatly interested" in Poor's perihelion paper. He had tried to work out a solution for Mercury's perihelion motion based on Newtonian principles but had not gotten anywhere. Nonetheless he was "confident that the anomaly will yet be explained under Newtonian laws." Curtis enclosed an abstract of the papers that he and Burns would be giving at the coming meetings; "a good deal more work must be done, but we think our results negative the spectrum shift argument." Pursuing his empirical bent, Curtis mentioned that he had another idea, should Poor's project get off the ground. "I have been hoping to get at an attack on the star displacement argument, by laboratory methods, and by a scheme which may, after all, be a "wild" one, and lead to no results. If you get the money you are after, and have some to spare, it is possible that I could use a thousand or so for apparatus, optical parts, etc., for this experiment."[49]

Before the eastern meetings took place, Curtis and Poor were united in their efforts to undermine relativity on all fronts. Poor focused more on theoretical attacks while Curtis preferred experiment and observation. These differences would later become a source of disagreement.

Confrontation

The Lick strategy to quell the rumors instigated by Poor was to stay on the offensive and announce new results from their 5-foot camera observations. The program had consisted of three separate exposures during the eclipse with twin cameras, yielding six plates. The first exposure was made on plates that were already in the camera, having been exposed to the check field the night before; the second exposure on new plates, made during the middle of the eclipse, had no check field; the third exposure, made on a third pair of plates, stayed in the cameras until the following night, when they were exposed to the check field. Trumpler began work on the 5-foot camera plates immediately after completing the 15-foot camera phase of the project. First he had to design a special device to adapt the original measuring engine to the differential measurement of stellar photographs. Then he measured the four plates from the first and third exposures, which had the check fields on them, together with the four corresponding Tahiti comparison plates. He was able to complete the reductions in time for Campbell to announce them at the eastern meetings. The large field of the plates necessitated including third-order terms in the formula, making the reductions complicated and lengthy. Nonetheless, the probable error of a star displacement from one pair of plates was ± 0″.3, which was only 1.6 times larger than for the 15-foot camera measures, even though the scale was three times smaller.[50]

Campbell presented Trumpler's paper to the American Philosophical Society on 26 April in an afternoon session that he chaired as one of the society's vice-presidents. J. A. Miller, who was in the audience along with Curtis, Keivin Burns, and Charles St. John, received the definitive statement that he had sought regarding the Lick results. The plates showed between 400 and 500 star images each, but Trumpler selected for measurement only 135 to 140 of the best defined images suitably distributed over the plate. He used a mean of the four plates to obtain light deflections for each of these stars. He arranged the stars in the order of their distance from the Sun's center (ranging from $0°.5$ to $10°.4$) and divided them into seven groups of about twenty stars each. The group means of the deflections then only had a probable error of ± 0″.04. The seven group means therefore provided an accurate test of the distance law. Trumpler compared the seven mean values to the theoretical value derived from the formula

$$\text{Light deflection} = 1″.75 \; R/d,$$

where R is the Sun's apparent radius and d the angular distance of the star from the Sun's center. Trumpler reported that the observed group means represented this formula "very closely," with no group residual exceeding 0″.05.[51]

Trumpler also took the opportunity to dispose of a recent critique of the British 19-foot camera and the Lick 15-foot camera results by Franz-Joseph Hopmann at Bonn.[52] Hopmann claimed that by assuming a different scale value of the plates, he would get a constant star displacement with angular distance from the Sun for both the British and Lick data sets rather than Einstein's predicted decrease. Trumpler showed that Hopmann neglected to point out that on his calculation, the amount of the limb displacements from the British and Lick observations would be "entirely discordant" (1″.41 and 0″.68, respectively), where on the Einstein interpretation they gave 1″.98 and 1″.72.[53]

Trumpler reported that the 5-foot camera results "speak still more decidedly against" Courvoisier's yearly refraction effect.[54] Any attempt to fit the group means of the light deflections to the Courvoisier formula led to some of the groups being way off the curve. Trumpler had even tried a combination of the Courvoisier effect and the Newtonian half-effect, but this was "no more successful in satisfying the observations." His conclusion was emphatic: "The preliminary results of the observations made with the 5-foot cameras at the 1922 eclipse fully confirm the prediction of Einstein's Generalized Theory of Relativity as to the numerical values and as to the distance law of the light deflections in the Sun's gravitational field; at the same time they give strong evidence against their interpretation as Courvoisier effect."[55]

In the same session where Campbell spoke, Curtis introduced Keivin Burns who presented his paper on the Allegheny solar spectrum results. J. A. Miller introduced Charles St. John, who gave a paper on "Exploring the Solar Atmosphere."[56] It was only about a year since St. John had finally been able to come up with definitive explanations of his years of observations, and he now found himself drawn into a debate about his interpretation. The controversy was transported to the National Academy of Sciences meeting in Washington, where Curtis presented the paper "Allegheny Results on the Shift of the Solar Lines Predicted by the Theory of Relativity," and St. John presented his side of the argument in a paper, "Gravitational Influence of Spectral Lines."[57]

The disagreement revolved around line intensity. Einstein's theory predicted that the Sun's spectral lines should be shifted very slightly to the red, the shift being only about eight-thousandths of an angstrom. The Allegheny capability could measure wave-length variations ten or more times smaller than this predicted shift. Burns claimed that the observed shifts did not conform to "the simple and uniform amount predicted by the relativity theory. Instead of all the solar lines being shifted by an equal amount to the red, and that amount the quantity predicted by Einstein's theory, a very marked line-intensity factor is found. That is, for the very faint solar lines there is little, if any, shift, and the amount of this shift increases as the wider and stronger lines are used."[58] Curtis presented a table of line intensity versus redshift that showed an "unmistakable progression" from the weakest line which was shifted .002 A to the strongest line which was shifted .015 A. The relativistic prediction lay in the middle of the range at .008 A. Burns and Curtis contended that this progression "must be due to some factor or factors other than relativity, and it does not seem possible to reconcile these results with that theory."

> For the theory requires that all solar lines be shifted to the red by a certain amount, while our results show that the very weak solar lines are shifted only one quarter or less of that amount. That is, if the relativity prediction

is true, we must postulate some cause to shift the very weak lines back toward the violet. Now, while various causes may shift spectrum lines to the red, there is no known case of anything shifting them to the violet, except velocity, which seems untenable in this case. Accordingly the authors regard these results as a negation of one of the so-called proofs of the theory of relativity.[59]

St. John was well aware that there were variations of redshift with line intensity. He and Evershed had battled for years over redshifts of cyanogen lines. Evershed relied on the strong, broad lines, getting a shift of the order of Einstein's amount, and St. John preferred the weak, narrow lines, getting zero shift, even at the limb. When astronomers realized that solar atmospheric pressure is negligible and St. John moved to using metallic lines, he found the same intensity effect Burns and Curtis were reporting. His results for 330 iron lines showed a mean shift to the red agreeing with Einstein's prediction, but when he grouped them according to line intensity "the displacements increase progressively with intensity." However, rather than seeing this pattern as a blow to relativity, St. John had a simpler explanation that, when accounted for, was consistent with relativity. Adams described St. John's approach in a report on work at Mount Wilson: "Intensity is an indication of the level in the solar atmosphere at which the lines originate, and St. John's earlier work had shown, in the case of certain elements, at least, the presence of convection currents, directed upward at low levels, and downward at high levels, with velocities of the order of a few tenths of a kilometer per second. The application of the correction for these convection currents removes the progression in the differences and yields the predicted relativity displacement."[60] St. John supported his conclusion using eclipse data on titanium lines. He showed that ionized lines of titanium, which rise to higher levels in the Sun's atmosphere, exhibit larger redshifts (in excess of the relativity amount) than lines of normal titanium, which occur at lower levels, even though these lines were the same intensity.[61] Earlier research on the nature of the circulation in the solar atmosphere had indicated upward currents occurring over the hot, bright granules of the Sun's surface and downward ones over the larger and cooler interspaces. These motions would yield the Doppler shifts required to explain the apparently divergent results for lines at different levels.[62]

St. John explained the extra redshift at the limb, which had also puzzled Evershed, by molecular scattering occurring as the light has to pass through a longer path in the solar atmosphere. Willem Julius's anomalous dispersion called for more refraction in the red end of the spectrum, which would widen the lines on the red side, producing the extra redshift.[63] St. John's conclusion was that three factors contributed to the observed line displacements: (1) the slowing down of the atomic clock in the Sun to the amount predicted by relativity, (2) radial velocities in the solar atmosphere, and (3) differential scattering in the limb values due to longer paths traversed by the light coming from the Sun's edge. Burns and Curtis simply glossed over velocity effects as "untenable" in explaining the intensity effects. For St. John, who had done extensive studies on motions in the solar atmosphere, they were an integral part of the phenomenon.

Just as Campbell was able to "put the quietus" on the rumors concerning the Lick eclipse results, this new area of controversy erupted around the other empirical test of relativity. Samuel Alfred Mitchell typified specialist opinion in his updated remarks in the second edition of his *Eclipses of the Sun*, published in 1924 after the eastern meetings in April. He insisted that the Allegheny results gave shifts that agreed "very well" with those obtained at Kodaikanal and Mount Wilson. "The real question is one of interpretation." He judged that St. John's picture of rising and ascending gases at different levels in the solar atmosphere had been "abundantly verified by very careful researches," concluding that the redshifts at the center of the Sun "appear satisfactorily explained." Mitchell was not happy, however, with St. John's explanation of the excess shift at the solar limb.

> This explanation of the limb effect seems to be a makeshift and no great confidence can be felt in it until verified by more complete observations. Unfortunately, the necessity of assuming such an effect enormously weakens all of the arguments tending to prove that the Einstein prediction has been verified to exist in the solar spectrum. Unquestionably the wave-lengths in the Sun are greater than those in the terrestrial laboratory, unquestionably the chief cause of the greater wave-lengths is the slowing up of the atomic clock in the Sun,—but there are minor differences in wave-length for which there is now no adequate explanation. Apparently there is nothing to do but wait until these discordances are explained.[64]

While professional opinion largely sided with St. John, the attack from Curtis and Burns kept the debate alive, especially in light of the unresolved limb effect.[65] As St. John and the Lick astronomers continued to face public criticism from the "irreconcilables," they began to coordinate their counterattacks to defend their professional integrity and scientific reputations.

Meanwhile, Poor's antirelativity campaign began to encounter difficulties. At the National Academy meeting in Washington, Curtis received a telegram from Poor urging him not to mention anything about the New York Academy's possible financial support of his antirelativity research project. Curtis kept the matter "entirely confidential" and he heard from Poor six weeks later that the plan had fallen through. "Something happened at the Washington meetings, or very shortly thereafter, which put a spoke in my wheels. The parties, with whom I was negotiating for funds for the Einstein investigation, suddenly cancelled appointments, and have since declined to put up the money."[66] Whatever had occurred to cause this sudden upset, it signaled the beginning of Poor's eclipse. The antics of T.J.J. See in the West and some of Poor's public statements in the East began to cast a pall on the antirelativity campaign. Poor was increasingly associated with the likes of See. Curtis even joked to him that he was afraid "that See may 'adopt' me" on account of his redshift results.[67] Others made the same connection. Several months later a correspondent asked Curtis about the Allegheny work: "So far, I have seen no signs of the monkey wrench you promised, except when Dr. See turns loose some of his characteristic talk."[68]

Poor began to feel increasingly isolated and felt there was a conspiracy to keep criticisms of relativity out of the scientific journals. When he saw the published abstract of Burns' work in *Science*, he wrote Curtis: "The report of your address must have been greatly emasculated in the process of editing, for no 'forbidden' matter appears. This does not surprise me in the least, for it is the definite policy of the editors of practically all scientific publications to decline to print any matter which is not 'a contribution to the theory.' "[69] Poor aired his conspiracy publicly in a letter to the *New York Times*, likely in frustration because his papers were being rejected for publication in the scientific literature. Years later the editor of *Science* wrote that "when we [he and Poor] were colleagues and friends he placed me in an embarrassing position by submitting papers that we were unable to accept."[70]

Curtis had more stature in the scientific community than Poor. He was a high-profile observer who had made his name while at Lick where he pioneered observations of the nebulae and was now director of one of the major eastern observatories. Nonetheless, his stature as a world-class observational astronomer could be reduced by emphasizing his relationship with Poor, and by association, with See. Conversely, the antirelativists could use Curtis's stature to bolster their position. As observational evidence in favor of relativity accumulated in the following years, the position of critic would become more difficult.

A NEW LINE OF EVIDENCE TO TEST EINSTEIN

The 1920s saw explosive growth in astronomers' understanding of the physics of stars. Much of the theoretical work underlying this revolution was due to Arthur Eddington. During the war years, he showed that stars are masses of gas in equilibrium. Outward radiation pressure plays a major role in balancing the inward force of gravitation, especially for giant stars. In 1920, Eddington used his theory to predict the angular diameter of several red giant stars.[71] He obtained the largest angular diameter for the star Betelgeuse—$0''.051$. This implied a volume so great that the orbit of Mars could fit inside it. The density would be so low that many physicists were dubious.[72]

Eddington's theoretical calculations faced the test of observation at Mount Wilson. Francis Pease and John Anderson were working with A. A. Michelson to design and build an interferometer that could measure stellar diameters with the 100-inch telescope. When Hale saw Eddington's published calculations, the estimated angular diameters were "so large that we are planning to measure Betelguese with the 100-inch as soon as possible." On 13 December 1920 the Mount Wilson astronomers found that the star was indeed "a whale of a thing."[73] Eddington first heard of the observations from an American sports reporter, "who, for once, was very welcome since he brought me more information than I could give him."[74] The Mount Wilson discovery was a "beautiful confirmation of [Henry Norris] Russell's theory, proving the existence of giant stars of enormous size, and of mean density about one-thousandth that of atmospheric air."[75] It also verified Eddington's theoretical prediction based on his theory of giant stars.

In 1924 Eddington made an even more astounding prediction. He realized that at the high temperatures and low densities in giant stars, gases would behave like a perfect gas. He could therefore use standard laws from the physics laboratory to calculate how temperature, density, and pressure are related in giant stars. He derived a simple relation showing that the intrinsic luminosity of a giant star depends solely on its mass. To test his equation, Eddington plotted data from a small number of actual giant stars with known masses and luminosities. They fell very close to the theoretical curve. Eddington then tried some dwarf stars to see how far they would be from the theoretical curve. He expected some deviation, because astronomers believed at the time that dwarf stars were too dense to obey the perfect gas laws. He was shocked when they fell onto the curve! Eddington concluded that at stellar temperatures atoms could be packed together much more closely than ordinary matter without any breakdown of perfect gas conditions. In principle, this meant that stars could be compressed to incredible densities. "Stars like the Sun are actually in the condition (have the compressibility) of a perfect gas and are therefore still a long way from maximum density. Then . . . on physical grounds . . . the maximum density [of a star] would be enormous."[76]

One possible candidate for an extremely dense star was the companion of the bright star Sirius. The spectra for both Sirius A and B are similar, indicating the same temperature. Since the companion is so much fainter than the primary, then it should be much smaller. Such very hot, faint stars are nicknamed "white dwarfs." Sirius B was classed as spectral type F, implying a temperature of about 8,000 degrees. Eddington used the temperature and luminosity to calculate its surface area and diameter, "which comes out not very much larger than the Earth; its mass is about 4/5 that of the Sun, so that the density would be enormous—about 50,000 gm. per c.c. That argument is well-known, but I think most of us have mentally added 'which is absurd.' According to the present conclusions, however, it is not absurd; so that, unless the spectroscopic classification has deceived us, the companion of Sirius may be an actual example of how disrupted atoms may pack together to a far higher density than ordinary matter."[77] Eddington figured that if Sirius B were such a highly dense object, then according to general relativity its spectral lines should be shifted to the red by a large amount. He asked Walter Adams at Mount Wilson if he might detect this redshift with the 100-inch telescope: "I have lately been wondering if you would find it possible with your great instrument to measure the radial velocity of the companion of Sirius—with a view to determining its density by means of the Einstein shift of the spectral lines (compared with Sirius.)"[78]

Eddington noted that Sirius B is ten magnitudes fainter than Sirius, but the same spectral type, so that "it should theoretically have a surface 10000 times smaller and a radius 100 times less." He calculated that the Einstein shift should be about fifty times its value for the Sun. He gave a rough estimate of about 30 km sec and a more precise prediction of 28.5 km sec. "Of course this involves a density of about 100,000 x water which seems incredible, but have recently been entertaining the wild idea that it may be just possible." Eddington explained to Adams that the high temperature in the star would strongly ionize the atoms in the stellar interior, stripping away the outer electron shells and effectively shrinking the diameter of each atom. "Owing to the high ionisation in the stars the atoms are almost certainly stripped down to the K-ring and I really do not see what is to prevent them packing close enough to give this density—if the pressure is high enough, as it would automatically become." Eddington emphasized the theoretical importance of his suggestion: "Should scarcely venture to suggest your following this wild idea, if I did not regard a negative result as also of interest. Suppose you prove that the Einstein shift is say under 3 km.[79] That gives a lower limit to the radius of the star and therefore an upper limit to the luminosity per unit area, and you have a very definite challenge to thermodynamicists—how with so low a rate of radiation it shows the A type spectrum."[80]

Adams was in a unique position to appreciate the importance of Eddington's request. As the outgoing president of the Astronomical Society of the Pacific, he had recently given an address to the Society's Board of Governors on the occasion of presenting its Bruce Medal to Eddington.

Adams had chosen to include some technical details of Eddington's achievements, since Eddington would not be present and "the printing is undoubtedly more important as Eddington himself will see it."[81] Adams wrote his remarks before he knew anything about Eddington's prediction regarding Sirius B. He outlined three areas of research for which the British astronomer merited the Society's highest award: "These are: his studies of stellar motions and the problem of the stable character of our universe of stars; the powerful support which he has lent to the theory of relativity, and the service which he has rendered through its interpretation and application; and finally, the series of remarkable researches on the physical conditions in the interior of stars which, for the first time, have given us an adequate explanation of why stars exist in the forms in which we find them, and how they pass through their various stages of evolution."[82] His talk included a detailed summary of Eddington's theory of giant stars, his result that "the total radiation of a giant star depends upon its mass alone, being independent of temperature or density," and its extension to dwarf stars.[83] The arrival of Eddington's letter several weeks after the Bruce Medal celebrations made a tremendous impact.

Adams, who had been the first astronomer to obtain a spectrum of Sirius B,[84] explained to Eddington that the observational problem was "very troublesome." Even under the best conditions "we are never able to obtain its [Sirius B's] spectrum free from that of Sirius. . . . The result is that our plates show the broad spectrum due to Sirius itself and superposed upon it a narrow darker strip due to the companion. I have looked at these plates many times with a view to measurement but have always laid them aside." Given the importance of the problem, however, he and his colleagues decided to try measuring the plates on the Koch registering photometer. This instrument shines light through the glass plate and converts the light intensity to a voltage that determines the deflection of a pen, recording a line tracing of the spectrum on a moving strip of paper. An accurate graph of intensity versus position results, with a dispersion greatly exaggerated compared to the original lines on the plate emulsion. Adams told Eddington that they had already made direct enlargements of the plates and would be doing the measurements soon. "Of course, our results for the companion will be integrated [with Sirius] to some extent, but I believe we may be able to get results of interest."[85]

By early March Adams could report results (table 10.2). "The curves were drawn in sets of three: First, on Sirius, then on the superposed spectrum of Sirius and the companion, and finally on Sirius again. The curves were measured by referring the vertex of the companion to the vertices of the two curves of Sirius, so the measurements were entirely differential. We used four lines and made four sets of curves for all except one of them.

TABLE 10.2.
Adams's Preliminary Sirius B Results

Line	No. of Curves	Displacement mm	Km/s
Hγ	4	+0.004	+9
4404	3	+0.010	+26
4481	4	+0.006	+16
Hβ	4	+0.009	+31

The slit of the microphotometer was shifted for each set so as to reduce the effect of silver grains on the enlargement."[86]

Adams computed the simple mean to be +20 km, "or, if Hβ is given double weight, being by far the best line, the mean is +23 km." He also felt that the result "ought to be multiplied by a factor to allow for the super-position of the spectra of Sirius and the companion. I think that a factor of 1.5 appears very reasonable, which would give a result of not far from 30 km. This is in agreement with your prediction." Adams admitted that the probable error was large, but he saw no reason why there should be a systematic error. "The values given by the individual lines show no greater discordance than we frequently meet in direct measurements of spectra with poor lines. I might say that the probable error for a single determination in the case of the first three lines is about 1/4 of the result. In the case of Hß it is considerably less." Adams was astounded. "I think the most extraordinary features about these results, if taken at their face value, is the evidence they yield for the existence of densities of ten to the fifth power or ten to the sixth power."[87]

Meanwhile, Eddington had revised his prediction for the expected Einstein shift. The change came from a revision of the spectral class of the companion from an A-type star to an F type, which is a little cooler.

> When I wrote to you I took the spectrum as Type A; shortly afterwards I found that later observations at Mount Wilson classed it as F0, so that the Einstein effect would be a little smaller. Taking F0 to mean an effective temperature of 8200° the Einstein effect would be 19 km. per sec.; it may, however, be a little larger, because in a star of this kind (with very large value of gravity at the surface) a higher temperature would be necessary to bring out the F spectrum than in ordinary stars. So I would conclude that 20–25 km per second is the most likely value.

Nonetheless, the measurement matched the prediction and Eddington was amazed at the result. "It is curious to think that the companion of Sirius is only about 20,000 km. in diameter—not very much larger than the earth—rather a contrast to Betelgeuse." Eddington told Adams that he had just corrected proofs of a "rather lengthy paper" which gave "all the evidence which has led me to the view that stellar matter can obtain very high density." He enclosed a diagram of his theoretical mass-luminosity relation (plotted as absolute magnitude as a function of mass). He noted that except for adjusting the vertical displacement of the curve to fit the observed values for the star Capella, "otherwise it is pure theory." He explained the key point about the stars behaving like they were constituted of "perfect gases" and the conclusion that enormous densities must be possible.

> I feel pretty well convinced that higher densities are *possible*, but whether we have been lucky enough to find an actual specimen is another question. It seems almost too good to be true. It looks all right, but owing to the difficulty of the measurement and the rather large probable error I regard your result with some caution—as I think you would wish, and shall not say anything about it publicly at present. I do not know whether you are proposing to take further spectrograms, or measures before you publish it. It is evidently of exceptional theoretical importance besides being a great practical triumph over difficulties.[88]

Adams wrote back that he was "glad" that Eddington took the results as "entirely provisional." Though he saw no reason that the measurements with the Koch microphotometer should be subject to systematic error in a differential comparison of the star, he felt that they "require confirmation" and he preferred "to publish nothing until we can secure this." The star had gone out of reach of the telescope by that time, and all Adams could do was "hope that next year we can undertake the observations with apparatus better suited for the purpose." Adams found Eddington's theoretical results of "extraordinary interest," and he looked forward to reading his paper. "If by chance the companion of Sirius should contribute a final bit of evidence to confirm your theory, it would be worth almost any trouble and difficulty."[89]

PART FOUR

1925–1930
FINAL ACCEPTANCE

Chapter Eleven

RELATIVITY TRIUMPHS

Relations between Poor and J. A. Miller had been friendly since their initial contact to plan for the 1923 eclipse. When Edwin B. Frost of Yerkes Observatory nominated Poor for membership in the American Philosophical Society, Miller and Curtis supported his application.[1] Poor's lunar test at the 1923 eclipse yielded preliminary results favoring an expansion of the Moon. Though Miller played it down until he could confirm his findings, the positive result encouraged Poor. Miller submitted a paper on the results in July 1924, but it was not published until the spring of 1925. Relations between Poor and Miller were to sour by the time it appeared.

The 1925 eclipse attracted intense public attention. The eclipse path embraced large urban centers in the eastern United States, including six observatories.[2] It would be the last total eclipse of the Sun visible in the United States until 7 March 1970. Despite unfavorable conditions, with a midwinter Sun low in the sky and a short duration of totality, astronomers organized many expeditions, especially in the East.[3] At Lick, Aitken joked to Miller that if he could guarantee "beautiful weather" he would love to go, but he decided to "let you Eastern people look after this eclipse without help from us. If you do have good luck you will have good reason to crow over us in the matter of climate."[4]

Miller and Curtis decided to locate at New Haven. Using almost the same equipment as they had in 1923, their program consisted of coronal direct photography, the Einstein tests, and the flash spectrum.[5] Though Miller's paper on the 1923 eclipse had not appeared, everyone in the know was aware that his party had "two Einstein cameras of 15-foot in order to try out some ideas of Chas. Lane Poor that there is no Einstein effect. Poor expects that the measured diameter of the *Moon* will determine this problem."[6] Curtis and Miller had considered not running Poor's test at New Haven because of the unfavorable conditions, but they decided to do it just for the practice. Curtis told Paul Merrill at Mount Wilson: "He [Miller] will run his Einstein's, but isn't calling them that as about all he wants is the practice and data as how best to do it next time."[7]

The publicity for the event was unparalleled. Full-page newspaper spreads on the eclipse plans appeared in the cities where the eclipse would be visible. Poor used the opportunity to describe the lunar test to reporters. He expressed his hope that it would disprove Einstein's theory. The *New York Times* declared: "If Dr. Poor fails in this [lunar test], his experiment will probably go down in astronomy as a sort of Custer's last shot against relativity, which now appears to have an abundance of confirmation." Miller told reporters that he doubted anything would be discovered at the eclipse to disprove Einstein's theory.[8]

Eclipse day in New England was clear and Miller was able to "crow" to Aitken: "The eclipse conditions were surprisingly good. The plates are better than one would anticipate with the Sun at such a low altitude. We got a good many satisfactory photographs." This judgment, however, held only for the coronal plates. Though he obtained excellent short exposure photographs of the Moon, he wasn't able to capture enough stars necessary to determine the scale of the plates. As Miller had not been expecting anything from this eclipse, he was not disappointed, but Poor had to backtrack with reporters. He claimed that the tests had been "hindered" and that he was not looking for conclusive results. Since the temperature was about 5 degrees above zero (Fahrenheit), he argued, the drop in temperature as the shadow passed was only about a tenth of a degree. Such a small temperature change could not produce the "abnormal refractions" he claimed caused the light bending. In previous tests, all done in hot climates, the cooling effect was about 2-3 degrees. His explanations were irrelevant since the real reason for no results was that the conditions had not been adequate. One wonders what he might have said had the plates been good.[9]

Curtis continued to be open to Poor. At New Haven he told the Columbia professor about the paper he had read at the National Academy the previous spring, describing the confrontation with St. John. Later he sent Poor a manuscript of the address. He included a description of an interferometer he had built "to get at another line of investigation bearing on the 'deflection proof.' " Poor replied enthusiastically. "You have to my mind completely disposed of the so-called third proof of relativity and I wish to congratulate you most sincerely on your results: also upon the clear cut way in which you state your position." Poor could not find "the slightest support for the theory" in the work at Mount Wilson. "I gather that St. John himself does not believe in relativity, that he has been obliged to hedge for political reasons." Poor asked Curtis if he had ever looked into how "to destroy the theoretical basis for this displacement, as well as the actual observational results?" He told Curtis that he had successfully attacked the other two predictions: "I can derive his equations for the motions of the planets and for the deflection of light-rays directly from his fundamental assumptions as to 'variable time,' without the use of any 'fourth dimension,' 'curvature of space,' or 'new' law of gravitation." He also claimed that he could show how Einstein had introduced the "elusive two" that made his 1916 prediction of the light bending double his 1911 prediction. Poor hoped that if Curtis would handle the theoretical side of the third test, then the two of them would have all three predictions covered.[10]

Curtis replied warily that "my point of view is somewhat different from your own: I have nothing to support my view-point, and am entirely ready at any time to admit it wrong, and may do so after seeing your papers." He argued that to try to discredit Einstein's mathematics missed the point. If the predicted phenomena were there, and they could not be explained in another way, then Curtis insisted: "We are 'goners.'" He had never gone into the theoretical basis for the third test, he told Poor, but only the observational side through Burns. "The shift, as Einstein predicted it, simply isn't there," he asserted. As for the other effects, the anomalous motion of Mercury's perihelion "appears to be really there." Trying to show that relativity did not predict it "does not satisfy my pragmatic bent." He outlined what he would try were he a good enough theoretician, including checking for its existence by "the tremendous task of doing it by mechanical quadratures" instead of by perturbations as Leverrier and others had done. He also mentioned trying to derive "an effective ellipticity" of the Sun due to its rotation. He told Poor that he had spent "one heart-breaking year" measuring Einstein plates for the 1918 eclipse, and having seen the British and Lick results from the later eclipses, felt that "they are good," and that the effect likely exists. "Now, again pragmatic, Einstein's factor of two does not bother me very much, not so much as the fact of a deflection."[11]

Poor responded by informing his observational colleague that he had been doing much of what Curtis suggested. He told him about his early attempts to explore oblateness of the Sun to explain Mercury's perihelion motion. He had tried to embark on the very task (mechanical quadratures) that Curtis had mentioned: "This was one of the problems I had in mind when I tried to raise money a couple of years ago, or was it only last year. I failed to raise the necessary money and have had to stop work." He also reminded Curtis of his Moon test for atmospheric refraction, and that Miller had apparently found the effect on the Mexican plates. "I have heard nothing from him as to the New Haven plates." Had Poor stopped there, he might have resonated with Curtis; but then he launched into a long explanation of his derivations of the Einstein equations "without any of his 'window-dressing' of warped space, fourth and fifth dimensions, or new law of gravitation." He further insisted on describing his discovery of how Einstein had come up with the "elusive factor of two." Curtis replied briefly that "I shall be very glad to see your 'showing up' of the gradual growth of Einstein's hypothesis, and hope that the day may come when you can get the money to go ahead with the investigations you had to drop."[12]

Miller's article appeared about this time, provisionally confirming Poor's belief that an expansion of the Moon had been observed during the 1923 eclipse. However, Poor's relations with Miller had deteriorated. Not aware of the experimental nature of the New Haven tests, he was still impatiently awaiting results. The next eclipse was to occur in January 1926 and would be visible from the islands stretching from East Africa to the Philippines. The maximum observable duration would take place on the west coast of Sumatra. Poor asked Curtis whether he had made any plans. If Curtis were going and was amenable to taking along his instrument, then Poor would be happy to make arrangements to cover expenses. He complained to Curtis that Miller had "bungled" two chances with the instrument, "and I do not want him to have a third chance." Curtis tried to placate Poor, suggesting he was being unjust to Miller and assuring him that the measures had been "carefully carried out." He reminded his theoretical colleague that weather conditions had prevented the taking of check fields. "The continuous rain and clouds for the last nine days formed about the most exasperating and heart-breaking condition I remember in any of my seven eclipses," he recounted, typically emphasizing the observational side.

Poor had specific complaints about the New Haven eclipse. It was almost four months after the event had taken place, yet he had still heard "nothing as to the results." Miller had taken over the final monetary arrangements, which Poor had initiated with the New York Academy of Sciences for the expedition. Now Poor felt out in the cold. "All I know is that the time limit for reporting to the Academy has long since expired, and nothing definite can be obtained from Miller. I can not even find out whether the agreed upon program was carried out at New Haven." Considering that Poor was at New Haven during the eclipse, we can guess the extent to which Miller had now divorced himself from the Columbia professor. Curtis explained Miller's decision to run the Einstein cameras and Poor's camera "more for practice than for anything else." "We both felt that it was absolutely hopeless to secure any results of any value whatever on such difficult problems with the Sun but 17 degrees above the horizon. As a matter of fact, we were so afraid that some of our astronomical colleagues might laugh at any attempt at that altitude that we called his instruments the twin camera and yours the Noble camera, i.e. not Einstein." Curtis told Poor that if he were running his own party he would be "glad" to run Poor's instrument for him. But he was going with Miller again. "He is not only a mighty good scout, with a combination of good judgment and hard work, but he is a good eclipse man to tie up with."

Curtis advised Poor to provide his instrument with a focal plane shutter to make the exposures short enough, should he decide to send it with Miller. Poor did not.[13]

The Relativity Debate circa 1925

The middle of the decade saw a swing toward acceptance of relativity, largely due to the efforts of Lick and Mount Wilson astronomers who continued to produce supporting evidence. The directors of these renowned observatories made no pretense of understanding relativity. Yet both contributed significant resources to the empirical testing of its predictions. Campbell's international reputation as an eclipse observer and director of one of the top U.S. observatories had initially induced Erwin Freundlich to seek him out. It was natural for Campbell to take up the problem, see it through to some definite conclusion, and to defend his results once they were made public. Mount Wilson was an acknowledged leader in solar spectroscopy, and it was as appropriate for Hale and his team to take up the search for the gravitational redshift as it was for Campbell to take up the Einstein eclipse problem. In both cases, the relativity test fit into research that had been going on before. Just prior to the announcement of the British eclipse results of 1919, both investigations were indicating a negative conclusion regarding the relativistic predictions. After the events of 1919 catapulted relativity into the limelight, Lick and Mount Wilson redoubled efforts to obtain definitive results. Their reputations as world-class research institutions benefited from the world's fascination with judging relativity. Both observatories announced positive conclusions about the same time in 1923, attracting attacks from critics. The nature of the ensuing debates forced Lick and Mount Wilson astronomers to go beyond defending their observations. They ended up supporting the theory.

Critics' motives were suspect. By 1924 T.J.J. See had claimed credit for a breathtaking array of discoveries from the origin of sunspots to the cause of Earthquakes. His tirades against relativity helped create a crackpot aura around the antirelativity group. In October 1924 he delivered a public address to the California Academy of Sciences in San Francisco, claiming that Einstein had made an error. Again, he based his claim on the misinformation being propagated in Germany about Soldner's 1801 paper. In earlier incidents, the press had given See a large hearing with little input from others. This time reporters were sure to solicit other opinions. The Princeton physicist Luther Eisenhart, Frank Dyson from England, and Arthur Eddington all made strong statements contradicting See's claim. Eddington, who was visiting California at the time to give a series of lectures on relativity, judged See's criticisms as "all bosh and nothing to it."[14]

Poor's antirelativity motivation stemmed from his past researches on the motion of Mercury's perihelion. He had made vain attempts to get observational colleagues to look for solar oblateness to explain the effect. After the British eclipse announcement in 1919, the relativity explanation for Mercury's perihelion overshadowed all others. Poor's first instinct was to approach the light-bending effect in the same way he had attacked Mercury's perihelion problem. He made theoretical calculations along classical lines to account for the light bending and suggested observational tests to search for solar and atmospheric phenomena that might explain the effect. Curtis became a strong supporter in these efforts, facilitating the running of his lunar test. When Poor failed to obtain financing for his antirelativity project, he turned to attacking "relativity mathematics." In this effort, he would follow in the footsteps of others in making erroneous statements regarding Einstein's "factor of 2." These errors and his blatant publicity seeking began to tar Poor with the same brush as See.

Curtis's motivations were similar to Poor's, though as an observer and an astronomer of international reputation, he did not feel isolated. He harbored lingering resentments against Campbell's handling of the 1918 Goldendale results, although he accepted the Lick 1922 results, whereas Poor and Keivin Burns did not. Nonetheless he encouraged Poor's attempts to discredit the Lick observations, ignoring his questionable practice of suppressing all discussion of probable errors. Curtis had no idea how general relativity predicted the various effects being tested, and he continued to try to find classical explanations for them. Poor's theory and Curtis's observation constituted a marriage of convenience. Except for alluding to an occasional unconventional idea to test the light-deflection prediction in a laboratory, all Curtis did was repeat the eclipse test with Miller. Upon assuming the directorship of Allegheny Observatory, he turned more and more to instrumental design and construction. After initiating the solar wavelength project with the Bureau of Standards, he had nothing to do with it except to build some of the apparatus.[15] It came as a pleasant surprise when the research opened up a line of investigation that seemed to negate the relativistic interpretation of solar line displacements. Curtis took on the debate with enthusiasm.

Burns had met Curtis in early years spent at Lick. He took a position at the Bureau of Standards in 1913 and became friendly with Curtis during his wartime work there. Burns left the Bureau in 1919 and was volunteering at Lick when the relativity news came from across the Atlantic. A convinced skeptic then, his position never wavered. After floating around Lick and Mount Wilson for a year, Curtis took him on as astronomer at Allegheny, one of his first acts as the new director. To him, Burns had a keen mind and was a top-notch spectroscopist. When the relativity by-product emerged from solar spectroscopy, it was natural that Burns would be delighted, as he shared Curtis's antirelativity sentiments.

William Meggers's attitudes differed from those of his astronomical colleagues on several levels. Curtis objected to the influx of Europeans into American scientific institutions. He maintained a strongly anti-German position after the war, especially regarding the entry of German scientists to the newly formed International Astronomical Union. In contrast, Meggers complained in 1921 to the secretary of the American section of the Union:

> I venture to add that the most interesting developments in modern science are the theory of relativity and the quantum theory, both of which have tremendous importance in Astronomy and Astrophysics. Most of the developments in these theories have been made by the very people which are at present not invited to cooperate with the Union. It would certainly not promote the work of the Union to ignore the investigations of Einstein, Planck, Sommerfeld, Franck, Schwarzschild, Küstner, Ehrenfest, Epstein, Kayser, Konen, Runge, Paschen, Eder, Vaenta, Exner, Haschek, etc., etc.[16]

Einstein visited the Bureau of Standards during his American tour in 1921. Meggers met the founder of relativity, and related the experience to Burns: "I had a pleasant visit with Professor Einstein last week and had no great difficulty in conversing with him in German. . . . By the way, have you seen in [Comptes Rendus] for March 7 in which Perot confirms the Einstein shift for the b group of magnesium?"[17] Curtis also met Einstein in Washington. Contrast his description of the famous scientist:

> He surely looks like the fourth dimension! Face is somewhat sallow and yellowish, redeemed by very keen bright eyes. But wears his hair a la Paderewski in narrow greasy curls of small diameter and four or five inches long. Came over for the Zionist movement and is apparently somewhat astonished and bothered by the fuss made over him and by requests for lectures, which are not so numerous now that some of his prices have become known. He wanted (doubtless for the Zionist fund) $1500 for a single lecture at Cincinnati, and a member of the Academy told me that he had asked $15,000 for a series of five lectures at the U. of Min[nesota]!! (Did not get it, however!)[18]

On the strictly scientific side, Meggers differed from Curtis and Burns as well. When Burns sent him a manuscript of some of the results that had accumulated by the fall of 1925, Meggers admitted he had been hoping to explain the intensity effect in a way conducive to relativity: "Einstein and St. John will not find any pleasure in these remarks, but there is apparently nothing else to do at present except call attention to the many factors involved in the line shifts. I was hoping that so-called anomalous dispersion would explain everything." Meggers had initially appealed to scattering to explain the intensity differences. The accepted scattering law would give "shifts in the right direction and of the correct order of magnitude for lines of different width," he told Burns, but "the effect in the red should be much smaller than in the violet . . . whereas the gravitational shift is directly proportional to wave length." He asked Burns: "Is it probable that the gas absorbing strong red lines descends faster than the gas absorbing strong blue lines?" Several years later, St. John proposed a different mechanism involving motion of gases that achieved the result Meggers was seeking.

Meggers told Burns that he was "ready to accept your statement that it does not seem wise at this time to form a judgment concerning the nature of the red shift." A copy of Burns's paper was mailed to Einstein in December and Meggers referred to it in a letter to Einstein a couple of months later: "I regret to state that the solar wave lengths behaved in such unaccountable manner that we were forced to suspend judgment on the gravitational shift."[19]

Meggers stuck by his position of suspending judgment, even as others began to accept the gravitational redshift as verified. Upon receiving a copy of a report from a physics colleague at Berkeley in September 1926, Meggers wrote: "I do not know who wrote the paragraph on relativity in last year's report, but the red displacement of solar lines was not discussed and consequently the results of Burns and Meggers . . . were not quoted. St. John's explanation appears to us to be somewhat artificial and we do not believe that he has settled the question." Meggers asked his colleague to add the statement: "Burns and Meggers have concluded from interferometer comparisons of solar and vacuum-arc-wavelengths that it is impossible at present to say whether or not the predicted red shift resulting from gravity, is present." He also requested that the conclusion in the report be changed to read: "The present status of this whole subject seems to be that, aside from [Dayton C.] Miller's own work, the displacement of solar lines, all of the experimental tests of relativity are reasonably well confirmed."[20]

J. A. Miller also became involved in the antirelativity activity through his association with Curtis and Poor. He had decided to conduct Einstein tests in an initial rush of enthusiasm. The 1922 eclipse in Australia was beyond his financial reach so he waited for the next one in 1923. During the planning stage, Curtis joined him and got him involved with Poor's lunar test designed to disprove relativity. While Miller was reducing the lunar observations taken at the 1923 eclipse in Mexico, he saw the newspaper reports of Poor's public activity directed against the Lick results. He never explicitly mentioned Poor when he asked Campbell to speak at the American Philosophical Society's 1924 annual meeting. Whether his association with Poor made him feel awkward or whether he was simply being discreet and reserving judgment on the matter is not clear. All evidence suggests that he accepted Campbell and Trumpler's result, while following through with his plans to try the test again at the 1925 eclipse. Miller continued to carry out the Einstein tests at subsequent eclipses with funding from parties hoping to disprove relativity. He would encounter a great deal of frustration over the light-bending test, harkening back to Campbell's early troubles with his 1918 observations.

D. C. Miller used the intense interest in relativity to get funding and support for the continuation of the ether-drift experiment he had abandoned after Morley's retirement. His association with Hale at Mount Wilson could only be advantageous. For Hale, it was natural to take on the ether-drift experiments of Miller and Michelson. They further highlighted the importance of his observatory for leading-edge research on the great problems of the day. When Ludwik Silberstein became involved, his theoretical prowess attracted Hale. For Silberstein, the collaboration enhanced his credibility and reputation. The Americans also provided him with the means to continue his search for observational proof of his beloved ether.

As tests continued and more results apparently contradicting relativity appeared, Lick and Mount Wilson took a more aggressive supporting stance for the theory. They could repel Poor's theoretical attacks by exposing where he made mistakes in interpreting Einstein's original predictions, and question his motivation for the assault by association with T.J.J. See. The entry of Curtis, Burns, and Meggers into the fray contradicting the Mount Wilson redshift results lent more credibility to the critics. Their work succeeded in convincing some scientists to suspend judgment on Einstein's "third test." The resurrection of ether-drift experiments satisfied some of the older generation of ether enthusiasts. As the decade progressed, however, it ultimately provided another avenue of support for relativity.

ANNOUNCEMENTS FOR AND AGAINST THE ETHER

In July 1924 D. C. Miller brought his modified equipment back to Mount Wilson. He was able to resume observations in March 1925 and continued until about the middle of April. He was still getting a slightly positive result. His stay on the mountain brought him into contact with Gustav Strömberg, who began to help him develop an acceptable interpretation of his results. Strömberg had been working for years deriving the solar motion with respect to stars and clusters. By grouping the stars according to their mean velocity, he had discovered a remarkable pattern. The low-velocity stars had a small range of velocities centered on the mean, while the high-velocity stars had a wide range of values around a different mean. The interesting feature was that the mean for the high-velocity stars was close to that of the spiral nebulae. The spirals had the highest mean velocity of any celestial object with the highest range of velocities. These facts led Strömberg to posit a double velocity distribution for all celestial objects. One corresponded to local motions (low-velocity stars) and the other to motions in a fundamental system or "world frame" (the reference frame of the spirals.)[21] Strömberg pictured the spirals as ships moving at random on the sea. Our vantage point is that of a passenger on one of the ships. The range of velocities of the other passengers (low-velocity stars) is small, as is their mean velocity. Sea gulls (high-velocity stars) can move fast in all directions, and they form a group with a large velocity range. Their mean velocity is almost at rest in the air; relative to the observer's ship, their mean velocity is high. The other ships have the highest velocity range, and their mean velocity is almost the same as that of the gulls.

To complete the picture, Strömberg had to explain why the velocity distribution in the "world frame" was not larger than observed. In principle, velocities for the spirals much greater than the observed mean would be possible. Strömberg had to postulate some restriction mechanism by which velocities of objects moving randomly in the "world frame" were limited. He took his cue from scientists who were objecting to Einstein's suggestion that matter completely determines the metrical properties of space. Strömberg suggested that space or space-time might have fundamental properties independent of the existence of stars, properties only modified in the neighborhood of matter. The limiting of cosmical velocities might be connected to his fundamental property of space, "which prevents the material cosmic system from 'evaporating,' and gives the stellar frame a certain degree of rigidity." Strömberg's idea led him to consider the ether as real and to search there for his velocity-limiting mechanism.

Of course, this notion was antithetical to the spirit of relativity. As early as 1922, the Mount Wilson astronomers realized that Strömberg's work was challenging Einstein. Strömberg was near the end of an extended investigation of the space-motion of the A stars as a class. He had found that the direction and speed of the Sun relative to stars of high velocity agreed with the solar motion derived from the spirals. Walter Adams reported to Hale that Strömberg was searching for a mechanism by which velocities would be restricted in some world frame. "He is going to write to Lorentz asking his opinion on the possibility of a system of coordinates in "inertial space" which would make velocities a minimum, and so account for the peculiar distribution of high velocity stars. It would hit the relativists to find nature preferring any special system of coordinates."[22]

By 1924, however, Strömberg could call for support from Einstein himself for his appeal to the ether. "The theory of relativity does not compel us to deny the existence of such a medium; on the contrary, the four-dimensional space-time has been identified by Einstein with the ether, which he regards as necessary in order to understand the existence of 'absolute' rotation and acceleration."[23] Strömberg was referring to the inaugural lecture Einstein gave on 27 October 1920 to launch his tenure as visiting professor at the University of Leiden. Lorentz created the position to bring Einstein regularly to Leiden. Einstein chose the problem of the ether in the theory of relativity as his topic, likely in deference to his mentor Lorentz. Einstein evolved the idea in correspondence with Lorentz. He equated the geometry of space, expressed as the metric tensor that describes the gravitational field, as the "ether." His conception was not the traditional ether that can be conceived as a manifestation of absolute space. Rather, he used the term to describe the gravitational field. His friend Besso teased him that he was being kind to Lorentz: "You have endowed the word with its only possible meaning in the new sphere, to prevent the people who believe in it, especially Lorentz, from being further alarmed by apparent deviations." Besso was right. Einstein was simply using the ether as a way of denying the idea "that empty space has no physical qualities whatsoever." Einstein never used the term 'ether' in any of his technical papers, although the Leiden lecture was reprinted many times over his lifetime.[24] His Leiden lecture and subsequent publications caused much confusion. His critics used it to attack him. Others, like Strömberg, used it to support their own ideas.

By 1925, D. C. Miller had dramatically changed his approach to his ether-drift experiment as a result of his association with Strömberg. He related this change in his presidential address to the American Physical Society later that year. "Previous to 1925 the Michelson-Morley experiment has always been applied to test a specific hypothesis" from a stationary ether through which the Earth moved, to effects of magnetostriction on the interferometer. "Throughout all these observations, extending over a period of years, while the answers to the various questions have been 'no,' there has persisted a constant and consistent small effect which has not been explained." Now Miller had decided to look on his interferometer as an instrument suitable for determining the relative motion of the Earth and ether. He then set himself the task of indicating the direction and magnitude of the absolute motion of the Earth and solar system in space, "independent of any 'expected result.' "[25]

Miller determined the magnitude and direction of the maximum displacement of the fringes as the interferometer rotated on a bed of mercury. The direction of maximum displacement varied periodically throughout the day, averaging about 45 degrees west of north. The amount of displacement also varied periodically, with a maximum of about 10 kilometers per second.[26] Strömberg helped Miller undertake graphical and numerical analysis to find a theoretical motion of the Earth and solar system that would fit the data; but the direction of maximum fringe displacement should have varied around north rather than 45 degrees west. If Miller ignored this western shift, he derived a motion of the Sun and solar system toward a point in the constellation Draco[27] at a velocity of 10 kilometers per second. Miller's results were independent of the time of year, which he interpreted to mean that the Earth's orbital motion of 30 kilometers per second must be imperceptible. Miller figured that if his equipment could not detect a velocity of 30 kilometers per second, then it was reasonable to suppose that a larger cosmical motion, about 200 kilometers per second, was showing up as a relative motion of Earth and ether of only 10 kilometers per second at Mount Wilson. "In order to account for these effects as the result of an ether drift it seems necessary to assume that in effect, the Earth drags the ether so that the apparent relative motion at the point of observation is reduced from two hundred, or more, to ten kilometers per second, and further that this drag also displaces the apparent azimuth of the motion about 45 degrees to the west of north."[28] Miller credited Strömberg with helping him with the "extensive calculations" of the solar motions. With Miller's new interpretation of his results, he could move beyond being the "lone critic" of the famous Michelson-Morley experiment, to being a contributor to the growing study of cosmical motions. His association with Mount Wilson further enhanced his credibility.

Meanwhile A. A. Michelson had completed the experiment in Chicago that Silberstein had convinced him to perform. Early in 1925 he announced provisional results at a public lecture hosted by University of Chicago president Ernest De Witt Burton. His results confirmed Einstein. A *New York Times* reporter quoted Michelson: "There is no question that the tests furnished another striking confirmation of his brilliant work." The reporter confused the story, however, writing, "Professor Michelson said . . . that the Einstein theory, as well as the ether drift theory, was correct."[29] By April, final results were in. Arthur Compton read Michelson's paper in his absence (due to illness) at the April 1925 annual meeting of the National Academy of Sciences. The paper, entitled "The Latest Test of the Einstein Theory," reported results on the rotation of the Earth that were consistent with general relativity.[30]

Miller presented his preliminary results the next day in a paper, "Report on Ether Drift Experiments." He claimed that he had obtained a positive result in his recent runs of the famous Michelson-Morley experiment conducted at Mount Wilson. He had definitely observed a positive displacement of the interferometer fringes that would indicate a relative velocity between the Earth and the ether of about 10 kilometers per second. "There are no corrections of any kind to be applied to the observed values.

In the work so far, every reading of the drift made at Mount Wilson has been included at its full value. No observation has been omitted because it seemed to be poor, and no 'weights' have been applied to reduce the influence on the results, since no assumption has been made as to the expected result." This dig clearly echoed Poor's public attacks on Campbell and Trumpler's eclipse data as well as earlier criticisms of the British decision to exclude their Sobral astrographic results. Miller interpreted his result as a partial drag of the ether by the Earth that decreases with height above the Earth's surface. He also expressed the belief that this interpretation would explain a reexamination of the earlier Cleveland observations done on the hill. Miller predicted that his work "will lead to the conclusion that the Michelson-Morley experiment does not and probably never has given a true zero result." He commented to reporters at the meeting that "it cannot be considered as basic experimental evidence of the Einstein theory of relativity."[31]

The reaction to Miller's announcement was dramatic. Ludwik Silberstein immediately proclaimed that relativity was dead. Curtis was at the meeting. He told Poor, "Silberstein acted as chief mourner, for he admitted that Miller's results appeared to shoot the theory of relativity full of holes." Silberstein later told the McGill University physicist, Louis King, "I had . . . the honour of killing officially Einstein's Relativity of 1905." In the pages of *Nature* he triumphantly asserted that the partial drag theories of the ether had been vindicated. Eddington immediately engaged him in a heated exchange. Sir Oliver Lodge, the great defender of the ether, did not appear so sanguine about Miller's effect. "The history of science has constantly shown that small residual effects *may* contain the germ of important discoveries. I hope that it may turn out so in the present instance, though I cannot say that I hope it with any confidence." Walter Adams wrote Miller: "Your values are very convincing, especially so in the matter of the constancy of the direction indicated. I think everyone will agree that the continuation of this work is one of the most vital in modern physics."[32]

Miller succeeded in putting the ether back on the research agenda and gave the antirelativity faction a tremendous boost. Scientists around the world immediately began making plans to redo Miller's experiment.

Announcement of the Sirius B Results

While D. C. Miller was at Mount Wilson resurrecting the ether, Walter Adams was sitting on his dramatic Sirius B results, awaiting another favorable appearance of Sirius and its companion. During the fall of 1924, Eddington had a chance to visit California for an extended visit. The British Association had its annual meeting in Toronto, Canada, that year, which brought Eddington across the Atlantic. Many scientific interests in the United States vied for him to visit. The University of California succeeded in attracting him to Berkeley for ten weeks, where he gave two courses in the Department of Physics, including an advanced course on the mathematical theory of relativity. He also offered a series of six non-mathematical public lectures on relativity. During his Pacific stay, Eddington also visited Mount Wilson for several days. He gave a few talks for the observatory staff and members of Caltech.[33] While there he undoubtedly had a chance to examine Adams' plates of Sirius B and D. C. Miller's interferometer.

Eddington's visit to the Pacific coast greatly advanced astronomers' understanding of relativity and stellar astrophysics; and the British theoretician witnessed firsthand the thriving centers of research there. He also had the dubious pleasure of commenting on the latest of T.J.J. See's antics at the California Academy of Sciences about his "discovery" of Einstein's errors. On his return trip in December, Eddington gave various lectures in the East.[34] The episode served to give him a high profile in America several months before Adams would be ready to announce results of his Sirius B observations.

While waiting for the Earth to move around its orbit and get Sirius onto its night side, Adams devised a new arrangement for photographing the spectrum. Supports for the auxiliary mirrors produced diffraction rays that interfered with the spectrum. Adams used diaphragms with circular apertures to reduce this interference and found a "marked improvement."[35] Sirius was again accessible to the 100-inch telescope in 1925. The observations and reduction of the data would take several months. As the annual meeting of the National Academy of Sciences in Washington approached, Hale and Adams were anxious to announce some results—even if Adams could not prepare a paper in time. The academy had arranged for a special public session on relativity and on the recent eclipse of January 1925. Remembering the previous year's relativity confrontations, the Mount Wilson team wanted to be able to contribute their new results. The home secretary of the academy, too, wanted some fireworks. He wrote Curtis several weeks before the meeting: "Your paper is certain to be of unusual interest. Further, you know how to talk intelligibly to the layman. Can you not introduce the Eclipse-Einstein group of papers?" Curtis replied that his own paper "contains some 'new stuff,' and yet it is of a nature rather more suited to one of the regular sessions than to an evening public session." Nonetheless, he was willing to cooperate.[36]

At Mount Wilson, the time to leave for the meetings arrived, and still no definitive answer was available. Hale went off to Washington while Adams stayed behind. On the eve of the meeting, Adams sent a telegram to Hale at the National Research Council: "Hardly possible wire full statement Sirius results for Academy but glad to have announcement if suitable. Displacement for Beta twenty nine kilometers Gamma eleven other lines intermediate. Spectrum Sirius bluer than companion hence superposition gives smaller displacement for violet lines. Correction for density factor gives values in good agreement with Beta where spectrum is nearly pure. Consider best value twenty five kilometers. Eddington gives twenty to twenty five kilometers."[37] Adams wired a similar message to Eddington in Cambridge. The International Astronomical Union was holding its triennial meeting there, and Eddington was slated to give a lecture on relativity. He was able to make the dramatic announcement that he had just received a telegram from Adams. The 100-inch telescope had revealed a displacement of the spectrum of Sirius B in agreement with Eddington's prediction based on general relativity.[38]

In Washington, the "Einstein-Eclipse" session turned out to be tame. Curtis had no new results and J. A. Miller had not yet measured his plates from the 1925 eclipse. Curtis gave his opening address on "Solar Eclipse Problems" as planned, but his other paper, coauthored with Burns, dealt with the infrared flash and coronal spectra of the New Haven eclipse. Hale's news of a startling verification of the gravitational redshift and the existence of a "white dwarf" must have made an impression on the assembled scientists, but the American press did not pick it up. The only relativity fireworks involved ether-drift experiments. The press ran with Michelson's and D. C. Miller's papers. First they announced that Michelson had verified relativity. The next day they proclaimed that Miller "Strikes a Blow at Einstein Theory."[39] Immediately upon hearing the news of D. C. Miller's ether-drift result, Science Service in Washington cabled Eddington for his opinion. Eddington's reply sent Edward E. Slosson scrambling for news on the latest development from Mount Wilson. He cabled Adams: "You may be interested in Eddington's reply to our cable asking opinion of your and Millers experiments. Quote it is difficult to reconcile Millers results of the Earths velocity with Michelsons latest experiment on the Earth rotation. I must await details of the experiment. Adams has accomplished a striking new test of general relativity that demonstrates the usefulness of this theory as an aid to astronomical progress and confirms an astronomical hypothesis previously doubtful."[40]

Eddington's remarks highlighted an important fact that would greatly affect astronomers' attitudes toward Einstein's theory. In addition to being a further test of the theory, the Sirius B result was the first time that relativity had shown itself to be a useful theoretical tool for astronomers. The latter half of the 1920s would further demonstrate that general relativity could assist astronomers interested in the large-scale structure of the universe. The course of these later developments was set at the April

1925 National Academy meetings. At one of the sessions, the Harvard astronomer Harlow Shapley introduced a young nonmember of the academy, Edwin Hubble from Mount Wilson. His talk, "Spiral Nebulae as Stellar Systems," resolved once and for all the long-standing debate on the nature of the spiral nebulae. For a long time the question had been whether or not the spiral nebulae are "island universes" similar to our Milky Way stellar system and external to it. Hubble showed that the spirals are outside the Milky Way—enormous stellar systems like our own. Within another four years, Hubble's work would open up a new observational field in astronomical research that verified cosmological predictions from general relativity. It would also help to create a new discipline in astronomy—relativisitic cosmology.[41]

Shortly after the academy meetings, Adams showed the Sirius B material to Henry Norris Russell. The Princeton theoretician agreed that the evidence was "amply sufficient to warrant publication." Adams submitted a paper to the National Academy and sent a copy to Eddington. "The results . . . have come out much more consistently than I had dared to hope." However, he was still dissatisfied with the need to apply a correction factor. "I wish that it were unnecessary to apply any correction factor for the superposition of the spectrum of Sirius. The results used, however, cannot be seriously in error. It should be possible, with our present knowledge of the problem, to secure somewhat better plates in the future and perhaps to avoid the necessity for such corrections as have been applied."[42]

When Adams photographed the spectrum of Sirius B, scattered light from Sirius superimposed its spectrum onto the plate as well. The intensity of the scattered light from Sirius was greatest at short wavelengths, while that of the companion peaked at longer wavelengths. Adams used the continuous spectrum to determine his correction factor. He compared the intensity of the companion with that of Sirius at five regions in the continuous spectrum. At these five regions, he determined the ratio of the photographic density of the companion to Sirius. He then applied these ratios to the hydrogen lines. The assumption was that for both the companion and Sirius, the relation of line intensity to that of the continuous spectrum is the same. At the wavelength of the hydrogen line $H\beta$, the interference from Sirius was a minimum and required no correction. For $H\gamma$, the ratio of intensities was nearly unity so Adams applied a correction factor of nearly 2. For other lines, he used an approximate formula. Since the relative intensity of the scattered light from Sirius was greater toward shorter wavelengths, the superposition of its spectrum on that of the companion tended to reduce the line displacement toward the red (longer) wavelengths. To correct for this effect, Adams multiplied the measured dis-

TABLE 11.1.
Final Means for Adams's Sirius B Measures

	Km./Sec.
Hβ	+26
Hγ	+21
Additional lines	+22
Mean	+23

placement by the correction factor. For Hγ, the final result would be almost twice the measured displacement.[43]

Adams's final results were based on four plates, measured in three different ways. He made eight determinations from all four plates using Hβ, six from three plates using Hγ, and one or two plates for eight other lines in the spectrum (table 11.1).

This result of +23 kilometers per second was a differential measure of the radial velocity of the companion relative to Sirius. The computed radial velocity for Sirius was 1.7 kilometers per second. Thus the displacement of the lines for the companion was +21 kilometers per second, or +0.32 angstrom. Interpreted as a relativity displacement, this result yielded a diameter of 18,000 kilometers for this strange object. Its density would be 64,000 times that of water.[44] Adams compared this result to the theoretical prediction:

> Eddington had calculated a relativity shift of 20 km./sec. on the basis of a spectral type of F0 and an effective temperature of 8000° for the companion. The resulting density is 53,000 for a radius of 19,600 km. Although such a degree of agreement can only be regarded as accidental for observations as difficult as these, the inherent accord of the measurements made by different methods, and in particular with the registering microphotometer, is thoroughly satisfactory. The results may be considered, therefore, as affording direct evidence from stellar spectra for the validity of the third test of the theory of general relativity, and for the remarkable densities predicted by Eddington for the dwarf stars of early type of spectrum.[45]

Adams wrote Eddington: "My principal feeling regarding the results of this work is a strong one of pleasure that they confirm so remarkably your beautiful prediction regarding the density of matter in the white dwarf stars." Eddington, of course, was "very delighted," and replied to Adams that he was deeply interested in the details of his paper. "But it is very staggering that it should turn out like this."[46]

When Adams's paper appeared in July, the press picked up the double story that a new test of relativity had emerged from Mount Wilson and that it also confirmed Eddington's theory of white dwarfs.[47] For profes-

sional astronomers, the importance of general relativity in verifying a startling result of the new discipline of theoretical astrophysics dramatically
vindicated relativity. In 1926 Eddington published his book *The Internal
Constitution of the Stars*, which summarized the theoretical work in this
growing field of research from 1916 to 1926. The book sold more than
1,200 copies in the first four years and became a standard text for astrophysicists. After presenting the details of Adams's investigation, Eddington concluded: "This observation is so important that I do not like to
accept it too hastily until the spectroscopic experts have had full time to
criticize or challenge it; but so far as I know it seems entirely dependable.
If so, Prof. Adams has killed two birds with one stone; he has carried out
a new test of Einstein's general theory of relativity and he has confirmed
our suspicion that matter 2000 times denser than platinum is not only
possible, but is actually present in the universe."[48]

JOHN A. MILLER AND THE ECLIPSE TESTS

While the climax of empirical verifications of relativity was building in
the western United States, parallel events were unfolding in the East. Despite Poor's withdrawal from his association with Miller and Curtis,
Miller carried out the lunar test during the Sumatra eclipse of 14 January
1926. He borrowed a 9-inch aperture lens of 63 feet focal length from
Indiana University for the purpose. Two of the expeditions that went to
observe the event were to tackle the Einstein problem: Miller's party and
one led by none other than Erwin Freundlich. Lick decided against sending
an expedition, and though the British dispatched an expedition, the program was solely to study the corona.[49] Unfortunately, the weather in Sumatra was cloudy at several places yielding mixed results. The British had
the best luck. Miller, not far away, had thin clouds floating over the region
of the Sun during the eclipse. Though Curtis was not able to obtain flash
spectrum photographs partly because of the haze, they secured successful
direct photographs with the coronal and Einstein cameras. Freundlich got
photographs, but being near Miller's group, he also took them through
haze.[50]

Miller wrote Aitken several months after the eclipse. In retrospect, he
judged the weather "very favourable" and claimed that the direct photographs of the corona were "very good." He was also pleased with the
Einstein photographs. "We have four Einstein plates made with six and
three-quarter inches [aperture] fifteen foot focal length. These plates contain thirty eclipse stars, a part of which are rather faint, but most of which
are very good images which can easily be measured. They are also well
distributed, nearly all within two degrees of the Sun. On two of the plates

there is a night field, and on the other two plates there is also a night field and a *day* field made during the time of the eclipse. I believe we have a pretty good set of plates."[51] Curtis's contribution to the project was to build a large precision-measuring engine for Miller to use on the plates. Miller did not want to have to rely on Schlesinger's equipment in Yale for this important work. Curtis started plans for the engine in June 1926 while Ross Marriott from Swarthmore went back to Sumatra to take the comparison plates.[52]

The Germans were less fortunate. Freundlich and his team left their instruments at Sumatra and went to California in April. Freundlich visited Mount Wilson for a few weeks while his associate, H. von Kluber, went to Mount Hamilton to examine his two Einstein plates from Sumatra. His job was to decide whether they should bother to go back and take comparison plates. After they had come and gone, Aitken told Miller confidentially that the Germans had been "a little disappointed" with their plates. In the end, nothing came of these observations. S. A. Mitchell related later in his eclipse book that the Potsdam observers "on account of haze secured few photographs of value." Miller and Curtis were left in possession of the field.[53]

By September, Curtis had made some progress on the measuring machine. Miller hoped to get it "as early as possible." "I gather from what I hear from relativitists that it does not make much difference whether the plates are measured or not. Russell has already told us what the results would be provided the plates were good, but the non-relativitists are anxiously awaiting for those plates to be measured."[54] An extensive delay in Curtis's work on the measuring engine aggravated the impatience of those awaiting Miller's results. Besides underestimating the amount of time he needed, Curtis came down with a serious case of pneumonia during the winter of 1927. It was only in the middle of May that he could begin to contemplate getting back to work. He apologized to Miller, but was inclined to blame it all on his illness: "Now, any cussing out that you can give me on the matter of this apparently ill-fated engine will be taken in good part. But this last delay is beyond me. I was too near death to be anything but reasonably careful now; no one, in fact, thought that I would be here at the present juncture. Get a pneumonia germ under the microscope over at the biology dept. and torture it; give it a few for me, too, for I think you would have had that machine in March but for one of said germ's brothers." Apologizing further, Curtis admitted that other elements were involved in the delay, including "a woeful underestimate of the amount of work necessary." Miller decided not to push Curtis. He told him that he would use Schlesinger's machine at Yale.[55]

Late in June, Miller wrote Curtis that Marriott was beginning to measure the Einstein plates and that some idea of the results would be available in about ten days. Unfortunately, the trip to New Haven to use Schlesinger's machine turned out to be a major obstacle. After their first measures in June, he and Marriott only went back once in November. By early December, Miller was nowhere near the end. He told Curtis that it was difficult for him to continue running to Yale to do the measuring. He would prefer to use the machine that Curtis was still constructing—more than six months after he had told Miller he was so close to completion.

> We took one of the plates that we made, one of the last pair, to New Haven and measured it, along with the diameters of Marriott's Moons but we could not be away from the Observatory too long, nor could we impose upon Schlesinger too long and I think we were a little hurried in our measurement of that plate. At any rate, I would very much rather measure my plates in my own Observatory. Schlesinger has been very good to us and I think would allow us to measure the remaining plates at his Observatory, but it is expensive from the standpoint of time and money to go up there and during the time when those plates are being measured we are knocking him out of his own problem.[56]

The delay in getting out the Einstein results was affecting other plans. Another total eclipse had taken place in June 1927, visible from England and the Scandinavian countries. It was a brief one, and despite popular enthusiasm in England, observers went merely to record the event and try for coronal photographs and spectrum data.[57] No one attempted the Einstein test. Miller did not even go, being busy with the 1926 measures. Miller wanted to go to the next eclipse, in May 1929. Sumatra and the Philippines were again the best sites. Miller wrote:

> We are beginning to think of 1929 and when I have gone to my friends regarding money for an expedition the first question I have been asked is, "What did you find out on the Einstein plates?" I have apologized for saying nothing about those Einstein plates so much that I am really ashamed of myself and I think that our failure to have measured those plates will have considerable delatereous [sic] influence in getting funds for the next expedition. But aside from that, I believe that I owe it to Astronomical science to at least measure and reduce those plates. Now I wish you would tell me exactly what the state of affairs is regarding the engine.[58]

He told Curtis that the coming Christmas vacation was the last time that he would have to spend a stretch of uninterrupted time measuring the plates. The president of Swarthmore College was leaving toward the end of December and Miller was to take charge until the following June. "I think it is pretty evident that I will not have any time much to work at those plates after the third day of January. . . . I would give a lot to be able to sit down at my own engine in my own observatory and measure those plates during Christmas vacation. Now, please tell me whether or not there is any ghost of a show to do that and I would appreciate it if you would tell me about what you think we ought to do. I can't very well go to New Haven during that vacation because Christmas is in there and it would be all broken up and all in all the matter is rather embarrassing."[59]

Curtis's response came quickly. He had been having trouble with the driving screw that moved the plate back and forth to set the cross-hairs on a stellar image. He could report that tests made to date "appear to indicate that the screw is essentially OK." He was still having a lot of trouble with what he called "creep"—if he came rapidly to a setting (for example, when changing from one star to another) the first measurement would be too high. He wanted a few more days to fix up another nut arrangement before shipping the machine. He included results of tests that he had taken for uniformity of the screw along with other instructions. "I keenly regret the delay and the embarrassment to which you have been put in this wild goose chase, but still feel that it's not entirely my fault, but that of the pneumonia germ." Curtis needed "a few days" for the "creep" problem. "Barring unforeseen snags I think I'll ship that —*** engine to you in time for the holidays, and the Lord have mercy on your soul."[60] In the end, the completed engine did not arrive on time. Curtis shipped it to Miller almost four months later on 23 March 1928.[61]

With Curtis's engine unavailable, Miller could do nothing more. He and Marriott used their vacation to prepare a paper based on their work to date. During their two visits to Yale in June and November they had completed measures for the diameter of the Moon. In January 1928 they submitted a paper giving results. It appeared in April, and the conclusions were definite:

1. That it is possible to obtain from the measures of short exposure photographs of the *Moon* made at time of total solar eclipse a reasonably accurate diameter of the *Moon*.
2. That no measureable effect has been found that could be interpreted as accounting even in part for the deflection of light from stars apparently near the limb of the *Sun* observed at the time of total solar eclipse.[62]

Miller's and Marriott's lunar results from the 1926 eclipse were particularly significant because Poor had designed the test to refute Einstein. The goal was to find refraction effects that would compromise the relativistic interpretation of the 1919 and 1922 eclipse results. But the effect did not show up.

A summary of Miller's 1926 eclipse expedition also appeared in the Astronomical Society of the Pacific's *Publications*. This paper listed all the items on the 1926 eclipse program, including "Photographs for testing the Einstein theory for the deflection of light," but only the coronal and spectroscopic studies were discussed.[63] Miller never published results on the light-deflection test from the 1926 eclipse. One can only speculate that when he sat down at his own measuring machine in March with the Einstein plates, he found problems like those that the Potsdam observers had encountered. Nonetheless, Miller did get his funding for the 1929 eclipse. Presumably his Moon results satisfied his funders. Curtis, too, recognized that the Einstein theory still figured prominently in eclipse fund-raising. As a declared antirelativist he felt the pressure even more. As the 1929 eclipse approached, he asked Miller if he could join his expedition again. "Have been thinking a lot about the eclipse and had almost decided to raise enough here, if possible, to go over by my lonely. But knew it might be a bit embarrassing because I could not get enough to run an Einstein program, and they might wonder why I did not."[64] In fact Curtis had lost interest in the Einstein eclipse test. He really wanted to get a flash spectrum in the red end of the solar spectrum, which he had tried to get at the New Haven eclipse. He also had plans for interferometric observations and extrafocal photometry on the corona. As an eclipse observer, Curtis had much wider interests than the Einstein test. He was obviously ready to move on to his own problems.

The 1929 eclipse happened to be especially favorable for the Einstein light deflection test. Though Lick was not planning to send an expedition, Trumpler published an article drawing attention to the fact that the eclipse field would include stars very close to the Sun. From Sumatra observers could photograph a star only 8″4 from the Sun's limb, giving a theoretical displacement of 1″14. From the Philippine Islands, a star 6″7 from the Sun's limb would have a theoretical displacement of 1″23. At the 1922 eclipse, Trumpler noted, "very few, and mostly faint stars were found within 40′ from the Sun's limb, and the results . . . mainly concern the light deflections of stars between 1 degree and 9 degrees from the Sun's center. Further observations of bright stars in close proximity to the eclipsed Sun seem to be the most desirable addition to the data at present available."[65]

Many astronomers decided to repeat the Einstein test. Early in 1929 there were reports that two German expeditions, two British, and one Australian, would try their luck.[66] Almost as many parties would be looking for light bending as had made the observations in 1922. Yet much had changed since then. As the 1929 eclipse approached, Miller was the only American interested in testing Einstein. Freundlich also went to Sumatra, determined to get definitive results after two failed attempts and over fifteen years of work on the problem.[67] His results would end up pitting him against Campbell and Trumpler at Lick.

Although antirelativity interests in the East continued to fund Miller's attempts to do the Einstein test again, he never was able to obtain definitive results. The Lick astronomers eventually moved away from this work. They applied the equipment and technical knowledge elsewhere, adapting the Einstein cameras to other projects just as they had tailored the Vulcan cameras for the Einstein problem. In the fall of 1929, after the Sumatra eclipse had come and gone, Charles E. Adams of Wellington, New Zealand, approached Aitken, who was now Lick director. He asked whether he would be repeating the Einstein program at the eclipse of October 1930. If not, could the New Zealanders borrow the Einstein cameras? Aitken answered that they were planning to use the equipment in observations of the asteroid Eros for the determination of the solar parallax.[68]

As far as the Lick was concerned, relativity had been verified and there were other problems to tackle. However, others persisted in their attempts to disprove or circumvent Einstein, forcing Lick and Mount Wilson to defend their work confirming relativity.

DAYTON C. MILLER AND THE ETHER DRIFT

D. C. Miller's new results from Mount Wilson made a strong impression on scientists. He received the AAAS prize of $1,000 for his paper on the "Significance of Ether-Drift Experiments" presented to the American Physical Society in December 1925.[69] Researchers in Europe and America quickly made plans to test his results.

In Germany, Einstein was besieged with requests from reporters to comment on Miller's preliminary results. On 19 January 1926 he published a statement, "My Theory and Miller's Experiment."

> If the results of the Miller experiments were to be confirmed, then relativity theory could not be maintained, since the experiments would then prove that, relative to the coordinate systems of the appropriate state of motion (the Earth), the velocity of light in a vacuum would depend upon the direction of motion. With this, the principle of the constancy of the velocity of light, which forms one of the two foundation pillars on which the theory is based, would be refuted. There is, however, in my opinion, *practically no likelihood* that Mr. Miller is right. . . . If you, dear reader, use this interesting scientific situation for placing a bet, then you better wager that Miller's experiments will prove to be faulty, i.e., that his results have nothing to do with an "ether wind"! I at least would be very ready to make such a bet.[70]

Einstein told an American correspondent several months later: "I don't have a high opinion of the reliability of Miller's results. . . . But today is not the moment to discuss this issue. The completion of those experiments which are entirely in keeping with Miller's method will have to be awaited."[71]

At Caltech, Millikan encouraged one of his staff, the physicist Roy J. Kennedy, to test Miller's conclusion with an interferometer that Kennedy had recently developed. The instrument was easy to control for temperature stability and made it simple to observe the fringes. Kennedy obtained "perfectly definite" results at Caltech. "There was no sign of a shift depending on the orientation." He repeated the experiment on Mount Wilson to check the idea that an ether-drift might depend on altitude. "Here again, the effect was null."[72]

Meanwhile, Michelson was on Mount Wilson preparing to measure the velocity of light using mirrors 90 miles apart. The word went out that he was planning to repeat the Michelson-Morley experiment to check Miller's results.[73] Walter Adams, now observatory director, urged him to do it. He believed the ether-drift experiment was "more important" than the velocity-of-light work and that "what the scientific world wants is *your* final word on the subject." Michelson acquiesced and started planning the work with his technical assistant, Fred Pearson, and Mount Wilson's Francis G. Pease.[74] Hale wrote to Einstein: "As you doubtless know, Professor Michelson will soon repeat the Michelson-Morley experiment on Mount Wilson with an improved interferometer. In my opinion he is not likely to confirm the strange results of Professor Miller. Professor Millikan has probably written you about the purely negative results recently obtained by Dr. Kennedy in Pasadena and on Mount Wilson." Einstein replied that such attempts would be very worthwhile (*verdienstlich*), but that he was already convinced that Miller's results were due to some temperature effect.[75]

When Kennedy's results at Pasadena and Mount Wilson indicated no ether-drift, Charles St. John decided to organize a conference on the Michelson-Morley experiment. He invited theoreticians and experimentalists to discuss the question of an ether drift from all viewpoints. The timing seemed auspicious. Michelson was planning to repeat the experiment to check Miller's results, and H. A. Lorentz was coming to Pasadena early in 1927. St. John wrote to Armin Leuschner in Berkeley: "The assemblage of such a body of men seemed a very happy circumstance and offered an opportunity too good to be missed."[76] The conference took place on 4 and 5 February 1927. Miller was just concluding a new series of measurements on the Case campus. Michelson and Pease were testing the setup at Mount Wilson for the instrument they had designed, before putting it into final form. The main speakers at the conference were Michelson, Lorentz, Miller, E. R. Hedrick of the University of California at Los Angeles, and Paul Epstein and Roy J. Kennedy from Caltech. Toward the end of the conference, Gustav Strömberg outlined his reasons for believing in the existence of "a 'fundamental' reference frame, or 'medium,' or 'ether,' whatever we prefer to call it."[77]

In the theoretical discussions, Hedrick questioned the standard theory of ether drift. Lorentz and Epstein responded to some of his points, but the unequivocal statement that St. John had been looking for was not forthcoming.[78] Miller presented results that were essentially the same as he had given before, but with one difference. Now he was not claiming a dependence of ether-drift on height. "Several critics seem to be under the impression that the earlier Cleveland observations gave a real zero effect and that it is claimed that the present positive effect is due to the greater elevation at Mount Wilson. This is not true."[79] Miller now claimed that the positive effects at Cleveland and at Mount Wilson "are so nearly equal" that it was not possible to deduce an effect due to altitude. Miller's position had changed when he began to interpret his positive result in terms of cosmical motions of the solar system through the ether. His new hypothesis of a large ether drift being reduced to a smaller one near Earth's surface made it unnecessary for him to search for an effect due to height. Lorentz pointed out that the validity of Miller's new explanation depended on the interaction between matter and ether. He suggested that an irrotational ether, for example, might satisfy all the conditions. "I tell you all this only to show how numerous the different possibilities for the theory are, if we are compelled by new experiments to go back to the notion of a substantial ether."[80]

On the experimental side, Michelson announced his intentions to redo the ether-drift experiment and referred to Kennedy's recent test as an "excellent piece of work."[81] Kennedy reported his negative results from the experiment performed in the Norman Bridge Laboratory at Caltech. Paul Epstein summarized experiments done independently by Rudolph Tomaschek of Germany and C. T. Chase of Caltech to check the 1903 Trouton-Noble experiment; but results at this time could not decide either for or against Miller's results. Epstein related results of Auguste Picard's balloon experiments at Brussels, which had addressed Miller's original dependence of ether drift on height. The experiment would have negated such a hypothesis, but Miller disclaimed it at the conference.

The proceedings of the conference did not appear until late 1928, by which time the various experiments presented in early 1927 had come to fruition or been repeated. These updates were included in the published proceedings. A note added in April 1928 indicated that K. K. Illingworth at Caltech had continued the work with Kennedy's apparatus, using improved optical surfaces and method of averaging. He concluded, "no ether-drift as great as 1 kilometer per second exists."[82] Epstein reported that Chase had continued his work at Harvard, increasing the accuracy of his measurements about three times. "Within this accuracy his results were negative, thus giving strong support to the theory of relativity."[83] Epstein also noted that Picard had repeated his experiment at an altitude of 1,800 meters in Switzerland, in collaboration with Ernest Stahel. The results "were completely negative, being only one-fortieth part of that expected according to Miller."[84]

Michelson and Pease got results late in 1927. They were also negative.[85] Michelson presented preliminary findings at a special meeting held in his honor by the Optical Society of America in November 1928, fifty years after the publication of his first communication on the velocity of light. "In a lecture room which normally seats hardly more than 300, there had gathered to hear him probably about 500 people, while many more who came were unable to get in." D. C. Miller was also in attendance. Michelson announced that his repeat of the ether-drift experiment at Mount Wilson showed no effect whatsoever.[86] The press quoted the two main antagonists. Michelson emphasized that his most recent experiments "are again negative." Miller insisted that he had initially conducted his experiment "in the honest hope of arriving at negative results also," but that his results had nonetheless "been positive."[87]

By the end of 1928, experiments by Kennedy in Pasadena, Chase at Harvard, and Picard and Stahel in Switzerland had all found no ether drift, contradicting Miller and supporting relativity. Michelson's preliminary results at Mount Wilson also contradicted Miller's findings. J. A. Miller's negative conclusions from Poor's lunar test, refuting any refraction explanation for light bending, had also appeared in April of that year. Einstein would have a banner year in 1928.

THE 1928 CLIMAX: THREE MORE PRONOUNCEMENTS

Adams's measurement of the spectrum of Sirius called for independent corroboration. In 1926, Joseph Haines Moore at Lick stepped up to the plate. His instrument of choice was the 36-inch refractor on Mount Hamilton. Adams had used the 100-inch reflector on Mount Wilson, but there were reasons that a refractor might have some advantages. Scattered light from the primary star was a particular challenge in this research. Polishing creates fine scratches on the silver coat of most mirrors, exacerbating the problem. A good refractor lens would avoid this additional source of scattering. The supports of the auxiliary mirrors on a reflector also produce diffraction patterns, which contribute to scattered light. Adams had used circular diaphragms to minimize this effect, but a refractor eliminates this problem. The only real drawback with a refractor is chromatic aberration. The lens refracts different wavelengths coming from the star by different amounts, creating a blurred image of the star on the slit of the spectrograph. To avoid guiding problems, astronomers typically place a dense blue screen in the eyepiece to produce a clean image of the star on the slit. However, the screen cuts out a lot of light from the star. The companion of Sirius is feeble in the first place, and the resulting image was extremely faint. Moore found that this challenge might actually have been an advantage. Since he could only see the faint image in perfect seeing (atmospheric conditions are still and clear), its disappearance was the signal to close the exposure.[88]

Moore never published results from his initial attempts, but there was an early rumor that he was not confirming Adams's findings. In the fall of 1926, Adams wrote to Aitken:

> I heard indirectly through St. John that Dr. Moore has been getting some fine spectra of the companion of Sirius and has not confirmed the relativity displacement. Of course I wish that we might have agreed, but I have the greatest admiration for Dr. Moore's ability in getting spectra of this object. I tried repeatedly last winter to secure spectra in the red without finding conditions good enough, and we are after it again now. I have great faith in the microphotometer measures of this star, but of course Dr. Moore may demolish them and we all want to know the truth.[89]

Aitken wrote back that "Dr. Moore's plates of the companion of Sirius, do not, as a fact, confirm the relativity displacement, but on the other hand, they do not deny it." Moore had indeed succeeded in getting spectra with a clear separation between that of the companion and Sirius itself, extending all the way to $H\beta$. Aitken explained, however, that "the spectra are too narrow to make it possible to decide the question one way or the other." Moore was hoping to get wider spectra and then to be able to make a definitive statement. Aitken told Adams that "we do not feel that his data warrant a statement in either sense."[90]

Moore obtained more spectrograms in 1927, but changed the focal length of the camera he used with his one-prism spectrograph, from 12 inches focal length to 16 inches focal length. He obtained four spectrograms in the winter of 1928 with the new camera. The plates were "so much more satisfactory for measurement on account of the wider star spectrum and higher dispersion" that he rejected the previous plates and only used the 1928 ones for his final measurements.[91] He announced his results on 15 June 1928 at a meeting of the Astronomical Society of the Pacific.

Moore measured four spectrograms of the companion on the Hartmann spectrocomparator, using the spectrogram of Sirius as the standard reference plate. He used the hydrogen lines $H\gamma$ and $H\beta$ as well as several others. Moore judged that the spectrum classed the companion to be about A5, and definitely "not as late as F0." Table 11.2 presents his re-

TABLE 11.2.
Moore's Sirius B Results

Date 1928	Relative Velocity Companion—Sirius	Number of Lines	Relative Density at Hγ Companion/Scattered Light
Feb. 13	+22 km/sec	7	3.7
Feb. 20	(+10)	4	1.2
Feb. 27	+29	6	10.0 underexposed
Mar. 20	+21	4, 9	2.8 mean of two
Mean	+24		

sults. He did not try to introduce a correction factor to determine the means, as Adams had done. He chose instead to use the uncorrected values, and to exclude from the mean the value for 20 February. On that night, light clouds were continually floating over the star, causing a lot of scattered light. The spectrum of Sirius was only "slightly fainter" than that of the companion.

During the winter of 1928 when Moore made the observations, the companion was receding from the primary at 5 kilometers per second. Subtracting this value from the mean, the resulting line displacement for the companion due to the relativistic effect was + 19 kilometers per second, or + 0.29 angstrom. As a further check, Moore calculated the observed displacements using Adams's procedure of a correction factor. He obtained a mean from all four spectrograms of + 26 kilometers per second. This value yielded a gravitational displacement of + 21 kilometers per second, or a shift of 0.32 angstrom. Moore concluded: "The results obtained from the four spectrograms at Mount Hamilton thus seem to afford additional evidence to that obtained by Adams, for the existence of a gravitational displacement of the lines in the spectrum of the companion of Sirius, of the order predicted by the general theory of relativity."[92] By 1928, both Mount Wilson and Lick had confirmed the gravitational displacement of the spectral lines of the companion of Sirius.

In the same year, a forty-five-page paper by Charles St. John gave the final results of his extensive measurements of solar spectral lines to determine whether the Einstein shift exists in the Sun.[93] He had completed the paper the previous year. "I have after much toil and moil got the Einstein paper off to the Press," he had remarked to Hale. "It makes rather a formidable document as the data are given in detail. It seems to me to put the case rather strongly."[94] St. John's conclusion supported his preliminary announcement made in Pasadena in 1923 that an Einstein displacement was present.

St. John had measured wavelengths of over 1,500 spectral lines at the solar center and 133 at the edge. He used 586 iron lines to obtain the main result for the center of the Sun. He found an average displacement of ± 0.0083 angstrom. The theoretical Einstein displacement is + 0.0091 angstrom. St. John found that the level in the solar atmosphere from which the various lines originated affects the observed displacements. He had to account for this effect when examining the result. The mean displacement for lines of medium level (520 km) was ± 0.009, exactly as relativity predicted. For lines of higher level (840 km) the mean displacement was 0.0027 angstrom greater, and for low-level lines (350 km) it was 0.0026 angstrom less. Measures of many different spectral lines confirmed these general results: 6 lines of silicon, 18 lines of manganese, 402 lines of titanium, and 515 lines of cyanogen. St. John interpreted his findings as due to a mixture of two effects that depend on level (height) in the Sun's atmosphere. Low-level lines are produced near the photosphere where upward currents added a Doppler shift to the violet due to motion toward the Earth. This violet shift is superimposed on the relativity redshift to produce the smaller displacements toward the red. For the higher-level lines, St. John appealed to a new theory by Edward Arthur Milne and Charles James Merfield to explain the greater redshifts. They proposed that upward-moving atoms, which would absorb radiation from the violet edge, tended to escape, leaving a greater number of atoms that absorbed from the red edge of the line. This effect should increase with height. At higher levels it would add an extra redshift to the shift caused by the gravitational field.[95] With these two other mechanisms at play, all the observed shifts would conform to the relativity prediction.

At the limb, St. John could combine the results for low and high levels in the atmosphere. Ascending and descending motions are across the line of sight and do not contribute to line shifts. He found that the mean of his 133 iron line displacements at the limb was 0.0015 ± 0.0004 angstrom greater than that calculated from the theory of general relativity. "This small residual, if real, is a true limb-effect."[96] This excess redshift at the solar limb became the object of study by solar spectroscopists for years. Other investigators confirmed its existence in different regions of the solar spectrum. A satisfactory explanation of this "limb excess" emerged only in the 1960s.[97] St. John did not elaborate this finding in his 1928 paper, preferring to leave it unsettled. The limb excess actually disappeared if he used only lines of very low level. For these lines he obtained the relativity prediction. St. John actually interpreted the increase in wavelength at the limb as corroborating his belief that upward currents reduced the wavelength for the low-level lines at the Sun's center.*

St. John attributed Burns's intensity dependence to effects of level, neatly disposing of any antirelativity implications. "Lines of different ele-

---

* At the limb, the upward movement of gases would be across the line of sight.

ments of very different intensities, but at the same level, give equal red displacements; while for lines of the same solar intensity, but at widely different levels, the lines of higher level give the greater red displacements. This points to level of origin rather than line-intensity as the controlling factor in line-displacement."[98] St. John also set the record straight about his early negative result, based on measures that he had made during the war years. He had based this study on lines of cyanogen rather than iron. In fact, most of the early investigators had used cyanogen, including Evershed, Schwarzschild, and Grebe and Bachem. St. John explained why: "The choice of lines for this purpose was made at a time when the pressure in the Sun's atmosphere was thought to be of the order of 5–7 atmospheres. As band lines show no appreciable pressure shift, their use seemed to eliminate one variable. High pressure in the Sun was then the accepted interpretation of the displacements to the red, now attributed to the Sun's gravitational field."[99]

This choice was "unfortunate" because in such bands the number of lines is large, resulting in crowding, overlapping of line series, and a high probability of undetected blends. St. John's strategy had been to select a small number of lines that looked appropriate for the task. "My original investigation was confined to some 40 lines and gave negative results. In view of later work on the complete band, these lines might be called the 'Forty Thieves.' "[100] The complete cyanogen band consists of 515 lines. St. John now used all of them in his study. He assumed that random errors due to faulty measures, blends, and overlapping series "are as likely to be positive as negative, and . . . their effect will be practically eliminated from the mean." As a check on this assumption, he sent a spectrogram to the physicist Raymond T. Birge of the University of California. He asked Birge to examine the structure of the band "with special reference to the overlapping of series." Birge selected a list of 184 lines which he considered especially suited to measurement.[101] St. John presented results for all 515 lines and for Birge's selected list of 184 lines (table 11.3): The influence of the forty-three lines that St. John used in his original study[102] was counterbalanced in the final mean, which was based on the "far greater number of lines." By adding 0.0026 angstrom to the results for the center of the Sun, St. John got the limb result. This amount (0.002 A) agreed with the mean limb-minus-center displacement for the cyanogen lines found by Adams and independently by St. John. For cyanogen, then, "the displacement at the limb is of the sign and approximate magnitude required by the theory of relativity."[103]

St. John's final conclusion was definite: "This investigation confirms by its greater wealth of material and in greater detail the conclusion announced in the Symposium on Eclipses and Relativity at Los Angeles, September 17, 1923, that the causes of the differences at the center of the Sun between solar and terrestrial wavelengths are the slowing up of the atomic

TABLE 11.3.
St. John's Final Solar Redshift Results for Cyanogen (1928)

(Unit = 0.001 A)			
Mean for 515 lines (center)	4.6	Mean for 184 lines (center)	5.0
Mean for 515 lines (limb)	7.2	Mean for 184 lines (limb)	7.6
Relativity shift	8.1	Relativity shift	8.1

clock in the Sun according to Einstein's theory of general relativity, and radial velocities of moderate cosmic magnitude and improbable directions, or equivalent conditions whose effects vanish at the edge of the Sun."[104]

In the same year, 1928, Trumpler finally published the main results from his complete measures of the Lick 5-foot camera plates from the 1922 eclipse. The six plates gave a mean limb deflection of +1″82 ± 0″15. Trumpler combined this value with the one he obtained from the 15-foot cameras (4 plates, +1″72 ± 0″11), giving double weight to the 15-foot camera results. The final result was a light deflection at the Sun's limb of +1″75 ± 0″09, "which agrees exactly with the prediction of Einstein's Generalized Theory of Relativity and has a probable error of only 5 per cent."[105]

In a later, more detailed publication of their 5-foot camera results, Campbell and Trumpler addressed Poor's claim that the Lick 15-foot camera data yielded a limb displacement of 2″05. They emphasized that the larger displacements "were merely given for an estimate of the possible influence of certain systematic errors." They noted that the higher displacements had "seemed, *a priori*, open to doubt," and that "the observations of the second instrument now available speak decidedly against it."[106]

In his summary paper, Trumpler also reported that the law of displacement with angular distance from the Sun obeyed the Einstein prediction. He illustrated the data graphically, judging that the fit of the theoretical curve was "very satisfactory." He displayed the deflections for the group means in a table. The fit with the theoretical values at the same angular distance was close. Trumpler noted "any arbitrary smooth interpolation curve drawn close to the plotted group means does not deviate from Einstein's curve by more than 0″03–0″04."[107]

The data also contradicted the Courvoisier "yearly refraction." Trumpler concluded, "having no theoretical basis, it is probably due to some systematic error peculiar to meridian circle observations."[108] As to abnormal refractions, he stated that a study of such effects in the Earth's atmosphere caused by the Moon's shadow "leads to the conclusion that these could not have had any appreciable influence on the eclipse measures."[109] Trumpler concluded: "The results of the two instruments, derived by the measurement of more than 3000 star images (87,000 bi-sections) confirm Einstein's prediction concerning the light deflections in the gravitational field not only in the amount of the deflections but also in the law according to which they decrease with increasing angular distance from the Sun's center, and Einstein's Generalized Theory of Relativity seems at present to furnish the only satisfactory theoretical basis for these observations."[110]

The year 1928 was a crowning year for the empirical verification of general relativity. There were three key results: the Lick 1922 light bending, the Mount Wilson measures of the gravitational redshift in the Sun, and the Lick result corroborating the Mount Wilson measurement of the gravitational redshift in the companion of Sirius. That year, J. A. Miller and Marriott published their lunar results discounting abnormal refractions as a mechanism for light bending. The results of the Michelson-Morley conference held at Mount Wilson and subsequent research contradicting D. C. Miller's ether-drift results also appeared. Taken together, these results sealed the verdict in favor of Einstein.

RELUCTANT ACCEPTANCE

As the decade ended, the empirical bent of American astronomers forced them to accept relativity, albeit reluctantly. Their reticence stemmed from an aversion to the mathematics and conceptual framework of the theory. As verifications of Einstein's predictions accumulated, this reluctance persisted. Early in 1927, Paul Merrill confessed to Aitken to having "a slight antipathy to relativity." Aitken answered that he was glad to know this "because I should hate to be the only one in that position. The trouble is that the arguments in favor of the theory seem to be growing steadily stronger."[111]

By this time, Aitken had come to realize that much of the resistance, even his own, came more from prejudice than a dispassionate assessment of the merits of the theory. He had also been party to some of the nastier opposition to relativity, including vitriolic attacks from T.J.J. See and attempts by Charles Lane Poor to discredit the Lick 1922 observations. Around the time he had his exchange with Merrill, Aitken published an excerpt from the reminiscences of the astronomer Simon Newcomb, one of the most revered American astronomers of the recent past. The passage had been published in 1903:

> Among the psychological phenomena I have witnessed, none has appeared to me more curious than a susceptibility of certain minds to become imbued with a violent antipathy to the theory of gravitation. The anti-gravitation

crank, as he is commonly called, is a regular part of the astronomer's experience. He is, however, only one of a large and varied class who occupy themselves with what an architect might consider the drawing up plans and specifications for a universe. This is, no doubt, quite a harmless occupation; but the queer part of it is the seeming belief of the architects that the actual universe had been built on their plans, and runs according to the laws which they prescribe for it. Ether, atoms and nebulae are the raw material of their trade. Men of otherwise sound intellect, even college graduates and lawyers, sometimes engage in this business.

Aitken remarked: "If in the foregoing excerpt, the word 'gravitation' were replaced by the word 'relativity,' Simon Newcomb's reflections concerning a certain kind of crank would be as apt today as they were twenty odd years ago."[112] This citation from Newcomb could not have been lost on Poor and the other antirelativity "irreconcilables."

Aitken's dig at the antirelativists was prompted in part by L. A. Redman, a lawyer in San Francisco. He sent a deluge of antirelativity letters to astronomers during the 1920s. This crank even plagued Poor, who corresponded with Aitken about how to handle him. Aitken had to decline to correspond with Redman after he published a personal letter from Aitken without his consent. Poor got rid of the pest by demanding a retainer's fee of $500 to check the man's statements, formulas, and calculations. Aitken told Poor that Redman looked upon him, Poor, as a supporter, but added diplomatically: "You are aware that I am not in agreement with you in a number of points relating to the Einstein Theory, but that has nothing to do so far as I can see with such writings as those of Mr. Redman."[113]

The almost universal inability to explain Einstein's theory to laypersons exacerbated astronomers' "antipathy" toward relativity. Their difficulty put them in an awkward position, so they concentrated on their competencies—observation. Aitken discovered in 1926 that even American astronomy's greatest theoretician could not explain relativity. He approached Henry Norris Russell at Princeton to write a short article on Einstein's theory. The Astronomical Society of the Pacific wanted to include it in its series of scientific leaflets for the general public, including potential benefactors of astronomical research. "What they want is a statement of the subject in a compass of fifteen hundred words that will enable a banker, a broker or a lawyer to understand what the theory is all about." "I knew of only one man in the country whom I could hopefully approach with such a proposition, and that man was you. If you think that it is at all a feasible thing, will you be willing to undertake the task and send an exposition of relativity in 'words of one syllable?' If you succeed, I think the astronomical world as well as the general public will take off its collective hat to you."[114]

Russell was unique among American astronomers in his international reputation as a theoretical astrophysicist in the same class as Eddington and Jeans. However, Aitken was to be disappointed. "Letters from you are always welcome," Russell assured his Western colleague, "but I am really somewhat horrified at the request of the ASP. . . . To write an account of relativity in fifteen hundred words is, I think, more than anybody can do. I tried to get the thing into five thousand for the Scientific American prize and couldn't do it,—being obliged to restrict myself to the old special theory. I do not really believe that it is possible to get a general discussion of the matter in so small a compass. I don't think even Eddington could do it."[115] And so the leaflet remained unwritten.

By 1928, all these experiences and insights came out in a luncheon address Aitken gave on the progress in research at Pacific observatories during the past year. He began lightheartedly: "A well known astronomer when asked recently to write an account of the theory of relativity for the general reader in about 1500 words, threw up his hands in horror. 'Absolutely impossible! It cannot be done!' was his response when he was able to speak. To recount the progress of astronomy in the Pacific Area in the past year in fifteen minutes, may, I am willing to admit, not be quite so hopeless an undertaking, but it is nonetheless a difficult one." Aitken then plunged into descriptions of research on various astronomical subjects, but his list brought him around again to relativity. "The theory of relativity continues to hold the interest of scientific workers as well as of the general public, and it is a coincidence worthy of note that two papers summing researches bearing upon astronomical tests of the theory appeared almost simultaneously a few months ago." He summarized Trumpler's recent paper on the deflection of light near the Sun, concluding that Trumpler's "masterly analysis not only shows conclusively that the observed value agrees with the theoretical one but also demonstrates that the variation in the amount of the deflection with increasing distance from the Sun's limb follows Einstein's formula or law. It appears further that the observations cannot be satisfied by any other theory which has so far been advanced."

Aitken then turned to St. John's paper on the solar redshift. He pointed out that it confirmed preliminary results announced in 1923. Though the problem was "even more complicated and difficult than the one Campbell and Trumpler faced," no one could read St. John's paper "without the conviction that he has solved it successfully, and that, over and above the displacements produced by all other causes, there remains in the observed values a displacement agreeing in amount and direction with Einstein's predicted value within the error of observation." Finally, Aitken added that Moore at Lick had just completed the measures and reductions of several one-prism plates of the spectrum of the companion to Sirius. Aitken reported that Moore's value "is practically identical with the one found by Adams" and that "the values are in substantial accord with the theory and that no other satisfactory explanation of the displacement is available. . . . Whether we like it or not, we are obliged to admit that in these three instances the Einstein theory has successfully stood the test of astronomical observations."[116]

In spite of a "slight antipathy" to the theory, the observations spoke for themselves. Relativity had been verified.

Chapter Twelve

SILENCING THE CRITICS

As 1930 approached it was clear that relativity had passed the three "classical tests" from astronomy set by Einstein. In the empirically oriented United States, the theory was accepted *because* it had passed crucial tests, despite a lack of theoretical understanding. The California astronomers had not taken up the research to pronounce judgment on the underlying theory. They adapted ongoing observational techniques to search for specific predicted effects. Critics seized upon their initial negative findings to denounce the theory, while supporters tried to circumvent them. After Lick verified the British eclipse result, ensuing debates seemed to cast observational astronomers as supporters of the theory. Yet it was actually the reputation of astronomers and their host institutions that drove the relativity debates of the 1920s. The issues at stake were the validity of the astronomers' work and the legitimacy of their specific skills within the professional scientific community.

As the 1920s drew to a close, Lick and Mount Wilson were forced to make a concerted effort to counter attacks from the "irreconcilables" and pronounce a final verdict in favor of Einstein's theory.

Charles Lane Poor versus the Lick Observatory

Charles Lane Poor, the Columbia celestial mechanics professor and ardent antirelativist, was chief catalyst for a concerted Lick and Mount Wilson effort to counter the critics decisively. Poor had found a sympathetic ear in Heber Curtis, another "irreconcilable" who had pursued antirelativity research at solar eclipses with Poor and with Keivin Burns at his own observatory. Poor had a falling out with Curtis and John A. Miller after they disproved his hypothesis that the diameter of the Moon expands during an eclipse due to atmospheric refraction. While Curtis encouraged Poor's anti-Einstein stance, he ultimately rejected many of his specific ideas. One of Poor's lines of attack involved abnormal refraction in the Earth's atmosphere as the cause of stellar displacements observed during an eclipse. Other astronomers adopted this argument and promoted it. In April 1927, Cincinnati astronomer J. G. Porter published a critique of relativity and the preliminary Lick results from the Australian eclipse, using a similar analysis to Poor's based on nonradial directions of some of the stellar displacements. Porter's purpose was to mount a "vigorous protest" against the fact that in several contemporary astronomy texts "relativity is treated as a theory practically proved." Praising Poor's treatment in his 1922 book attacking relativity, Porter analyzed in detail the directions of the stellar displacements in the Lick data. He concluded that only a quarter of the stars "deviate from this [radial] direction by 10 degrees or less." He called upon refraction by the solar atmosphere and abnormal refraction in the Earth's atmosphere due to cooling in the eclipse shadow cone as the primary causes for the displacements. "That the refraction at the sun, combined with the refraction in the earth's atmosphere will best account for the irregular deviation actually observed, is the opinion of many who have examined the subject. Here again relativity both fails to explain the observed facts, and is besides totally unnecessary."[1] Curtis was "greatly pleased" with Porter's note "because it coincides so closely with my own view of the subject." He reiterated his belief that Burns had found evidence for a solar redshift mechanism that varied with intensity and wavelength, contradicting the relativity explanation. Nonetheless, he cautioned Porter against appealing to abnormal refraction to explain the stellar displacements around the eclipsed Sun. While he lauded Porter's critical stance against relativity, he presented cogent, quantitative arguments why refraction due to cooling of the Earth's atmosphere in the eclipse shadow cone could not play a significant role in the stellar displacements. "I have never been able to see how we could get an effect symmetrical about the Sun from refraction near the center of the swiftly moving shadow cone." Porter appreciated Curtis's letter but was unmoved. "I see the point of your criticism, and I believe it is well taken," he replied. "But, of course, there would be more or less abnormal refraction due to the disturbed condition of the atmosphere; and I think that is plainly shown on the plates."[2]

Curtis, the inveterate observer, had also not been sympathetic to Poor's attempts to attack relativity mathematics. Poor persisted, however, and presented his "exposé" at a meeting of the American Astronomical Society in Philadelphia on 29 December 1927. Some Toronto members of the Royal Astronomical Society of Canada were sufficiently impressed with Poor to make a special request to the editor of their society's journal to publish Poor's critique.[3] Poor divided his paper into an observational critique and a theoretical one, but he emphasized theory. He set out to present relativity in terms of Newtonian physics, classical optical theory, and, if necessary, special relativity. Two misconceptions made this possible. The first was a belief that Einstein's light-bending prediction hinged on the hypothesis that "light has weight," a conception that stemmed from Eddington's earlier popular expositions. The second was his interpretation of the principle of equivalence in terms of the Lorentz transformations from special relativity: "The effect of gravitation upon ideal 'clocks' and 'measuring-rods' at rest at a given point in a gravitational field is identically the same as that caused by a motion of the 'clock' and 'rod' through free space with a velocity equal to that which they would have acquired had they fallen, under the action of gravitation, from infinity to that point." According to Poor, these effects of gravitation could be calculated "under the theory of relativity . . . by the Lorentz formulas."[4]

Einstein's 1911 calculation of the bending of light had been based on the slowing of a clock in a gravitational field. He deduced this result from considering the propagation of energy from one place to another in the field.[5] Poor interpreted this "assumption" or "new tenet of relativity" in Newtonian terms as a repulsion due to the gravitational field of the Sun. He obtained the amount of retardation of the clock by using his peculiar notion of the equivalence principle and the Lorentz transformations.[6] He was able to derive the same formula for the variation of the speed of light in a gravitational field as had Einstein. His analysis "worked" because the Newtonian law of gravitation is implicit in Einstein's 1911 calculation.

Einstein's 1916 paper had derived an expression for the velocity of light as a function of position in a gravitational field. Using this equation, Einstein followed the path he had taken in 1911 to calculate the light bending from the new formula. Poor performed mathematical sleight-of-hand to show how Einstein had arrived at double the value for the light bending—the "elusive factor of 2." He wrote down the two expressions for the velocity of light from the 1911 and 1916 theories, using polar coordinates.

$$\gamma = c/c_0 = 1 + \phi/c^2 \qquad\qquad (1)$$
$$\gamma = 1 + \phi/c^2 \, (1 + \sin^2 \theta) \qquad\qquad (2)$$

c_0 = the speed of light in empty space
c = the speed of light in the gravitational field
ϕ = the gravitational potential
θ = the angle between the radius-vector at any point of the path of the ray, drawn from the gravitating body to the path[7]

In 1911, Einstein wrote equation (1) in similar form, but in his 1916 paper he did not use the polar form. If integrated, equation (2) gives the same result for the light bending that Einstein obtained. However, Poor got a different result.

Poor "explained" how Einstein got double the amount of light-deflection as follows. A simple inspection of equation (2) shows that for a transverse ray ($\sin \theta = 0$), the formula reduces to equation (1). For a radial ray ($\sin \theta = 1$), the coefficient of ϕ/c^2 becomes 2, or double the 1911 value. Of course, for an arbitrary ray, one obtains the deflection as it passes the

Sun by integrating equation (2) over the entire light path. One does not simply consider that part of the ray immediately tangential to the Sun or strictly radial to it (as it approaches from a distance). Poor knew this obvious fact, yet he insisted that Einstein had merely inserted sin θ = 1 into his equation to get the double value. "Einstein used in his calculation the value of this factor [sin θ] for a radial ray, not that of a transverse ray; and thus introduced the mysterious factor two (2). This, of course, was a plain, straight mathematical error:—the rate of variation in the speed of a radial ray has not the slightest thing to do with the 'bending' of a transverse ray." Poor claimed that when he did the integration properly, "one will then obtain for the full deflection according to Einstein's assumptions and basic formulas: 1."10 and not the 1."70 of Einstein's paper."[8]

When Poor sent a copy of his Philadelphia paper to Aitkin, the Lick director told Poor that he was aware that "you do not believe in this doctrine." Nonetheless, he felt the tide was turning in favor of relativity. "Doubtless there are others who fully agree with you, though the trend of modern astronomical thought seems to be in favor of the doctrine."[9] Poor disagreed. He told Aitken that the trend of astronomical thought "is decidedly away from the doctrine." "At first the doctrine was accepted 'on faith' by nearly every one. As it was investigated, however, astronomers found it illogical, mathematically wrong, and contrary to observed facts. Several prominent men have openly declared their doubts as to the doctrine: [Forest Ray] Moulton, [George C.] Comstock, Curtis, [Jermain Gildersleeve] Porter, and others. Many others have privately expressed their adverse views, but have neither written, nor publically [sic] spoken against the theory."[10]

The influential names on Poor's list highlight the antirelativity bent of the eastern astronomy establishment. For example, Forest Ray Moulton was a celestial mechanics specialist and cosmogony theorist at the University of Chicago. His collaborator on theories of cosmogony, the geologist Thomas Chrowder Chamberlin, had written a lengthy and favorable preface to Poor's 1922 book against relativity. Another Chicago man, celestial mechanics specialist William Duncan MacMillan (not on Poor's list), participated in a public debate on relativity at Indiana University, taking the antirelativity side against Robert D. Carmichael.[11] MacMillan had embraced Ludwik Silberstein's attack on relativity in the summer of 1921. Edwin B. Frost, director of Yerkes Observatory in Chicago, had nominated Poor for membership in the American Philosophical Society, recruiting Curtis to support the nomination. George C. Comstock, now retired, was a positional astronomer of the traditional school. He had held many prestigious positions, including director of Washburn Observatory, dean of the graduate school at the University of Wisconsin, president of the American Astronomical Society, and member of the National Academy.

J. G. Porter, director of Cincinnati Observatory and professor of astronomy, had just published his objections to relativity, arguing like Poor, that the Lick results indicated the presence of refraction, not a gravitational effect.[12] Philip Fox, positional astronomer and director of Dearborn Observatory, had supported Silberstein's critique of relativity in 1921. Later in the decade, he tried to test the light-bending effect near the limb of the planet Jupiter."[13]

The geographical concentration of these critics and skeptics in the eastern part of the country highlights tensions that existed in professional astronomy in the early twentieth century. The leaders in relativity testing were the western astronomers who were in the forefront of the newer field of astrophysics. With their advanced technology, they were in the vanguard of testing Einstein's revolutionary theory. By contrast, all the doubters mentioned above, except Curtis, were celestial mechanics specialists or positional astronomers of the old school, predating spectroscopy and astrophysics. This correlation suggests that conservatism played a large part in relativity denial. Poor's continuing attacks and the positive responses to them in eastern astronomical circles forced the western astronomers to prepare more detailed rebuttals. They defended their observations, and by implication, Einstein's theory.

In May 1928, when the Canadian journal appeared containing Poor's paper, he sent a copy to Aitken, who replied: "We cannot quite agree with you in the matter of your statement about the value of the Einstein prediction." Besides, Aitken noted, the observed value agreed with the one Einstein predicted.[14] Poor complained that Aitken's remarks were "apparently based on a most casual reading of the paper and without an attempt to check the facts as to the way in which the deflection $1''75$ is derived by Einstein, Eddington and others." He reiterated his arguments in great detail, concluding that if the Lick director would check the calculations in his paper, he would find that Poor was right.[15] Aitken had in fact checked—with Trumpler, who had told him "that he had carried the integration through, going back to the original data, and had arrived at the results of $1''75$." Aitken was "a little at a loss to know how to account for the different result you reach." He admitted candidly that his attitude was not based on firsthand knowledge. "On a question of this kind I . . . am forced to depend upon my good friends and my judgment as to their capacity for work of this kind." He told Poor that Trumpler was planning to publish "a little note" on the subject of the Einstein formulas. "When this note appears, perhaps you will be able to point out wherein his process differs from that from which you reach the results $1''10$. . . . While philosophically speaking I do not find myself able to follow or accept the conclusions that result from the relativity theory in its generalized form, I am reluctantly forced to conclude . . . that it has met all the observational tests so far suggested."[16]

Trumpler's "little note" appeared early the following year. His paper was the most complete treatment of general relativity that appeared in an American astronomical journal during the entire decade. Trumpler began by referring to Einstein's 1916 calculation of 1″.7 for the deflection of light passing the Sun's edge and Poor's claim "that the light deflection calculated according to Einstein's method amounts to only 1″.1." He remarked that it would be "perhaps not out of place here to give the calculation of the deflection of a light ray in full detail, in order to remove any doubt as to the correct value of the result."[17]

Right from the beginning, Trumpler avoided Newtonian language. He described how a local measuring system is defined at each point in the space-time continuum. Relative to these local systems, a particle moving in the Sun's gravitational field seems to follow its "natural path" (a geodesic), and no force is apparent. In the local measure, the speed of light is constant. For an observer not using local measure, and using a Euclidian coordinate system, the speed of light is variable. Trumpler followed Einstein's procedure in deriving the variation of the speed of light in a gravitational field. He obtained the formula directly from the components of the fundamental interval ds. He explained lucidly the difference between the 1911 and 1916 calculations. In the 1916 formula, he showed exactly where the light bending due to space measurements and time measurements occurred. The 1916 theory had introduced the space term, which was equal to the time term, hence doubling the total amount of bending.

Trumpler also discovered Poor's integration error. To perform the calculation, Poor had shifted from polar coordinates to rectilinear coordinates. In doing so, he had omitted an important term.

Finally, Trumpler used Newton's and Einstein's theories to calculate the light bending for light paths of different lengths, all just grazing the Sun. For short paths near the Sun, both theories yielded very nearly the same light path. As the path lengths increased, the theories yielded different paths, with different bending. At about a path length of five solar radii, the Einsteinian value was double the Newtonian. Beyond that distance, the light path was practically straight. Trumpler pointed out that the geometry of the situation in which light from a distant star is observed from Earth as it passes close to the Sun produces a light path from infinity to infinity. So only the double value can be observed.

Trumpler's analysis was important in several ways. His insistence on setting up the treatment in strictly general relativistic language increased astronomers' familiarity with it. He showed that Poor's own peculiar method of applying the Lorentz transformations and the principle of equivalence was not what Einstein had done. Trumpler at last fully explained the "elusive two," an issue that had plagued discussions of relativity for years. He also found Poor's mathematical error. Finally, Trumpler showed American astronomers that at least one of their number could handle the theory. The integrity of the Lick results was upheld. C. A. Chant of Toronto, who had originally published Poor's paper, wrote Trumpler that he was "pleased" to see his article. He asked his permission to reproduce it in the Canadian journal. "It will be a good companion-piece to Poor's article, which was printed at the special request of some of the R.A.S.C. members here. Of course I was pleased also to present Poor's side of the question."[18]

Soon after this exchange, *Popular Astronomy* published a scathing review by Poor of the Lick 5-foot camera results from the Australian eclipse. Poor aimed his assault at the Lick astronomers' methods of observation and data reduction. Trumpler responded, exposing several instances where Poor was deliberately quoting out of context and distorting the facts.[19] One example will suffice. Poor claimed that in the final reduction of the data, Trumpler had written the Einstein law into his final solution. To make this assertion, he quoted Trumpler out of context and omitted important information. As was normal procedure for the Einstein test, Trumpler had used check-field stars to determine all the constants of the reduction formula except one—the scale value of the plates. As he remarked in the Lick *Bulletin*: "*The scale of the eclipse field photographs is necessarily linked together with the light deflection*; the two cannot be separated unless some assumption is made concerning the law according to which the light deflections are related to the star's angular distance from the Sun's center."[20]

The Lick observers described in great detail how they obtained the displacements using the reduction formula. Then they used two different laws to determine the scale value and compare the reduced observations to the theoretical ones—Einstein's law of displacements, and then Courvoisier's "yearly refraction" law. The Einstein law fit the observations very well, while the Courvoisier law did not.[21] Poor ignored all this careful work as well as discussion of the two laws that Trumpler used to evaluate the data. Poor quoted the italicized phrase above out of context and glibly asserted: "The Einstein law of decreasing size for the deflections was thus apparently written into the methods of determining the so-called 'observed deflections,' and then these very deflections, so determined, were compared with the assumption from which they were derived."[22] Poor never mentioned the Courvoisier work, nor the fact that in both cases, the law of displacement versus angular distance was used to determine the scale value. Einstein's law fit the observations; Courvoisier's law did not. Trumpler used this example to "illustrate the unreliability of Professor Poor's statements." Poor's transgressions were so blatant and numerous that Trumpler pulled no punches in his conclusion: "The whole review has so many mistakes and misstatements that it is not possible to take all of them up individually. It looks as if Professor Poor either did not read the Bulletin with any care and understanding or that he wrote the review mainly to vent his personal feelings and prejudices against anything connected with Einstein's Theory."[23]

Shortly after the publication appeared, Chant sent Trumpler a copy of the Canadian journal with his "factor of 2" article. By now he no longer felt the need to give both sides equal time. He told Trumpler that his article "deserved a place in the Journal to balance the original article [by Poor]." "When Poor was here a month ago we treated him with all courtesy, but I do not think he showed himself a very powerful destroyer of Einstein. I note also that you have crossed swords with him in [*Popular Astronomy*]. You are doing excellent service, which few people could undertake."[24]

Diehard critics persisted nonetheless. J. G. Porter published a short piece in the Astronomical Society of the Pacific's *Publications* challenging Trumpler's contention in his critique of Poor's "relativity mathematics" paper that only relativity "is at present able to account for the numerical values of the observed star displacements."[25] Porter described how he used a protractor to examine the direction of stellar displacements on the Lick diagram in their preliminary publication of the 1922 eclipse results. He claimed that "less than a quarter" of the stars exhibited radial displacements, most deviating more than 45 degrees and some more than 90 degrees from the radial direction. Porter asserted that abnormal refractions must be the cause of the stellar displacements, not relativity.[26] Porter was making exactly the same argument he had made two years earlier, that Curtis had privately warned him was unconvincing. Aitken used the opportunity to insert an Editor's Note after Porter's paper. The editors were "glad" to print it "because he [Porter] is not alone in his views." Porter's analysis was based on preliminary data published in 1923, however, and Aitken noted that "his figures would be rather different if he had studied the definitive results of the eclipse observations . . . issued in March, 1928. After careful consideration of the points raised by Professor Porter, by Professor Charles Lane Poor and others, the astronomers at the Lick Observatory still maintain that the observed light deflection at the Sun's limb at the Australian eclipse is, in its amount and in its character, in agreement with the predictions made on the theory of relativity."[27]

ANTIRELATIVISTS RALLY IN THE EAST

In October 1929, the antirelativity forces organized a symposium designed to knock down the observational underpinnings of relativity. The Optical Society of America provided the venue at their annual meeting, one year after they had fêted Michelson and heard from him that he had found no ether drift. The event took place at Cornell University in Ithaca, New York. Floyd Karker Richtmyer from the physics department was on the program committee. He explained to William Meggers at the Bureau of Standards that the Program Committee wanted to solicit material "for a *critical* discussion of the experimental facts supporting or contradicting the theory." Meggers preferred to let Keivin Burns discuss the Allegheny-Bureau data on the solar redshifts. Burns accepted and suggested Curtis and John A. Miller for material on the eclipse test.[28] Miller and Curtis had successfully obtained photographs at the eclipse of 9 May 1929 in Sumatra, where two British and one Australian expedition to the Malay peninsula had failed.[29] The program committee invited Curtis to speak, but ended up with Poor. As Curtis told Miller: "They asked me to handle the eclipse results, but I declined, pointing out that the time was not yet ripe. So they put in Poor instead, who will not do a great deal of real damage."[30]

Curtis hoped that his and J. A. Miller's 1929 Sumatra plates would cast doubt on the Lick verification. Several weeks before the Optical Society meeting, J. A. Miller wrote him that he was "pleased" with the Einstein plates. The seeing had not been as steady as he had hoped, but the drive was good and the images were "round and well-defined." To his friend Karl Lampland at Lowell Observatory he wrote: "I think we secured the best set we have ever made. We have about 65 stars on the eclipse field on the very best plate. I think it is possible, however, that some of those are a little too faint to measure. We also have impressed upon the plate a field for comparison that we made at eclipse time, shifting the camera in declension [*sic*] by 25 degrees and also a field that we made a few nights before the eclipse. The pair of plates made at beginning of totality are not so good as those made at mid-totality and the sky was evidently not quite so clear."[31]

Curtis asked Miller if "there would be any objection, if called upon, to my telling of the favourable preliminary examination of your Takengon plates?" Miller saw "no reason at all why you should not" but he insisted that Curtis not get anyone's hopes up:

> I don't want to raise the expectation of the astronomical public too high until after we have achieved. My experience in measuring Einstein plates makes me feel chary about what we will find on those plates. We have not yet mastered the idiosyncrasies of the measuring engine and when the plates are examined critically under the measuring engine the percentage of stars may not be so large as I believe now it will be. I still stand by the letter I wrote you, which was not an over-statement and I believe that we have got an awfully good set of plates. . . . All of the plates have a sufficient number of stars on them to measure but none of them are quite as good as the one I

described as the best plate. I think the next best plate had 47 stars on it and the short exposure plates . . . have about 30. Now, I think it worth while for the Society to know what we have done but don't make it too glowing.[32]

In the end, Miller's worst fears were realized. His plates turned out to be useful only for coronal studies.

The final Optical Society program consisted of Poor on the light bending, Burns on the solar redshift, Herbert Rollo Morgan of the U.S. Naval Observatory on Mercury's perihelion, and Dayton C. Miller on the ether drift. Meggers appeared after all to talk on the redshift of solar and stellar lines and attack the observations of Adams and Moore on the spectrum of Sirius B.[33] The organizers were able to solicit papers on all the tests of relativity. Every presentation attempted to show that relativity had not been verified. The society published three of the presentations in full in its *Journal*.[34]

The Optical Society's editorial board harbored a strong antirelativity sentiment. The editor in chief was Paul D. Foote of the Bureau of Standards in Washington, where Meggers worked. Richtmyer from Cornell was the assistant editor in chief and business manager. He was also president of the society in 1920. Richtmyer wrote what would become a standard textbook, *Introduction to Modern Physics*. He first published it in 1928. The first and second editions had only one reference to relativity— the variation of mass with velocity—which Richtmyer claimed was surrounded with "controversy and much weighing of both theoretical and experimental evidence." It was not until the third edition in 1942 that a chapter treated special relativity, eight years after Richtmyer's death in 1934.[35] Among the associate editors were the following: Henry G. Gale, University of Chicago (1925–26), Dayton C. Miller (1924–25), Heber D. Curtis (1925–26), and Ludwik Silberstein (1926–27).

D. C. Miller took up the antirelativity cause with gusto. A month after the Optical Society symposium, he presented his paper to the National Academy of Sciences at their fall meeting in Princeton, New Jersey.[36] In addition to presenting his ether drift experiments, Miller repeated Poor's assertions that the Lick eclipse data called for a displacement greater than Einstein's prediction.

While Curtis was disappointed that J. A. Miller obtained no results from Sumatra, he got satisfaction from an unexpected source. The only other successful expedition to Sumatra had been a German-Dutch party led by Erwin Freundlich. Einstein's protégé and former champion eventually published results claiming that a deflection larger than the Einstein amount had been detected. It was the very conclusion that Poor had tried to extract from the Lick 15-foot camera results. Freundlich's announcement enjoyed considerable publicity in Germany and the news filtered to the United States. The Potsdam observers also claimed that their reduction of the Lick 1922 results led to higher values for the light deflection, in line with their own 1929 results.[37]

The affair was put to rest by Trumpler, who was vacationing with his family in Switzerland for several months in the summer and fall of 1931. Trumpler visited Potsdam to discuss the matter with Freundlich and his associates. He remeasured their plates himself and obtained results that closely agreed with the Lick values. He also showed that the Potsdam observers had used a faulty weighting system that had yielded the higher values.[38]

THE FINAL SHOWDOWN

Continuing attacks and misinformation from the few vocal critics compelled Lick and Mount Wilson astronomers to go on the offensive. Charles St. John took up the cause at the April 1930 meeting of the National Academy of Sciences, where he presented a paper on "The Michelson-Morley Experiment and Two Predictions of General Relativity." In addition to reporting that no ether drift had been found at Mount Wilson, he reviewed the Lick eclipse results and his solar redshift conclusions. At the same meeting, St. John heard D. C. Miller's paper on his ether-drift results, and his repetition of Poor's attack on the Lick eclipse results. He noted with alarm that scientists in the East were receptive to Miller. He wrote Campbell that they should take further steps to silence the critics.

> I have given the Einstein resume including your deflection results and called attention to the number of lines [sic] [he meant "stars"][39] 62–85 with the 15 foot [camera] and 185–140 with the five foot [camera] as indicating the weight of your results. Miller spoke his old piece that he has given dozens of times but criticized your eclipse work saying that though the numbers of stars were large, many of them were rejected, some giving negative values. I think you ought to know of his remarks upon that subject, and that he ought to be packed up. The paper appealed strongly to those who ought to know, like [W. F. G.] Swann [of Yale], [Oswald] Veblen [Princeton], the Comptons [Karl and Arthur], Stewart of Illinois, [Frank] Schlesinger and others. I found the ground was ready for the good seed.[40]

About a week later, St. John wrote again from Washington. "I omitted one remark of Miller." He told Campbell that Miller "said that the actual value given by your observations is 2″06. I am not sure upon what he based this value." He also mentioned that Ernest Merritt, head of physics at Cornell, had been at the Optical Society meeting and had heard Poor's paper attacking the Lick results. The paper was now available in the society's journal. "You have seen Poor's paper in the Jour. Optical Soc. April. This is what he gave at the Ithaca meeting when all observations were knocked out of the running. I gather from Merritt that Poor made a strong impression at Ithaca, and that Miller also, was accepted. We will take care of Miller and turn Poor over to your tender care." He told Campbell how he planned to tackle Miller:

> As to the Michelson-Morley observations our results show nothing more than the small probable errors of 0.0015 fringe. These seem to be accepted by men who count: [George W.] Burgess [head of the Bureau of Standards], [Ernest W.] Brown, Swann, Schlesinger, Stewart, Veblen, [George] Birkhoff and others. It is evident that a systematic error is hidden in the crude mercury floatation system which Miller has never changed from the old rough apparatus. The Mount Wilson apparatus is a real piece of apparatus, worked surfaces of tank and float are true to 0.001 inch, and the centering exact. When this is done everything that Miller found disappears. The feeling among the people I mentioned is that Miller is in a tragic position, and that the only thing for him to do is to reconstruct his instrument and find out for himself that he has been wrong, but I doubt his doing this.[41]

Campbell was grateful for the intelligence. He handed St. John's letters over to Trumpler. "If any man is sure of his ground," he told St. John, "that man is Dr. Trumpler." "It is to be regretted the poor man who lives in New York should be able to mislead those men of science who are not very familiar with the operations involved in applying the Einstein eclipse test. On my own account I shall attempt no answer, but it is quite likely that Trumpler will, if he can only get hold of what the poor man said at Ithaca."[42]

Meanwhile, Miller went on with experiments at Cleveland during 1930. As late as 1933 he presented a paper at the National Academy claiming a positive result. He presented the same paper to the American Physical Society in the same year.[43] Poor also soldiered on, publishing an article in the November 1930 issue of *Scribner's Magazine*. Robert Aitken told Henry Norris Russell that Poor's article "excited my indignation and I was moved to make a short comment, confining myself entirely to the statements relating to the results of the Australian eclipse expedition in 1922." Russell replied: "I agree entirely with you about Poor's article in the Scribners'. They oughtn't to have published it. But I *won't* write anything for them. To argue with Poor is a waste of time. I don't think he is mentally sound when it comes to this topic. But you are absolutely right when you call attention to his misrepresentation of the Lick observatory data."[44]

Despite continued appearances of D. C. Miller's and Poor's works in print, the main trend of opinion was that astronomical research had amply verified relativity. Nonetheless, St. John went ahead with his coup de grâce. It did not appear until October 1932. In his lengthy article, St. John summarized results on the Michelson-Morley experiment and the three astronomical tests of general relativity.[45]

St. John described D. C. Miller's main results and then turned to the Mount Wilson experiments by Michelson, Pease, and Pearson. St. John capitalized on the fact that Miller had often emphasized how difficult it was to make his observations. For example, in his AAAS prize paper that he had presented to the American Physical Society in December 1926, Miller had elaborated the point as follows:

> I think I am not egotistical, but am merely stating a fact when it is remarked that the ether-drift observations are the most trying and fatiguing, as regards physical, mental and nervous strain, of any scientific work with which I am acquainted. The mere adjustment of an interferometer for white-light fringes and the keeping of it in adjustment, when the light path is 214 feet . . . requires patience as well as a steady "nerve" and a steady hand . . . , the observer has to walk around a circle about twenty feet in diameter, keeping his eye at the moving eyepiece of the attached to the interferometer which is floating on mercury and is turning on its axis steadily . . . ; the observer must not touch the interferometer in any way, and yet he must never lose sight of the interference fringes . . . ; these operations must be continued without a break through a set of observations, which usually lasts for about fifteen or twenty minutes, and this is repeated continuously during the several hours of the working period.[46]

By contrast, the Mount Wilson setup was designed to avoid the difficulties that Miller had faced: "Increased precision was obtained by arranging for the observer to make the micrometric settings of the interference fringes from a fixed position . . . such that he was free from the exacting conditions described by Professor Miller, who says . . ." At this point St. John quoted the first sentence from Miller's paper, cited above.[47]

St. John also reported independent results by Roy Kennedy at Caltech and Georg Joos at Jena in Germany. Like the Mount Wilson observers, both had found no ether drift. The frontispiece of St. John's article showed the Miller interferometer and the Jena instrument. The captions under the double photograph spoke eloquently:

a) The Miller Interferometer, carried on a wooden float in mercury, centered by a pin, and kept in rotation by slight impulses. The observer followed the rotating instrument and read the position of the fringes as he walked around.

b) The Jena Interferometer, mounted on ball bearings and motor driven. The axis of rotation was adjusted within 1″ of the vertical. Fringes were photographed and their positions measured with a micrometer.[48]

St. John included a short section on the excess over Newtonian theory of the observed motion of Mercury's perihelion. At the Optical Society meeting in October 1929, Herbert R. Morgan had listed possible sources of uncertainty in the observed motion of the perihelion to justify his assertion that the excess should be higher (50″9) than the value usually quoted (43″). St. John displayed results from three highly respected astronomers, Leverrier (1859), Simon Newcomb (1886), and Charles Doolittle (1912), who had calculated the Newtonian contributions to Mercury's perihelion motion. Comparing these figures with the observed value, the discrepancy came to 43″49, whereas general relativity predicted 42″9 per century.[49]

In his section on the deflection of light, St. John discussed Freundlich's recent results from the Sumatra eclipse. The Germans had found a limb deflection of 2″24 or 28 percent in excess of the theoretical prediction. They had used an absolute method of measurement, using photographic copies of a *réseau* impressed upon the eclipse and night comparison plates for the plate corrections, including scale factor.[50] Trumpler had criticized their method, challenging the scale determination from the réseau in face of temperature changes between night and day. St. John reported that Trumpler had remeasured the Potsdam plates, and calculated the deflections using the same techniques that he had used with the Lick data. He obtained a limb deflection of 1″75. St. John presented a table of accumulated results from all eclipse expeditions to date.[51] The agreement with general relativity was excellent. St. John concluded this section with a discussion of refraction due to cooling in the shadow cone. He disposed of this suggestion, citing J. A. Miller and R. W. Marriott's paper on the 1926 eclipse test for the Moon's diameter.

Finally, St. John turned to the redshift of the Fraunhofer lines in the Sun. He presented a detailed description of his convective motion explanation of the limb-center shifts. He showed how this could account for the smaller differences observed for low-level lines than those relativity predicted. For high-level lines, he appealed to the effect due to Milne and Merfield. St. John then moved on to a summary of the results on the spectrum of the companion of Sirius. He showed in detail how the scattered light from Sirius was practically eliminated from the spectrum of the companion for frequencies around the $H\beta$ line.

After his review of the observational evidence in favor of relativity, St. John drew on the credibility of one of the great British theorists, Hale's old friend Sir James Jeans: "In this brief review of the observational results the cumulative effect of the experimental evidence gives strong support to the statement of Sir James Jeans, who says: 'The general theory of relativity has long passed the stage of being considered an interesting speculation . . . and has qualified as one of the ordinary working tools of astronomy.' "[52]

It took roughly two decades for Einstein's jury of astronomers to pass judgment on his theory of relativity, by focusing on the three "classical tests." The jury's verdict was largely based on work that grew out of existing lines of research—eclipse photography, solar and spectral spectroscopy—that observers adapted to the problem. During the latter half of the 1920s, the astronomical community shifted from being Einstein's jury into witnesses on his behalf as astronomers gradually began to use relativity to advance their discipline.

THE EMERGENCE OF

RELATIVISTIC COSMOLOGY

After the Great War, the Allied countries created the International Astronomical Union (IAU) to rebuild disrupted international collaborations. At the inaugural Brussels conference in July 1919, the founders created thirty-two standing committees. The very first committee, Commission 1, was on relativity,[1] reflecting the intense interest in Einstein's theory. Four months later, the British verified Einstein's light-bending prediction. Enormous publicity focused attention on the astronomical tests. Eddington became the first chair of the IAU's relativity committee, due to his prominent role in explaining the theory and testing its predictions. Despite his penchant for theory, he set down strict, observational guidelines for international cooperation: "I assume that our committee is concerned chiefly with the astronomical aspects of the subject; and although no strict line can be drawn, it is not intended that it should deal with the more purely mathematical and physical developments. We are thus concerned primarily with the three 'crucial tests.' "[2]

At the first meeting in 1922, the commission's discussions centered on astronomical tests.[3] Eddington decided "at the present stage progress must be made chiefly by individual effort, and . . . there is not much scope for research organized on a large international scale."[4] By the time the second meeting of the IAU came around in 1925, Lick and Mount Wilson had verified the two tests they and others had been investigating for more than a decade. The members of the relativity commission decided not to meet, and the executive committee recommended "its re-appointment was unnecessary."[5] Research on the astronomical tests of relativity shifted into commissions that dealt with other areas of the discipline. For example, the Einstein eclipse problem went to Commission 12 on solar physics.[6] To this day, Commission 1 of the IAU is dormant.

This short-lived appearance of relativity as a special subdivision of the overall astronomy discipline reflects the increased acceptance of Einstein's theory of gravitation and its gradual transformation into a working tool for astronomers. When it was a controversial theory requiring sophisticated astronomical observations to test it, general relativity deserved to stand apart as a topic. As the theory passed its tests, astronomers began to use it to assist in other problems and explore new realms. Schwarzschild's solution of Einstein's field equations for a mass point would ultimately lead to the theory of black holes. Eddington used general relativity to test his theory of stellar interiors with observational assistance from Adams. St. John found it helpful in explaining the complex mixture of effects acting on the solar spectrum. In this context, it was natural that work on relativity would be incorporated into different divisions of the astronomy discipline.

The most dramatic and far-reaching connection between astronomy and relativity emerged in the latter part of the 1920s, as Einstein's theory continued to pass astronomers' tests. It led to the emergence of a new astronomical discipline—relativistic cosmology. The subject grew out of research on the stellar system and the nebulae, and of cosmological considerations of Einstein's field equations. The observational side of the problem came primarily out of the Pacific observatories in the United States. De Sitter, Einstein, and others pursued the theoretical side in Europe. Later, others took it up in America.[7]

Vesto Melvin Slipher of Lowell Observatory in Flagstaff, Arizona, had pioneered the measurement of radial velocities of the spiral nebulae. He published his first results in 1913. Slipher enjoyed a virtual monopoly on radial velocities of spiral nebulae and globular clusters well into the 1920s.[8] It was not until 1916, when plans for the 100-inch reflector at Mount Wilson were underway, that Hale began to develop a "nebular campaign" for his observatory.[9] By the second half of the 1920s, Edwin Powell Hubble and his colleague Milton Humason were using the 100-inch to get spectra of the faint nebulae that Slipher could not reach with his 24-inch refractor. By that time, the only telescope in the world that could obtain the redshifts of the most distant nebulae was the 100-inch reflector at Mount Wilson.

Hubble heralded the new field of relativistic cosmology in January 1929. He announced startling new results from Humason's spectroscopic observations of spiral nebulae with the 100-inch telescope. Hubble had found a linear relationship between radial velocities and distances among extragalactic nebulae. The farther away a spiral nebula is, the faster it recedes from us.[10] In 1917, Willem de Sitter had published his static solution of Einstein's field equations for an empty universe with the feature that particles would have a "tendency to scatter." He had speculated that Slipher's early results for spiral nebulae might be due to such an effect. Einstein's own static solution for the universe as a whole had no such feature, because he had incorporated a cosmological constant to keep his universe from expanding.[11] Hubble was not a theoretician, so he did not elaborate "the obvious consequences of the present results." Nonetheless, he concluded that its "outstanding feature . . . is the possibility that the velocity-distance relation may represent the de Sitter effect."[12]

About a year after Hubble's announcement, Eddington discovered two mathematicians' nonstatic solutions to the Einstein field equations. Alexander Alexandrovich Friedmann of Russia and Georges Lemaître of Belgium had developed these solutions years earlier, but no one had noticed them. Eddington published a paper on the dynamic solution and informed de Sitter, who also published a paper for astronomers.[13] From then on, the velocity-distance result for the nebulae was widely interpreted as evidence for an expanding universe. The idea was bold and exciting. It caught the imagination of the public in much the same way that the verification of the relativistic bending of light had done more than a decade earlier. When the world learned that astronomers had discovered that the universe is expanding, they learned that Einstein's theory of relativity had predicted it and Mount Wilson astronomers had proved it.

The power of general relativity as a theoretical tool for astronomers studying the large-scale structure of the universe cemented its acceptance among astronomers and physicists. During the preceding decade, astronomical observations had been a tool to check the validity of the theory. After Hubble's announcement, the situation changed. The more relativity passed astronomers' tests, the more confident they were in using it for other problems. The more successfully it assisted other research agendas, the more confidence astronomers placed on it. On 19 June 1929, the astronomer Donald Menzel noted this shift in his review of progress in astronomy at the AAAS Pacific Division meeting in Berkeley: "It is very curious, but most of the spiral nebulae . . . appear to be receding from us, and observations recently made at Mount Wilson indicate that the further away these spirals are, the faster their motion of recession appears to be. . . . Before generalizing too hastily, we must subject our observations to rigid test, according to the theory of relativity." The theory had been verified "to such an extent" over the past two years that "most astronomers and physicists have been convinced that Einstein's theory—or at any rate some slightly modified form of it—is to be regarded as a fact."[14]

Menzel called on recent work by the Caltech physicist Richard C. Tolman on Einstein's theory to apply the relativity test to Hubble's observations. Tolman found that there were only three possible descriptions of the universe as a whole that could come out of a static solution of Einstein's field equations. The third solution was the one that de Sitter had found in 1917. Menzel summarized Tolman's description of the de Sitter effect, by which light received from very distant objects would appear reddened. Spectroscopically, this effect would be observed as a redshift that would be interpreted as a Doppler effect due to recessional velocity. "Not only does de Sitter's theory account for this apparent tendency of the universe toward expansion, but the astronomer's discovery of the effect is additional evidence that relativity is correct. Furthermore, Hubble has investigated the relation between the amount of shift and distance of the nebulae, and finds that the observations agree very well with the theoretical predictions."[15] Menzel emphasized the fact that relativity had passed all the tests astronomers could muster. Noting that "the equanimity of relativists was considerably upset" a few years earlier by D. C. Miller's ether-drift announcement, he reported that Michelson, Pease, and Pearson had been working on the problem at Mount Wilson "with the most sensitive apparatus yet devised for the study of the problem." Their final results, "gratifying indeed to the relativist," showed no effect.[16] Menzel also referred to the three investigations Aitken had announced the year before, verifying light bending, solar redshift, and gravitational redshift of Sirius B: "Though it is quite obvious, it does not seem to be universally recognized that these observations stand for themselves—that they are *independent of any theory.* Very often those who, for some reason, take issue with relativity, appear to forget that the ultimate theory (whether it be Einstein's or not) must be able to explain these observational results. That relativity does so is a strong argument for its reality; that it predicted many of them before any study had been made is even more convincing proof of its correctness." Of course, Menzel proudly noted that "a major portion of the work of verification has been done on the Pacific Coast."[17]

Two years later, Einstein came to meet the Pacific astronomers who had been in the front rank of verifying and incorporating his theory into their discipline. He spent January and February 1931 at Caltech, his first visit to California. On 15 January, the community held a dinner in his honor at the Athenaeum. It was a truly southern California affair. Russell R. Ballard, the president of the California Institute Associates, a group of promoters of southern California scientific and scholarly research, delivered the opening remarks. Ballard proudly offered to "call the roll" of Einstein collaborators present: in order of seniority, Albert A. Michelson had started it with the ether-drift experiment; Charles E. St. John had verified the solar redshift; William W. Campbell had verified the bending of light near the Sun; Robert A. Millikan had verified the photoelectric effect; Walter S. Adams had verified the gravitational redshift of the companion of Sirius; Richard C. Tolman had elaborated the theoretical side of the cosmological predictions for Hubble; and Edwin P. Hubble had determined the linear relation for nebular velocities and distances. (Fig. E.1) Except for Tolman, the entire California group was made up of observers—a powerful complement to the European theorist Einstein.[18]

In the final dinner speech, Walter Adams highlighted the problem of "the nature and structure of the universe." He announced that "Professor Einstein is now inclined to consider the most promising line of attack on this problem to be based on theories of a non-static universe, the general equations for which have been developed so ably by Dr. Richard Chase

Figure E.1. Einstein and the leading American relativity testers gathered in the Hale Library at the Pasadena headquarters of the Carnegie Observatories in January 1931. *Left to right*: Milton L. Humason, Edwin P. Hubble, Charles E. St. John, Albert A. Michelson, Einstein, William W. Campbell, Walter S. Adams. A portrait of George Ellery Hale is in the background. (Courtesy Carnegie Observatories, Carnegie Institution of Washington.)

Tolman, of the California Institute of Technology."[19] The Caltech event was a triumph for Einstein as well as for the California scientific community. Both had come of age. Hale and Campbell had put California on the world scientific map and Einstein was one of its new stars. Einstein's first visit to California heralded new research possibilities for relativity theorists and astronomers alike.

Hubble's discovery could not have come at a better time for Hale and the scientific community he was building in Southern California. The opportunities that the 100-inch telescope had provided for nebular research inspired him to push for the construction of an even larger instrument. His entrepreneurial genius and the successes that characterized his past efforts convinced the Rockefeller Foundation to finance Hale's latest dream. A year before Hubble's announcement, the California Institute of Technology decided to construct a new observatory, whose main instrument would be a 200-inch reflecting telescope. In addition to a remote site for the large reflector, the new observatory was to have an astrophysical laboratory and a graduate school of astrophysics located in Pasadena at Caltech.[20] Hale was the father of this ambitious scheme, and his colleagues at Mount Wilson were to play leading roles in the design and construction of the 200-inch reflector. Hubble's announcement came just as work on the new observatory began. What a powerful vindication of Hale's vision this was. Hubble's discovery opened a new and exciting field of research that could be pursued only with the 100-inch reflector at Mount Wilson and the proposed 200-inch. When the Mount Palomar Observatory opened, it dominated the field of observational cosmology and studies of the universe for over fifty years.

Hale's success in California was part of a general shift in astronomy after the First World War. The astronomical center migrated from Europe to the United States. This preceded the rise of fascism in Germany, which accelerated the migration of leading scientists to the "New World." Within a few years, Einstein would leave Germany for good and emigrate to the United States, where he would spend the rest of his days.

By the 1930s, general relativity had weathered the previous decade's storm of debate and had moved triumphantly into the new field of cosmology. There, fresh problems such as the timescale of the universe would yield new results—such as the big bang theory of creation—that would, decades later, capture the imagination of the public once again. General relativity also gave rise to new specialties within the astronomy discipline, such as relativistic astrophysics. The public would be startled yet again by the theory of black holes that emerged from this field. Einstein's theory came of age and has flourished to this day within the astronomy discipline. Returning to Donald Menzel's words, used in his address to Pacific astronomers in 1929: "If the physicist is to be considered the father, and the mathematician the mother of the Einstein theory, the astronomer is certainly the rich and doting aunt who adopted the child soon after its birth and undertook the strenuous task of raising it. Were it not for the careful nurturing that the infant theory had in astronomical observatory and laboratory, it would never have grown to be the husky youngster it is at present, now well able to take its own part."[21]

FINAL REFLECTIONS

How Scientists Accept Theories

Where does this story leave the twenty-first-century reader? For one thing, it liberates us from a rigid understanding of how the scientific community accepts scientific theories. It is a messy process. There has been a tendency to simplify the historical picture with relativity, largely because the 1919 British verification of Einstein's light-bending prediction launched Einstein and his theory to international fame. This exciting story has captured the attention of writers and historians interested in Einstein and relativity. All major Einstein biographies focus on the British eclipse expeditions and most ignore or pass quickly over other attempts to test light bending before 1919. None deal with attempts after the British success.[1] The effect of the British announcements have also fascinated historians and scientists interested in the larger impact of Einstein and his theory of relativity.[2] The British eclipse results and the ensuing explosion of publicity have overshadowed other historical realities. Ever since this dramatic episode took place, the prevalent view of astronomers' participation in the reception of relativity has been almost exclusively as an effort to prove Einstein right. The historical record is more complex.

Consider the testing of Einstein's astronomical predictions by Lick and Mount Wilson astronomers. It began before the war as the application of existing research results to look for a specific effect. As individual investigators became more involved with the work, they tailored the procedures more directly to the relativity test. Eventually they developed a bona fide "Einstein Problem" with its own techniques and instrumentation. Their challenge was to obtain unambiguous measurements of the effects that general relativity had predicted. At Lick it was determining accurate star positions during an eclipse. At Mount Wilson the challenge was accurate measurements of solar spectral lines and identification of various laboratory and solar phenomena that shifted spectral lines. The astronomers involved did not question relativity theory's validity at first, since they had no adequate understanding of the theory anyway: their skills were in precision measurement. Once specific results began to emerge from this specialized research, the participants began to view the whole enterprise in a different light. Those debating the validity of the underlying theory began to cast the astronomical work as determining the truth of a controversial theory. For the astronomers conducting the research, it was actually about precise measurement of astronomical phenomena.

Here, critics played a central role. They made emphatic pronouncements on the death of the theory in face of any negative result. Announcements favorable to relativity's predictions prompted alternative explanations from traditional theory. Positions polarized as people sought an
unambiguous statement for or against the theory. And so evolved the notion of "crucial tests." This picture has no room for those astronomers,
like Eddington and de Sitter, who found Einstein's theory useful, beautiful, and profound and were less concerned about its "reality." Despite a
few voices to the contrary, the public saw the issue in terms of relativity's
truth or falsehood. The passing or failing of crucial tests was something
they could understand.

While this picture did not always do justice to why astronomers were
conducting the research in the first place, it did serve a useful purpose.
The debates thrashed out differences between the old and new way of
looking at specific phenomena of interest to astronomers. The issue of the
"elusive factor of two" that surrounded the light-bending prediction is a
case in point. It culminated in Trumpler's masterly exposition of the general relativistic way of looking at the bending of light in a gravitational
field. However, the purely empirical nature of the research among the
influential American astronomers and others involved in the debate had
a longer-lasting effect. It helped to foster the belief that scientists accept
theories on the basis of crucial tests that exhibit the truth or reality of
the theoretical picture. This simplistic empirical view of scientific method
exists to this day.

In the face of a result that verifies a theory's specific prediction, what
makes one scientist adopt the theory into one's research repertoire and
perhaps one's system of beliefs, while another continues to search for
alternative explanations? For Eddington and de Sitter, the observations
vindicated their conviction that the theory was important and useful.
Mount Wilson and Lick astronomers continued the observational research because their expertise and prestige lay in those areas. Their pos-
ture in favor of relativity evolved gradually, first defending their own observations and interpretations of the data, and then actively debating
against the continued attacks of critics. For Aitken, who was not involved
in any of the research, his change of attitude was partly in recognition of
extrascientific factors motivating critics. He saw the anti-Einstein propaganda from Germany. He was also aware of prejudices in his own country: a nationalistic fear of overdependence on Europeans; a conservative
resistance to newfangled notions as opposed to hard-nosed "common
sense"; an empirically based resistance to accept the "metaphysical notions" of general relativity; and a concern for their own specialties of
celestial mechanics and positional astronomy. The fortunes of Charles
Lane Poor, for example, provide a case study of a professional who viewed

general relativity as a threat to his own status as expert. His eventual decline among his peers was due to his breaking certain codes of behavior—deceptive presentation of others' results, publicity tactics—and his committing technical errors in the course of his ubiquitous attacks, that is, the "elusive factor of two." These different and often conflicting motivations all operate concurrently. Over time, a theory becomes accepted or it does not. Einstein's theory of relativity eventually became part of the "working tools" of astronomers, physicists, and other scientists.[3]

ASTRONOMERS' RECEPTION OF RELATIVITY

We can identify three levels of reception that took place as astronomers judged the merits of Einstein's theory. The first is "exposition"—describing an important development and sharing it with colleagues. The second level is "empirical research"—testing specific predictions of a theory. The third level is "elaboration"—actively working with the theory and elaborating its implications.

Most astronomers could do their work related to relativity without making a judgment of the theory's usefulness or its future prospects. At the first level, nearly all astronomers presenting relativity to their colleagues before the war missed or evaded the more fundamental implications of Einstein's work (Curtis was the exception). The early astronomical publications on relativity implied no research commitment by the authors. They merely wanted to understand what was going on and how it might affect their discipline. After 1916, except for Eddington and de Sitter, many commentators tried to ignore the geometrical features of general relativity while explaining its chief astronomical elements.

At the second level, astronomers committing time and resources to test specific predictions of Einstein's theory did not need to accept the underlying theory. They did not even have to understand the theory, and most did not. Some even embarked on the enterprise hoping to disprove it. Most astronomers took on the work because famous physicists were heralding a revolution in science. It was a hot topic and promised rewards for those who could verify or disprove its predictions. Curtis, for example, could use Einstein's equation for light bending to determine the values he must look for, while believing the theory is nonsense.

Only on the third level, those who undertake the elaboration of a new theory are the most likely to confront its wider implications. By working with general relativity, Schwarzschild, Eddington, and de Sitter discovered the fecundity of its new ideas. The theory also proved to be useful in ongoing lines of research in astronomy, such as stellar interiors and cosmology. This increased the chances that the broader implications of

Einstein's theory would enjoy a wider examination. This level was almost entirely absent in the American community of astronomers. By the 1930s, the third level of reception found its complete expression in the emergence of relativistic cosmology. The Americans at Mount Wilson played an important role, but Hubble significantly referred to the new field as "observational" cosmology.

In the United States, the three levels of reception followed one another sequentially. Early expositions represented a "first look" at the new theory. Testing occurred next and took over a decade. Once the theory had been judged from the expositions as important and the tests to be accurate, astronomers could begin elaborating it for cosmological, and later astrophysical, purposes. However, this "evolutionary" picture of reception was unique to the United States. It did not occur everywhere. Schwarzschild, de Sitter, and Eddington in Europe, for example, were elaborating the theory long before the tests were well under way. The fact that astronomers in the United States went through these three levels of reception in a temporal sequence is strictly due to their completely empirical orientation. This historical circumstance partly explains today's common belief that relativity's acceptance was solely because it passed the three "classical tests" of astronomy. In fact, for many years it was general relativity's theoretical foundations that attracted physicists, mathematicians and astronomers to Einstein's theory, until other observable implications, both terrestrial and cosmological, were discovered in the 1950s and 1960s.[4]

Relativity and Us

The story of how astronomers received Einstein's theory helps us understand why relativity, despite its popularity today, is still not part of our general culture. In the period covered in this account, astronomers were like an educated public hearing about relativity for the first time. As they came into contact with it, they became involved in aspects that directly related to their concerns. Many of them never dealt with the deeper implications of Einstein's ideas. It was not necessary. The same is true for most people today. Scientists in industry apply the equations of special relativity daily, and Einstein's general theory is guiding cosmologists in their search for a deeper understanding of the cosmos. Yet apart from the few specialists who work the equations, we do not need to understand the underlying concepts, nor do many of the experts. It will take a few deep thinkers who can also communicate in a powerful and meaningful way to bring Einstein's ideas into our general consciousness. They will need to make relativity relevant to us.

Specialists often say that this is impossible. Special relativity deals with the high-speed world at the atomic and nuclear scale. General relativity deals with the cosmological scale. We humans are right in the middle. Yet the space program has opened up new avenues of research that explore the effects of gravity on the human scale. Orbiting spacecraft provide a free-fall environment that allows scientists to experience Einstein's equivalence principle in action and explore its consequences at the macro, human level. Perhaps, like Hale, we need to expand the multidisciplinary mix of people working together. Add to the astronomers, physicists, and chemists some biologists, neuroscientists, and other specialists in the life sciences and see what happens. It is my personal belief that Einstein's ideas about how we measure the world and how we move within it hold further secrets that are highly relevant to each one of us.

NOTES

PREFACE

1. J. Crelinsten (1981).
2. J. Crelinsten (1983).

CHAPTER ONE. EINSTEIN AND THE WORLD COMMUNITY OF
PHYSICISTS AND ASTRONOMERS

1. P. Forman et al. (1975), quotation on 56; S. Goldberg (1984), 185; C. Jung-nickel and R. McCormmach (1986).

2. For details of Einstein's life and career, see R. W. Clark (1971); B. Hoffmann and H. Dukas (1972); J. Bernstein (1973); A. Pais (1982); A. Fölsing (1997); and D. Overbye (2000).

3. A. Einstein, "Zur Elektrodynamik bewegter Körper," *Annalen der Physik, 17* (1905), 891–921; "On the electrodynamics of moving bodies," in A. Einstein et al., *The Principle of Relativity* (New York: Dover, 1952), sec. 6, 51–55 (English translation of Einstein's original 1905 paper).

4. S. Goldberg (1984), 185–203 (Germany), 205–220 (France), 221–240 (Britain), 241–263 (United States).

5. H. Minkowski (1908). Cf. L. Pyenson (1977); T. Hirosige (1976).

6. Einstein to C. Habicht, 18 or 25 May, 1905, *The Collected Papers of Albert Einstein* (hereafter *CPAE*), vol. 5, 31. English Translation, 20; A. Einstein, "Über einen die Erzeugung und Umwandlung des Lichtes betreffenden heuristischen Standpunkt" [On a heuristic viewpoint concerning the generation and transformation of light], *Ann. der Phys., 17* (1905), 132–184; "Über die von der molekulartheoretischen Theorie der Wärme geforderte Bewegung von in ruhenden Flüssigkeiten suspendierten Teilchen [On the movement of particles suspended in fluids at rest, as postulated by the molecular theory of heat], in *ibid.*, 549–560; "Ist die Trägheit eines Körpers von seinem Energieinhalt abhängig?" [Does the inertia of a body depend on its energy content?], in *ibid.*, 639–641.

7. A. Fölsing (1997), 229–230.

8. A. Einstein (1907); Einstein to Habicht, 24 Dec. 1907, quoted in Fölsing (1997), 231–32. See *CPAE*, vol. 5, 82, note 5.

9. A. Fölsing (1997), 250.

10. For a detailed look at the nature of the astronomical community in the early twentieth century, see J. Crelinsten (1981) 2–28, 450–520.

11. A telescope that uses a lens to focus the light, as opposed to a reflector that uses a mirror.

12. By photographing the spectrum of a star and comparing it to a spectrogram taken in the lab, one can determine the component of the star's velocity along the

line of sight (radial velocity). If the star is moving away, the spectral lines will be shifted toward the red. If it is approaching the Earth, the lines will be shifted toward the blue.

13. Curtis to Campbell, 22 Jul 1920 (LO). Parallax is the apparent shift of a nearby star's position relative to the distant stars as the Earth orbits the Sun. Astronomers take photographs of the stars six months apart and compare the two photographs. Precise measurements of the apparent shift of the closest stars gives an accurate determination of their distance.

14. Curtis to Campbell, 26 Jul 1921 (LO).

15. Curtis to Office of the Secretary, Carnegie Foundation for the Advancement of Teaching, 15 Dec 1921 (LO).

16. Cf. K. Hentschel (1994).

17. William Huggins in England and Samuel P. Langley in the United States were two of the earliest pioneers of astrophysics. Cf. W. Maunder (1913); W. Huggins (1897); H. Dingle (1972); S. P. Langley (1884); D. F. Moyer (1973. See also A. M. Clarke (1903); M. Rothenberg (1974), 222–233; R. W. Smith (1982), 6–7; L. Rosenfeld (1973), esp. 381–382; H. Dingle (1973); C. A. Whitney (1976).

18. Slipher to Hubble, 20 Mar 1953 (LA). Cf. V. M. Slipher (1913).

19. H. C. King (1955), 308–318. The Yerkes 40-inch refractor went into operation in May 1897. H. Wright (1966), chaps. 4–5.

20. H. C. King (1955), 311–312; W. H. Wright (1941); J. H. Moore (1939).

21. R. Trumpler (1938).

22. W. W. Campbell (1901), (1903), (1907). For Campbell's program on radial velocity, see H. Wright (1966), 44–47, and J. H. Moore (1939), 145–147.

23. W. W. Campbell (1911a), (1911b), (1913a); R. Trumpler (1935); W. de Sitter (1917), 26–27.

24. Campbell and Barrows (secretary of the Pacific Division) to Slipher, 24 May 1915 (SP).

25. W. H. Wright (1941).

26. H. Wright (1966), 71–72, 88, 115, 145–146.

27. Quoted in H. Wright (1966), 38.

28. H. Wright (1966), chapter 4 for Yerkes, chapters 7 and 8 for Mount Wilson, Halm quote on 215; D. Osterbrock (1988); W. G. Hoyt (1976), 27–54.

29. H. Wright (1966), 237–251; R. Kargon (1977a); J. R. Goodstein (1991).

30. E. C. Pickering (1914), 74.

31. Meggers to Alfred Fowler, 6 Nov 1921 (MC).

32. Meggers to Kayser, 11 Sept 1923 (MC).

33. H. A. Bumstead to Hale, 23 Jul 1918 (HM).

34. Hale to Campbell, 7 Mar 1919, 15 Mar 1919; Campbell to Hale, 30 Aug 1919 (HM).

35. Campbell had three sons in the fighting.

36. Campbell to Perrine, 23 Apr 1919; Perrine to Campbell, 18 Jul 1919 (LO).

37. Stebbins to Campbell, 29 Mar 1922 (AAS). See also B. Schroeder-Gudehus (1973), (1978).

38. D. H. DeVorkin, 36–37 and 39–43; H. Wright (1966), 260–261.

39. H. Konen (1912). Translation in P. Forman et al. (1975), 5, note 4.

40. *MNRAS, 71* (1911), 384.

41. Merrill to Campbell, 8 Feb 1915; Campbell to Merrill, 15 Feb 1915 (LO).

42. R. A. Millikan (1910), (1916).

43. Jewett to Millikan, 3 Oct 1923 (CA). For an interesting discussion of Millikan's style as a physicist, see R. K. Kargon (1977b); also J. R. Goodstein (1991).

44. Hinks to Hale, 8 Mar 1914 (HM). Hinks left astronomy and was permanent secretary of the Royal Geographical Society for the rest of his career. Cf. W. M. Smart (1945–46), 89. For Eddington's career, see A. V. Douglas (1956).

45. F.J.M. Stratton (1945–46).

46. J. H. Jeans (1919); Jeans to Hale, 11 Oct 1917; Hale to Evelina Hale, 5 Nov 1918; and Hale to Jeans, 9 Sep 1922 (HM).

47. S. Weart (1979), 299–300.

48. Meggers to Sommerfeld, 6 Jul 1926 (MC).

49. "Stellar spectroscopy in 1915," *MNRAS*, 76 (1915–16), 357; "Stellar spectroscopy in 1917," *MNRAS*, 78 (1917–18), 315–316.

50. Curtis to Campbell, 23 Dec 1920 (LO).

51. The 1927 edition of James M. Cattell's *American Men of Science* included 242 astronomers, representing the leading practitioners in American astronomy. Of these, 121 had a Ph.D. See J. Crelinsten (1981), 25, 27–28, 479–492.

CHAPTER TWO. ASTRONOMERS AND SPECIAL RELATIVITY:
THE FIRST PUBLICATIONS

1. These articles were the only ones written on the subject in English-language astronomical journals during the early period. J. Crelinsten (1981), 30.

2. J.M.A. Danby (1975); J. Crelinsten (1981), 31.

3. H. C. Plummer (1909), (1910). J. Crelinsten (1981), 31–39.

4. H. C. Plummer (1910), 252. In 1887, Michelson and Morley split a light beam, each half running at right angles to the other. Mirrors reflected them back to their starting point, creating observable interference fringes. If one beam ran parallel to Earth's motion, the other would run at right angles. Rotating the apparatus to exchange beam positions would shift the interference fringes if motion through the ether affected light speed. There was no shift! To explain this, Lorentz postulated that matter-ether interactions had shrunk the apparatus in the direction of Earth's motion, compensating for the changed light speed.

5. Ibid.

6. E. T. Whittaker (1953). For a detailed and convincing critique of Whittaker's viewpoint as expressed in his 1953 volume, see G. Holton (1960).

7. H. C. Plummer (1910), 258.

8. Ibid., 265.

9. R. McCormmach (1970a).

10. H. C. Plummer (1910), 266.

11. E. T. Whittaker (1910a). This article was listed in Maurice Lecat's bibliography of relativity, but it was listed under an anonymous author, because the council note was signed only as "E.T.W." The article is cited under Whittaker's full name in the literature of the day. J. Crelinsten (1981), 39–49. M. Lecat (1924).

12. D. Martin (1976); W. H. McCrea (1957), quote on 237.

13. Quoted in G. Temple (1956), 319.

14. E. T. Whittaker (1910b), (1953).

15. E. T. Whittaker (1910b), 363.

16. I am grateful to the late Banesh Hoffmann for this explanation.

17. Proper motion refers to the motion of a star across the line of sight. It is measured by photographing stars on different dates and measuring the two plates to see if the star has changed position relative to the distant stars. Proper motions are measurable only for nearby stars, relative to the fixed pattern of distance stars.

18. E. T. Whittaker (1910b), 363.

19. R. L. Waterfield (1938), 118.

20. A. V. Douglas (1956), 20; R. L. Waterfield (1938), 118.

21. A. Pannekoek (1961), 478–480; Waterfield (1938), 120–121, 341–352; Douglas (1956), 20–22, 25–28.

22. E. T. Whittaker (1910b), 365.

23. Ibid., 364.

24. Ibid., 363.

25. Ibid.

26. For the electromagnetic view of nature, see R. McCormmach (1970b), (1970a). For a discussion of the older mechanical worldview, see M. J. Klein (1972). For more on worldviews among scientists, see R. McCormmach (1975).

27. E. T. Whittaker (1910b), 364.

28. A. Einstein et al. (1952), 51–55.

29. L. Pyenson (1977); P. L. Galison (1979); T. Hirosige (1976), 74–76.

30. E. T. Whittaker (1910b), 365.

31. Ibid., 365–366.

32. Ibid., 366.

33. G. Holton (1973), 269–280. For a list of treatments of relativity, see J. Crelinsten (1981), 35–36, note 4.

34. E. T. Whittaker (1953).

35. Whittaker left Dublin in 1912 to take a mathematics chair at the University of Edinburgh. H. C. Plummer took over Whittaker's position in Dublin. J. Crelinsten (1981), 40, note 4.

36. G. Holton (1973), 326; E. T. Whittaker (1955).

37. See chapter 4, 103ff, and Epilogue.

38. Whittaker to McVittie, 13 Nov and 23 Dec 1951 (AIP).

39. Whittaker to McVittie, 14 Oct 1952 (AIP).

40. G. J. Burns (1910). The papers as cited by Burns were "Burnstead [sic, should read Bumstead] (1909); [D. F.] Comstock (1910); [R. C.] Tolman (1910); [G. N.] Lewis and [R. C.] Tolman (1909)." See S. Goldberg (1968), 375–421, esp. 381–384 on Comstock, 385–403 on Lewis and Tolman.

41. G. N. Lewis and R. C. Tolman (1909), 517; in S. Goldberg (1968), 398.

42. A. Einstein (1965), 266. Einstein originally delivered this address as the Herbert Spencer Lecture on 10 June 1933 at Oxford, England. It was first published in *Mein Weltbild* (1934).

43. A. S. Bucherer (1908); C. V. Burton (1910). S. Goldberg (1968), 315–318.

44. W. de Sitter (1911), 388.

45. Ibid., 389.

46. Ibid. H. Poincaré (1905); H. Minkowski (1908).

47. W. de Sitter (1911), 390.

48. Ibid., 408.

49. Ibid., 408–409. H. von Seeliger (1906a). L. Pyenson (1976), 115, note 15.

50. W. de Sitter (1911), 413.

51. Ibid., 414.

52. Ibid., 415.

53. Ibid.

54. Ibid.

55. The campus moved to Stockton in 1925 and the College of the Pacific became the University of the Pacific in 1961.

56. M. A. Hoskin (1971).

57. H. D. Curtis (1911).

58. Ibid., 219.

59. Ibid.

60. Ibid., 220.

61. Ibid.

62. Ibid., 221.

63. E. Wiechert (1911a, b). Wiechert believed in absolute space and the ether and opposed special relativity. Curtis cited part one of Wiechert's paper (1911a), and probably had not seen part two when he wrote his article. Wiechert developed his criticisms of the relativity interpretation more fully in part two (1911b), esp. on 747–749, 750–752, but he alluded to it in part one on 701.

64. H. D. Curtis (1911), 222–223.

65. Ibid., 224.

66. Ibid., 225–226.

67. Ibid., 226–227.

68. Ibid., 226.

CHAPTER THREE. THE EARLY INVOLVEMENT, 1911–1914

1. A. Einstein (1911), translated in Einstein et al. (1952), 99–108.

2. A. Einstein (1907).

3. A. Einstein et al. (1952), 99. All other quotations are from this translation; unless specified otherwise.

4. Translation by H. D. Curtis, in Curtis (1913), 78. The standard English translation of Einstein's paper, by W. Perrett and G. B. Jeffery, in Einstein et al. (1952), leaves out the conditional clause in the first sentence: "It would be a most desirable thing if astronomers would take up the question here raised.* For apart from any theory there is the question whether it is possible with the equipment at present available to detect an influence of gravitational fields on the propagation of light" (108). The asterisk* indicates where Curtis's translation included the extra clause, "although the foregoing reasoning may prove to be insufficiently founded or even entirely illusory." Why the other translators, who did their version in 1923, chose to omit this clause, is a mystery. Perhaps in the midst of controversy over the validity of the theory, the more enthusiastic translators chose to strike out Einstein's own admission, written more than a decade earlier, that his reasoning might be faulty.

5. A. Pannekoek (1961), 406; R. Waterfield (1938), 20ff; S. A. Mitchell (1932), 130–136.

6. E.G. Forbes (1975), 90.

7. D. E. Osterbrock (1979–80), 67.

8. G. de Vaucouleurs (1957), 88–90, 92; A. Pannekoek (1961), 359–363; R. Waterfield (1938), 31; W. Z. Watson (1969), ix, 94.

9. S. A. Mitchell (1932), 153.

10. Campbell to Mitchell, 19 Mar 1908 (LO).

11. H. von Seeliger (1901), (1906a), (1906b). See A. Wilkens (1927); L. Pyenson (1976), 115.

12. Campbell to von Seeliger, 6 Mar 1908, and Seeliger to Campbell, 8 Apr 1908 (LO): "Meine Arbeit ist von andrer Seite *nicht* kritisiert worden und ich habe vorläufig nichts hinzuzufügen. . . . dass man meine Arbeiten in England z.B. überhaupt zu ignorieren pflegt. Ich kann mir das nicht erklären. Dagegen freut es mich, dass die amerikanischen Collegen, besonders die vom Lick Observatory anders denken."

13. E. W. B[rown] (1910); Campbell to Brown, 13 Mar 1912; Brown to Campbell, 22 Mar 1912 (LO).

14. Brown to Campbell, 5 Mar 1912; Campbell to Brown, 13 Mar 1912; Brown to Campbell, 22 Mar 1912.

15. W. de Sitter (1911).

16. W. de Sitter (1913); W. W. Campbell (1913b); F. Wacker (1906). For Wacker's ideas, see L. Pyenson (1974), 47–50.

17. L. Pyenson (1989), 228; K. Hentschel (1997), 5–6.

18. L. Pyenson (1976), 105–106.

19. Ibid., 106. See also Einstein to Freundlich, 1 Sep 1911, in A. Beck and D. Howard (1995), 201–202.

20. Einstein to Zangger, 20 Sep 1911 and Einstein to Freundlich, 21 Sep 1911, *CPAE*, vol. 5, 325 and 326; English in A. Beck and D. Howard (1995), 207.

21. Freundlich to "Sir," 25 Nov 1911 (LO; HA; US; USNO); C. D. Perrine (1923).

22. E. C. Pickering to Freundlich, 12 Dec 1911 (HA).

23. Freundlich to Pickering, 22 Dec 1911, 20 Jan 1912; Pickering to Freundlich, 5 and 8 Jan 1912 (HA).

24. Freundlich to "Sir," 25 Nov 1911, and C. G. Abbot to Freundlich, 2 Mar 1912 (SA, Record Unit 45, Box 22); Captain Jayne, Superintendent Naval Observatory, to Freundlich, 9 Jan and 29 Mar 1912; and Freundlich to "Sir," 26 Jan 1912 (USNO general correspondence).

25. Freundlich to Pickering, 20 Jan 1912 (HA).

26. Einstein to Freundlich, 8 Jan 1912, quoted in K. Hentschel (1994), 150; Cf. A. Beck and D. Howard (1995), 246.

27. Freundlich to Campbell, 24 Feb 1912 (LO). The earliest correspondence between Freundlich and the Greenwich Observatory preserved in the Greenwich Observatory Archives dates from 1913 and pertains to the eclipse of 1914.

28. Campbell to Freundlich, 13 Mar 1912 (LO).

29. Campbell to Perrine, 13 Mar 1912 (LO).

30. Freundlich to Campbell, 3 Apr 1912. Freundlich's English.

31. Perrine to Campbell, 23 Apr and 26 Aug 1912; Campbell to Perrine, 9 May and 31 May 1912; E. C. Pickering to Campbell, 14 Oct 1912; Perrine to Campbell, 25 Feb 1913 (LO). Cf. C. Perrine (1923).

32. Campbell to Freundlich, 30 Apr 1912 (LO).

33. Undated fragment (LO, Curtis folder). Underlining in original. Cf. H. D. Curtis (1911).

34. Curtis to Freundlich, 6 Jun 1912 (LO).

35. Freundlich to Campbell, 11 Oct 1912 (LO); Freundlich to Abbot, 30 Apr 1912, and True to Freundlich, 15 May 1912 (SA); Freundlich to USNO, 22 Apr 1912 (US); E. F. Freundlich (1913).

36. Freundlich to Campbell, 6 Feb 1913 (LO). E. F. Freundlich (1913), 371–372.

37. Karl Kamper, an astrometrist at the David Dunlap Observatory, personal communication.

38. Einstein to Freundlich, 27 Oct 1912. *CPAE*, vol. 5, 503–504; English in A. Beck and D. Howard (1995), 5, 323.

39. A. Pais (1982), 212.

40. Einstein to Sommerfeld, 29 Oct 1912 (". . . und das ich grosse Hochachtung für die Mathematik eingeflösst bekommen habe, die ich bis jetzt in ihren subtileren Teilen in meiner Einfalt für puren Luxus ansah! Gegen dies Problem ist die ursprüngliche Relativitätstheorie eine Kinderei"), *CPAE*, vol. 5, 505; English in A. Beck and D. Howard (1995), 324.

41. A. Hermann (1979), 24. For a description of the Einstein-Grossmann collaboration, see A. Pais (1982), 208–223, and D. Overbye (2000), 235–241.

42. E. G. Forbes (1961), 130; C. Fabry & H. Buisson (1910).

43. E. G. Forbes (1961), 131.

44. H. Wright (1966), 229–230.

45. A. H. Joy (1958), 4–5, 7.

46. E. G. Forbes (1961), 134.

47. G. E. Hale (1903). H. Wright (1966) 150–151, 207–208, and note 19 on 446.

48. G. E. Hale (1908a). See also H. Wright (1966), 221.

49. G. E. Hale (1908b), 146, 150.

50. W. S. Adams (1910). Forbes (1961), 134–135.

51. F.J.M. Stratton (1956a). See also *Kodaikanal Report* (1907), 1.

52. *Kodaikanal Report* (1908), 4.

53. Ibid., 5.

54. J. Evershed (1909a).

55. J. Evershed (1909b).

56. His dividing line between the two groups was lines that are displaced by 0.025 angstrom under 10 atmospheres of pressure.

57. *Kodaikanal Report* (1912), 4.

58. Einstein to Julius, 24 Aug 1911, *CPAE*, vol. 5, 312–313; English in A. Beck and D. Howard (1995), 199.

59. Julius to Einstein, 26 Aug 1911, ibid., 315–316; English in ibid., 201–202.

60. Julius to Einstein, 17 Sep 1911, ibid., 323; English in ibid., 205–206. Einstein to Julius, 22 Sep 1911, ibid., 327–328; English in ibid., 208–210. Julius to Einstein, 27 Sep 1911, ibid., 329–330; English in ibid., 209–210.

61. Einstein to Julius, 15 Nov 1911 ("Es ware so interessant, wenn wir über den Grund der Sonnenverschiebung etwas Zuverlässiges erfahren könnten"), ibid., 347; English in ibid., 221.

62. Julius to Einstein, 20 Nov 1911, ibid., 355; English in ibid., 225. Einstein to Julius, ibid., 357; English in ibid., 226.

63. Einstein to Freundlich, 8 Jan 1912, ibid., 387–388; English in ibid., 246.

64. Abraham to Schwarzschild, 13 Oct 1912 (SM).

65. K. Schwarzschild (1914). E. G. Forbes (1961), 140.

66. J. Evershed (1913).

67. E. Freundlich (1914).

68. J. Evershed and T. Royds (1914). I am grateful to Dr. M.K.V. Bappu, director of the Indian Institute of Astrophysics, for sending me a copy of this paper and several other *Bulletins*.

69. Ibid., 71.

70. Ibid., 77–80.

71. Ibid., 80.

72. C. E. St. John, "On pressure in the Solar atmosphere," read at seventeenth meeting of the AAS, 1914. See C. E. St. John (1918); C. E. St. John and H. D. Babcock (1914).

73. C. G. Abbot (1935), 274.

74. Ibid., 278. See C. E. St. John (1913a–d).

75. Freundlich (1913), 372: "Ich bin jederzeit bereit mit solchen Herren welche Aufnahmen wollen, zwecks genauerer Besprechung in Verbindung zu treten"; Freundlich to Campbell, 6 Feb 1913; Campbell to Freundlich, 6 Mar 1913 (LO).

76. Freundlich to Campbell, 28 May 1913; Campbell to Freundlich, 14 June 1913 (LO).

77. Freundlich to Dyson, 7 Feb 1913; Dyson to Freundlich, 16 Feb 1913 (RGO); JPEC, Minutes, 13 Dec (RAS). See also W. H. McCrea (1979).

78. W. W. Campbell (1913b).

79. Campbell to Freundlich, 27 May 1913; Freundlich to Campbell, 2 July 1913 (LO); W. W. Campbell (1923a), 16.

80. Campbell to Hale, 27 Aug 1913; Hale to Campbell, 19 Sep 1913 (HM). H. Wright (1966), 260–261, 300–301; D. DeVorkin (1981), esp. 36–40.

81. Campbell to Hale, 27 Aug 1913 (HM); W. W. Campbell (1913c); H. H. Turner (1913a), (1913b), (1913c).

82. Einstein to Freundlich, mid-August 1913, *CPAE*, vol. 5, 550; English in A. Beck and D. Howard (1995), 351.

83. A. Fölsing (1997), 327–329; D. Overbye (2000), 248–250.

84. A. Pais (1982), 239.

85. Nernst to Lindemann, 13 Aug 1913, in A. Fölsing (1997), 329; Einstein to Stern, in A. Fölsing (1997), 334.

86. Einstein to Freundlich, 26 Aug 1913, *CPAE*, vol. 5, 554–555; English in A. Beck and D. Howard (1995), 353–354.

87. R. W. Clark (1971), 162–163.

88. Einstein to Hale, 14 Oct 1913 (HM), *CPAE*, vol. 5, 559–560; English in A. Beck and D. Howard (1995), 356; printed in Hoffmann and H. Dukas (1972), 113, and in H. Wright, J. Warnow, and C. Weiner (1972), 66–67. See also L. Pyenson (1976), 108.

89. Hale to Campbell, 1 Nov 1913; Campbell to Hale, 4 Nov 1913; Hale to Einstein, 8 Nov 1913 (HM). The last letter is in H. Wright et al. (1972), 68–69, and *CPAE*, vol. 5, 566–567; English in A. Beck and D. Howard (1995), 361.

90. Einstein to Freundlich, 7 Dec 1913, *CPAE*, vol. 5, 581; English in A. Beck and D. Howard (1995), 369; L. Pyenson (1976), 109; K. Hentschel (1994), 170; A. Fölsing (1997), 319–320; C. D. Perrine (1923), 283–284.

91. H. D. Curtis (1913). Using a small guiding telescope mounted on the main instrument can help the observer adjust for drifting due to the Earth's rotation. A bright star in the field of view makes this task much easier.

92. A. Fölsing (1997), 356.

93. Campbell to Hale, 3 Mar 1914 (HM); Campbell to Byron S. Hurlbut, 15 Jan 1914 (LO).

94. Tucker to Campbell, 20 Jul 1914 (LO).

95. Campbell to Mr. Bruce, 28 Jul 1914 (LO).

96. Mrs. Campbell's diary notes; Campbell to "Consul for Great Britain and Ireland, 3 Aug 1914; Charles Wilson (American chargé d'affaires) to Campbell, 11 Aug 1914 (LO).

97. Einstein to Ehrenfest, 19 Aug 1914, *CPAE*, vol. 8, 56, and note 4, 57; English in A. Hentschel and K. Hentschel (1998), 42. C. D. Perrine (1923); Mrs. Campbell's diary notes (LO).

98. Hale to Campbell, 13 Oct and 5 Nov 1914; Campbell to Hale, 16 Oct 1914 (HM).

99. Campbell to Hale, 16 Oct 1914 (HM).

CHAPTER FOUR. THE WAR PERIOD, 1914–1918

1. Einstein to Ehrenfest, 19 Aug 1914 ("Ich döse ruhig hin in meinen friedlichen Grübeleien und empfinde nur eine Mischung von Mitleid und Abscheu"), *CPAE*, vol. 8, 56. English in A. Hentschel and K. Hentschel (1998), 41. For details of Einstein's marital breakup, see D. Overbye (2000) and A. Fölsing (1997).

2. Freundlich to Struve, 10 Mar 1915, and Struve to Freundlich, 12 Mar 1915, cited in *CPAE*, vol. 8, Doc. 54, notes 3 and 4, 89; Einstein to Freundlich, 5 Feb 1915 (". . . schimpfte er weidlich über Sie. Sie thäten nicht, was er von Ihnen fordere, etc."), ibid., 89; English in A. Hentschel and K. Hentschel (1998), 66.

3. Einstein to Otto Naumann, 1 Oct 1915, and Einstein to Freundlich, 24 Nov 1915, ibid., 178–179 and 203–204; English in A. Hentschel and K. Hentschel (1998), 133 and 149–150.

4. Einstein to Sommerfeld, 9 Dec 1915 ("Sagen Sie Ihrem Kollegen Seeliger, dass er ein schauerliches Temperament hat. Ich genoss es neulich in einer Erwi[d]-erung, die er an den Astronomen Freundlich richtete"), ibid., 217, and see Doc. 161, note 7, 218. English in ibid., 159.

5. E. Freundlich (1915/16a,b); W. W. Campbell (1911a). See K. Hentschel (1994), 155–159.

6. Einstein to Freundlich, Mar 1915, *CPAE*, vol. 8, 94–95 (quote on 95), and note 2. "*Es ist schade, dass Ihre Ausführungen nicht ausführlich genug sind, um die Unsicherheit abschätzen zu können, die Ihren Schätzungen anhaftet.* Deshalb kann sich ein Nichtfachmann über die Zuverlässigkeit Ihrer Rechnung kein Bild machen. Eine viel ausführlichere Darlegung wäre erwünscht. Am schlimmsten ist in dieser Beziehung die Angabe über die mittleren Dichten." Italics in original. English in A. Hentschel and K. Hentschel (1998), 70.

7. Einstein to Sommerfeld, 15 Jul 1915, *CPAE*, ibid., 147 ("sicher fundamental"); English in ibid., 111.

8. Freundlich to Struve, 7 Aug 1915, and von Seeliger to Struve, 15 Aug 1915, quoted in K. Hentschel (1994), 159, 162.

9. Cf. D. Overbye (2000), A. Pais (1982), and A. Fölsing (1997); Einstein to Walter Dällenbach, 13 Mar 1915, in A. Fölsing (1997), 369.

10. Einstein to de Haas, Aug 1915, in A. Fölsing (1997), 370.

11. Einstein to Freundlich, 30 Sep 1915, and note 3, *CPAE*, vol. 8, 178; English in A. Hentschel and K. Hentschel (1998), 132–133.

12. A. Einstein, *Sitzungsberichte*, Preussische Akademie der Wissenschaften (1914), 1030, cited in A. Pais (1982), 250–251. For a detailed discussion of Einstein's tortuous but ultimately successful path to his final equations, see ibid. For a less mathematical but equally thorough treatment, see D. Overbye (2000).

13. Einstein to Sommerfeld, 9 Dec 1915, *CPAE*, vol. 8, 217. "Das Resultat von der Perihelbewegung des Merkur erfüllt mich mit grosser Befriedigung. Wenn kommt uns da die pedantische Genauigkeit der Astronomie zu Hilfe, über die ich mich im Stillen früher oft lustig machte!" English in A. Hentschel and K. Hentschel (1998), 159.

14. Einstein to Ehrenfest, 17 Jan 1916, quoted in A. Pais (1982), 253. For a detailed account of Einstein's work during this period, see ibid., 243–261; A. Fölsing (1997), 369–378; D. Overbye (2000), 293–294.

15. Einstein to Ehrenfest, 26 Dec 1915 ("das Bezugssystem nichts Reales bedeutet"), *CPAE*, vol. 8, 228–229, on 228, and 5 Jan 1916 ("Dass Du die Zulässigkeit allgemein kovarianter Gleichungen noch nicht eingesehen hast, kann ich Dir nicht verübeln, weil ich selbst so lange brauchte, um über diesen Punkt volle Klarheit zu erlangen. Deine Schwierigkeit hat ihre Wurzel darin, dass Du instinktiv das Bezugssystem als etwas 'Reales' behandest"), ibid., 237–239, on 238; English in A. Hentschel and K. Hentschel (1998), 167–168 on 168, and 173–174 on 174.

16. Schwarzschild to Einstein, 22 Dec 1915 ("Wie Sie sehen, meint es der Krieg freundlich mit mir, indem er mir trotz heftigen Geschützfeuers in der durchaus terrestrischer Entfernung diesen Spaziergang in dem von Ihrem Ideenlande erlaubte"), ibid., 224–225, on 225, and note 3, 225; English in ibid., 163–164. Cf. A. Fölsing (1997), 384; K. Schwarzschild (1916a); Einstein to Schwarzschild, 29 Dec 1915 ("Die Theorie befriedigt mich sehr. Schon dass Sie die Newton'sche Näherung ergibt, ist nicht selbstverständlich; umso schöner ist es, dass sie auch noch die Perihelbewegung und die immerhin noch nicht genügend gesicherte Linienverschiebung liefert. Am wichtigsten ist nun die Frage der Lichtstrahlenkrüm-

mung"), *CPAE*, vol. 8, 231–232, on 232; English in A. Hentschel and K. Hentschel (1998), 169–170.

17. Schwarzschild to Einstein, 6 Feb 1916, ibid., 258–260; English in ibid., 190–192. K. Schwarzschild (1916b), 424–434. Cf. Fölsing (1997), 364–365.

18. A. Fölsing (1997), 385. See also J. Eisenstaedt (1982) for a discussion of subsequent work on Schwarzschild's solution during this period.

19. Einstein to Zangger, 26 Nov 1915, *CPAE*, vol. 8, 204–205; English in A. Hentschel and K. Hentschel (1998), 150–151.

20. Freundlich to Einstein, 17 Jun 1917, ibid., 469–471; English in ibid., 342–344.

21. Einstein to Sommerfeld, 28 Nov 1915 ("Nur die Intriguen armseliger Menschen verhindern es, dass diese letzte wichtige Prüfung der Theorie ausgeführt wird"), ibid., 208; English in ibid., 153.

22. Einstein to Naumann, 7 Dec 1915, ibid., 214–215; English in ibid., 157–158.

23. Struve to Naumann, 20 Dec 1915, cited in ibid., Doc. 160, n. 8, 216. Quotations in A. Fölsing (1997), 382, 383.

24. Freundlich (1915/16b), 21; K. Hentschel (1994), 159–160.

25. Von Seeliger to Struve, 12 Jan 1916, and von Seeliger (1917), quoted in K. Hentschel (1994), 161.

26. Einstein to Sommerfeld, 2 Feb 1916, *CPAE*, vol. 8, 255–257; English in A. Hentschel and K. Hentschel (1998), 188–189.

27. Einstein to Schwarzschild, 9 Jan 1916, ibid., 241 and note 4, 242; English in ibid., 176–177. Schwarzschild to Einstein, 6 Feb 1916, ibid., 258; English in ibid., 190. Einstein to Hilbert, 18 Feb 1916, in ibid., 264, and notes 2 and 3, 265; English in ibid., 195. Einstein to Hilbert, 30 Mar 1916, ibid., 277, and notes 2 and 3, 278; English in ibid., 205–206. Einstein to Struve, 13 Feb 1916, ibid., 261–262; English in ibid., 193.

28. Einstein to Ehrenfest, 26 and 29 Dec 1915; 3, 17, and 24 Jan 1916, ibid., 228–229, 237–239, 242–244, 249–253; English in ibid., 167–168, 173–175, 177–179, 182–188. Einstein to Lorentz, 1 and 17 Jan 1917, ibid., 232–234, 245–247; English in ibid., 170–171, 179–182.

29. Einstein to Besso, 3 Jan 1916, ibid., 234; English in ibid., 171.

30. Einstein to Lorentz, 17 Jan 1916, ibid., 245–247; English in ibid., 179–180.

31. Ibid., note 3, 247.

32. A. Einstein (1916a); issued separately as Einstein (1916b) and translated in Einstein et al. (1952), 111–164. See Einstein to Wien, 28 Feb and 18 Mar 1916, *CPAE*, vol. 8, 266–267, 274, and note 1, 275; English in A. Hentschel and K. Hentchel (1998), 197, 203–204.

33. M. J. Klein (1970), 297–302.

34. Eddington to de Sitter, 11 Jun 1916 (LC). That this is the paper de Sitter sent appears from Eddington to Dyson, 9 Dec 1916 (RGO).

35. Eddington to de Sitter, 4 Jul 1916 (LC). De Sitter to Einstein, 27 Jul 1916, *CPAE*, vol. 8, 323; English in A. Hentschel and K. Hentschel (1998), 239. Eddington to Wesley, RAS Secretary, 24 Aug 1916 (RAS); W. de Sitter (1915–16), (1916–17a), (1917–18). Karla and Franz Kahn (1975), 454, misrepresent Eddington's

initial reaction to Einstein's theory and to de Sitter's desire to advertise it as "somewhat cool"; but Eddington's letter emphasizes the difficulties about publishing in the RAS's *Memoirs* only to recommend the alternative of the *Notices*, not to discourage de Sitter.

36. Eddington to de Sitter, 11 Jun and 31 Dec 1916 (LC); A. S. Eddington (1916), (1916–17), (1918).

37. Eddington to de Sitter, 4 Jul 1916 (LC); W. de Sitter (1916).

38. W. de Sitter (1916), 418.

39. E.g., E. B. Wilson (1917).

40. T.J.J. See (1916).

41. H. Wright (1966), 118–119.

42. W. G. Hoyt (1976), 119–123.

43. Dooley (H. H. Turner) (1899). See also J. Lankford (1980).

44. J. H. Jeans (1917).

45. Eddington to de Sitter, 31 Dec 1916 (LC); A. S. Eddington (1917a).

46. Jeans to Lodge, 14 Aug 1917 (UC).

47. D. F. Moyer (1979), esp. 56–70, who, however, sometimes misinterprets Eddington's early writings. Cf. J. Crelinston (1981), 29–63, 100–101, 107–108, and S. Goldberg (1970), 89–126.

48. H. D. Curtis (1917), 63. Curtis adopted de Sitter's phrase "time-space."

49. Ibid., 64.

50. Ibid.

51. M. Born (1916); Einstein to Born, 27 Feb 1916, *CPAE*, vol. 8 266, notes 2 and 3; English in A. Hentschel and K. Hentschel (1998), 196.

52. E. Freundlich (1916). See the quotations in A. Fölsing (1997), 378, and K. Hentschel (1997), 17.

53. Foerster to Einstein, 25 Mar 1916, *CPAE*, vol. 8, 275, and note 2 (re Weinstein), 276; English in A. Hentschel and K. Hentschel (1998), 204.

54. A. Einstein (1917a). Cf. A. Fölsing (1997), 378–379.

55. Einstein to Weyl, 23 Nov 1916, *CPAE*, vol. 8, 365; English in A. Hentschel and K. Hentschel (1998), 265.

56. E. Gehrcke (1916); P. Gerber (1898), reprinted as Gerber (1917); Einstein to Wien, 17 Oct 1916, *CPAE*, vol. 8, 344, and note 2, 345; English in A. Hentschel and K. Hentschel (1998), 255.

57. M. von Laue (1917).

58. Einstein to Hartmann, 27 Apr 1917, *CPAE*, vol. 8, 439–440, and note 4, 440. English in A. Hentschel and K. Hentschel (1998), 320–321. E. Hartmann (1917).

59. See, for example, correspondence with Rostok philosopher Moritz Schlick in ibid., 388, 389, 417, 426, 450; English in ibid., 283, 284, 305, 312, 333.

60. Einstein to Hartmann, 27 Apr 1917 ("Bei der Lektüre philosophischer Bücher habe ich erfahren müssen, dass ich dastehe wie ein Blinder vor einem Gemälde. Ich begreife nur die induktiv zu Werke gehende Methode"), ibid., 440; English in ibid., 321.

61. Einstein to Besso, 28 Aug 1918, ibid., 864; English in ibid., 633.

62. Einstein to de Sitter, 22 Jun 1916, ibid., 302; English in ibid. 223.

63. Cf. A. Einstein (1916c); W. de Sitter (1916–17b). It was not until the 1960s that technology caught up to theory and gravity wave detectors started to be built. In 1978 their existence was indirectly demonstrated in a binary pulsar. See A. Fölsing (1997); C. Will (1986), 181–206.

64. Einstein to de Sitter, 22 Jun 1916, *CPAE*, vol. 8, 302, and note 7 for the de Sitter quotation; English in A. Hentschel and K. Hentschel (1998), 223.

65. Einstein to Schwarzschild, 9 Jan 1916, ibid., 240–241; English in ibid., 176.

66. Ibid.

67. A. Einstein (1916a), 112–113.

68. De Sitter to Einstein, 1 Nov 1916, *CPAE*, vol. 8, 358; English in A. Hentschel and K. Hentschel (1998), 260.

69. Ibid.

70. W. de Sitter (1916–17a), 155–184, esp. sec. 30; Eddington to de Sitter, 13 Oct 1916, cited in *CPAE*, vol. 8, doc. 272, note 5, 359. Einstein to de Sitter, 23 Jan 1917, ibid., 383; English in A. Hentschel and K. Hentschel (1998), 279–280.

71. Einstein to Besso, 31 Oct 1916, ibid., 347–348; English in ibid.,257.

72. Einstein to de Sitter, 2 Feb 1917, ibid., 385; English in ibid., 281.

73. Einstein to Ehrenfest, 4 Feb 1917, ibid., 386; English in ibid., 282.

74. Einstein to Ehrenfest, 14 Feb 1917, ibid., 390; English in ibid., 285. A. Einstein (1917b).

75. Einstein to Freundlich, 18 Feb 1917 or later, ibid., 393; English in ibid., 287.

76. Einstein to de Sitter, before 12 Mar 1917; de Sitter to Einstein, 15 Mar 1917, ibid., 411–413; English in ibid., 301–302.

77. De Sitter to Einstein, 20 Mar 1917, ibid., 414–416; English in ibid., 303–305.

78. Einstein to de Sitter, 24 Mar 1917, ibid., 421–423; English in ibid., 308–310.

79. De Sitter to Einstein, 1 Apr 1916, ibid., 427–428; English in ibid., 312–313.

80. W. de Sitter (1917–18), 3–28.

81. D. Overbye (2000), 325.

82. G. E. Hale (1915), 261.

83. Hale built the 150-foot tower telescope to obtain dispersions large enough to measure the Sun's magnetic field using the Zeeman effect on spectral lines. Construction of the 150-foot steel tower began in 1910 and the telescope was completed by 1912. See H. Wright et al. (1972), 45, 75, 257, 267.

84. C. E. St. John (1915a), (1915b); J. Evershed (1916a, b). See also E. G. Forbes (1961), 137–138.

85. C. E. St. John (1917a).

86. Ibid., 450.

87. St. John obtained numerical values from Eddington (1916–17).

88. C. E. St. John (1917a), 450.

89. Ibid., 452.

90. C. E. St. John (1917b), esp. 258–260. See also "The Equivalence Principle of Relativity," in G. E. Hale (1917), 209–210.

91. J. Evershed (1918a), 373.

92. Ibid. 373–374.

93. Ibid., 375.

94. Ibid.

95. Eddington to W. S. Adams, 28 Jan 1918 (HM).

96. A. S. Eddington (1918), 57.

97. Sommerfeld to Weyl, 7 Jul 1918, quoted in K. Hentschel (1994), 165.

98. Weyl to Einstein, 18 Sep 1918, *CPAE*, vol. 8, 877–879, and Einstein to Weyl, 27 Sep 1918, ibid., 893–894; English in A. Hentschel and K. Hentschel (1998), 644 and 655.

99. Evershed to Campbell, 14 Sep 1914 (LO). In practice, the Venus spectrum and an iron arc spectrum are taken on one plate, and compared to another plate with the sky spectrum and iron arc on it.

100. Campbell to Evershed, 23 Apr 1915 (LO).

101. Evershed to Campbell, 15 Jun 1915 (LO).

102. *Kodaikanal Report* (1916), 3.

103. J. Evershed (1918), 278. For the pole effect, see C. E. St. John and H. D. Babcock (1915a, b). At Kodaikanal, Thomas Royds was the one who studied this phenomenon extensively.

104. A. Pais (1982), 507.

105. Campbell to Perrine, 6 Aug and 20 Sep 1915 (LO); C. D. Perrine (1916).

106. R. G. Aitken (1917).

107. K. Bracher (1979).

108. Slipher to Lowell, 11 Feb 1918; Slipher to Miller, 19 Mar and 15 May 1918 (SP); J. A. Miller (1935).

109. K. Bracher (1979), 411.

110. Campbell to William M. Cake, 30 Mar 1918 (LO).

111. Undated ms., likely a news release, late Apr 1918 (LO).

112. Slipher to Miller, 18 Mar 1918 (SP).

113. W. W. Campbell (1918a) and (1918b), 229. The instrument arrived at Mount Hamilton on 21 Aug 1918, three months after the eclipse had taken place.

114. Curtis to Hale, 2 Jun 1918 (HM).

115. W. W. Campbell (1918c); Ethel Crocker to Mrs. Campbell, 6 Jul 1918 (LO).

116. W. W. Campbell (1918d).

117. W. W. Campbell (1918b), 230.

118. Campbell to Hale, 29 Aug 1917, and 22 Jul 1918, and Curtis to Hale, 2 Jun 1918 (HM); Burton to Leuschner, 18 Jul 1917, and Levy to Campbell, 3 Aug 1917 (LO).

119. Campbell to Secretary of the Navy, 8 May 1918 (LO).

120. Curtis to Hale, 27 Jul 1918 (HM).

121. W. W. Campbell (1918b), 230; Curtis to Campbell, 8 Jul and 5 Aug 1918, Campbell to Curtis, 10 Jul 1918, and Campbell to Pickering, 13 Aug 1918 (LO).

122. Campbell to Curtis, 14, 27, and 30 Nov 1918; Curtis to Campbell, 8 Dec 1918; Boothroyd to Campbell, 31 Dec 1918 (LO).

123. Einstein to Hilbert, 18 Feb 1916, and 30 Mar 1916, *CPAE*, vol. 8, 264, note 3, 265, and 277–278, quotation on 277; English in A. Hentschel and K. Hentschel (1998), 195 and 205–206, quotation on 205.

124. Einstein to Hilbert, 30 May 1916, ibid., 293; English in ibid., 216.

125. De Sitter to Einstein, 27 Jul 1916 (2 letters), ibid., 322–324, and note 4, 324; English in ibid., 238–240. Einstein to de Sitter, 2 Feb 1917, ibid., 385 and note 1, 386; English in ibid., 281. Einstein to de Sitter, 12 Mar 1917, ibid., 411, and note 4, 412; English in ibid., 301.

126. Freundlich to Einstein, 17 Jun 1917, ibid., 469–471, and note 1, 471; English in ibid., 342–344. Einstein to von Siemens, before 16 Dec 1917, ibid., 570–571, and note 4, 571; English in ibid., 478. Freundlich to Einstein, 6 Dec 1917, ibid., 563–564; English in ibid., 413–414. See also A. Fölsing (1997), 410–412; K. Hentschel (1997), 38–40.

127. Freundlich to Einstein, 17 Jun and 4 Dec 1917, *CPAE*, vol. 8, 469–471, 560; English in A. Hentschel and K. Hentschel (1998), 342–344, 411.

128. A. F. and F. A. Lindemann (1917).

129. Freundlich to Einstein, 17 Jun 1917, *CPAE*, vol. 8, 470; English in A. Hentschel and K. Hentschel (1998), 343.

130. Ibid.

131. Einstein to Freundlich, 3 Feb 1915, ibid., 88; English in ibid., 66.

132. Freundlich to Einstein, 17 Jan 1917, *CPAE*, vol. 8, 469–471; English in Hentschel and Hentschel (1998), 342–344.

133. Krüss to Einstein, 6 and 9 Jan 1917, ibid., 601, 603; English in ibid., 439, 440; Müller to Einstein, 9 Jan 1917, ibid., 603; English in ibid., 440–441.

134. Einstein to Krüss, 10 Jan 1917, ibid., 604–606; English in ibid., 441–442.

135. According to K. Hentschel (1997), 38, the plan fell through, but he mixes up the chronology of events somewhat. See *CPAE*, vol. 8, doc. 486, note 2, 684, for a reference to the fact that Freundlich was able to arrange a leave of absence from the Royal Observatory.

136. Einstein to Freundlich, 17 Jan 1917, *CPAE*, vol. 8, 608; English in A. Hentschel and K. Hentschel (1998), 444.

137. Freundlich to Klein, 11 Mar 1918, quoted in K. Hentschel (1997), 40.

138. Einstein to Freundlich, 17 Jan 1917 ("Vielleicht ist es besser, wenn ich mich persönlich für das Instrument verwende, da auf mir kein bellicoses Odium lastet"), *CPAE*, vol. 8, 608, and note 2, 609; English in A. Hentschel and K. Hentschel (1998), 444.

139. Schweydar to Einstein, 14 Apr 1918, ibid., 717, and notes 2 and 4, 718; English in ibid., 526. Planck to Einstein, 19 Mar 1918, ibid., 682, and notes 1 and 3, 684; English in ibid., 501.

CHAPTER FIVE. 1919: A YEAR OF DRAMATIC ANNOUNCEMENT

1. A. Pais (1982), 508.
2. J. Evershed (1919).
3. Ibid., 51.
4. Ibid., 52.

5. J. Evershed (1920a).

6. J. Evershed (1919), 52.

7. Ibid.

8. Adams to Hale, 7 Mar 1919 (HM).

9. Campbell to Curtis, 6 Jan 1919, and Curtis to Campbell, 13 Jan 1919 (LO).

10. Campbell to Curtis, 28 Feb 1919 (LO).

11. Ibid.

12. Curtis to Campbell, 9 Mar 1919 (LO). Scale errors refer to differences in scale between comparison and eclipse plates, taken months apart. Orientation errors are due to slight differences in the position of the plates in the plate holder, relative to the optical axis of the telescope. How Curtis corrected for these errors is described later when discussing his results.

13. Campbell to Curtis, 13 and 28 Feb, 3 Mar 1919 (LO); Curtis to Folks, 22 Feb 1919 (LO copy from MA).

14. Campbell to Curtis, 3 Mar 1919 (LO).

15. Campbell to Curtis, 17 Mar 1919 (LO).

16. A. V. Douglas (1956), 40.

17. F. W. Dyson (1916–17); W. W. Campbell (1918e).

18. A. V. Douglas (1956), 92–95, on 95.

19. Eddington's Notebook, quoted in ibid., 40

20. Curtis to Campbell, 22 Mar 1919, and same to same, 24 and 25 Mar 1919, plus enclosure, "Outline sketch of frame for measuring relativity plates" (LO).

21. Campbell to Curtis, 8 Apr 1919 (LO). When two of the larger steel rods were placed side by side with two ends in contact, the other ends stood as much as a quarter of an inch apart.

22. W. W. Campbell, "The Crocker expedition from the Lick Observatory, Jun 8, 1918. Some eclipse problems," manuscript of talk to American Philosophical Society, read 25 Apr 1919 (LO).

23. Hinks to Campbell, 1 Jan 1919, and Campbell to Hinks, 2 Jun 1919 (LO).

24. Ibid.

25. Curtis to Burckhalter, 16 Jun 1919 (LO); cf. W. W. Campbell (1923a), 18.

26. Curtis to Hale, 23 Jun 1919 (HM).

27. Curtis to Campbell, 23 Jun 1919 (LO).

28. Curtis, "The Einstein Effect: Eclipse of Jun 8, 1918 (preliminary paper)," 9 pp. Unpublished ms. (LO). The uninformative published abstract reads, in toto, "The result of measures on plates taken with the Vulcan cameras at Goldendale, Washington, at the eclipse of Jun 8, 1918." H. D. Curtis (1919).

29. Curtis, "The Einstein Effect" (LO). For Campbell's discovery, see W. W. Campbell (1911a,b).

30. Astronomers referred to this residual shift of spectral lines for B stars as the "K term." Commenting on a paper on radial velocities by Campbell for the Pasadena meeting, Curtis asked: "St. John found no 'relativity shift' to the red for the sun. Has any one yet discussed the possibility of the K term for the B and other early types being due to this?" Curtis, undated fragment (LO).

31. W. de Sitter (1915–16), 719–720; (1916–17a), 175–177; (1917–18), 26–27; A. S. Eddington (1917), 374.

32. R. J. Trumpler (1935).

33. Curtis, "The Einstein Effect" (LO). For Eddington's use of the phrase, see, for example, A. S. Eddington (1916–17) and (1919).

34. Curtis, "The Einstein Effect" (LO).

35. Ibid.

36. Ibid.

37. Ibid.

38. Ibid.

39. Curtis to Campbell, 13 Jan 1919 (LO).

40. Curtis, "The Einstein Effect" (LO).

41. Ibid.

42. See H. H. Turner (1912).

43. Curtis, "The Einstein Effect" (LO).

44. Hale to Curtis, 24 Jun 1919 (HM).

45. Campbell to Curtis, 22 Jun 1919; Curtis to Campbell, 23 and 24 Jun 1919 (LO).

46. "Record of meetings of the delegation to Brussels," 2 (NA, International Astronomy Union, American Section).

47. Eddington's Notebook, quoted in A. V. Douglas (1956), 40.

48. Joel Stebbins, "Report on the organization of the International Astronomical Union," 8 pp. (AAS).

49. *Observatory* (1919a), 298.

50. Ibid., 299.

51. Ibid.

52. See to Campbell, 3 Jul 1919 (LO).

53. Curtis to See, 5 Jul 1919 (LO).

54. Campbell to "Astronomer," 16 Jul 1919 (LO).

55. Curtis to Aitken, 16 Jul 1919 (LO). "Paris" referred to the *Carte du ciel* and "Boss" to a catalog of proper motions published by Lewis Boss of Dudley Observatory in 1904. See R. Waterfield (1938), 120.

56. Curtis to Aitken, 16 Jul 1919 (LO).

57. Campbell to "Astronomer," 21 Jul 1919 (LO).

58. Campbell, telegrams of 13 and 27 Aug 1919; Campbell to Cunard Steamship Company, 4 Sep 1919 (LO).

59. Freundlich to Einstein, 15 Sep 1919, and Einstein to Freundlich, 19 Sep 1919, *CPAE*, vol. 9, 156–157, 158–159; English in A. Hentschel and K. Hentschel (2004), 88–89, 89–90.

60. Lorentz to Einstein, telegram, 22 Sep 1919; Einstein to Pauline Einstein, 27 Sep 1919; Planck to Einstein, 4 Oct 1919; A. Moszkowski (1919); and A. Einstein (1919), all quoted in A. Fölsing (1997), 439–440.

61. Lorentz to Einstein, 7 Oct 1919, *CPAE*, vol. 9, 185–186; English in A. Hentschel and K. Hentschel (2004), 109.

62. J. Earman and C. Glymour (1980a), 73–76; A. Fölsing (1997), 441–442. Sixty years later, to mark the centenary of Einstein's birth (1979), the Sobral plates were remeasured using modern measuring and reduction techniques. The 4-inch plates yielded $1\rlap{.}''90 \pm 0\rlap{.}''11$, not much different from the $1\rlap{.}''98 \pm 0\rlap{.}''18$ that Dyson et al. got. The astrographic plates gave $1\rlap{.}''55 \pm 0\rlap{.}''34$, a significant improvement over the $0\rlap{.}''93$ obtained earlier, with an unspecified large error. The new measure

yielded a limb deflection of 1.″87 ± 0.″13, a result within one standard error of Einstein's prediction. See G. M. Harvey (1979). I am indebted to Don Fernie, director emeritus of the David Dunlap Observatory, for bringing this work to my attention.

63. A. N. Whitehead (1926), 13. See A. Fölsing (1997), 442–444; D. Overbye (2000), 361–363; J. Earman and C. Glymour (1980a), 76–78.

64. *London Times* (1919a); *New York Times*, 9 Nov 1919; J. Crelinsten (1980a); See to Campbell, 16 Nov 1919 (LO); *Observatory* (1919b); F. W. Dyson et al. (1920).

65. See to Campbell, 16 Nov 1919 (LO).

66. See to Campbell, 6 Dec 1919; Campbell to See, 8 Dec 1919 (LO).

Chapter Six. Men of Science Agog

1. *PASP* (1919). "According to press dispatches . . . the photographs obtained fully demonstrated the correctness of Einstein's theory."

2. Eddington to Einstein, 1 Dec 1919, *CPAE*, vol. 9, 262–263; A. Hentschel and K. Hentschel (2004), 158–159.

3. Einstein to Eddington, 15 Dec 1919, quoted in A. V. Douglas (1956), 41. Cf. J. Earman and C. Glymour (1980b), 199. See also *CPAE*, vol. 9, 304–305; English in A. Hentschel and K. Hentschel (2004), 184.

4. Dyson to Hale, 19 Dec 1919 (HM).

5. Dyson to Campbell, 29 Dec 1919 (LO); Schlesinger to Dyson, 16 Feb 1920 (SM).

6. Hale to Dyson, 9 Feb 1920 (HM).

7. Schlesinger to Dyson, 16 Feb 1920 (SM).

8. Dyson to Schlesinger, 18 Mar 1920 (SM).

9. *Observatory* (1919b), 394–395, 396–397.

10. *MNRAS* (1919), 111–112.

11. H. N. Russell (1920). Coelostats are flat mirrors driven by a clock to track the Sun and reflect the solar image into the main telescopes. The issue of nonradiality of stellar displacements and possible refraction effects continued to surround the light-bending test in America throughout the 1920s, but in Britain, they were eventually abandoned.

12. H. N. Russell (1921). Reprinted from *Princeton Lectures*, No. 2 (Princeton, 1 May 1920).

13. *Observatory* (1919c), 428–429.

14. *Observatory* (1919b) 395–397; *Observatory* (1919c); H. F. Newall (1919); *Observatory* (1920a); *London Times* (15 Nov 1919). See also J. Crelinsten (1981), 120–122.

15. Hale to Newall, 20 May 1920 (HM).

16. W. W. Campbell (1922), 19.

17. R. G. Aitken (1920).

18. Rutherford to Hale, 13 Jan 1920 (HM).

19. R. Clark (1971), 144.

20. Rutherford to Hale, 13 Jan 1920 (HM).

21. Wright to Hale, 27 Feb 1920 (HM).

22. Brown to Schlesinger, 7 Nov 1920 (SM).

23. Boothroyd to Slipher, 8 Mar 1920; Aitken to Slipher, 19 Mar 1920; Slipher to Boothroyd, 20 Mar 1920; Boothroyd to Slipher, 26 Apr 1920 (all in SP). See J. Crelinsten (1981), 177–178.

24. *New York Times*, 12 Nov 1919.

25. *New York Times*, 16 Nov 1919. J. Crelinsten (1980), 121–122.

26. See Chapter Nine passim.

27. Meggers to Schlesinger, 9 Jan 1920, and Schlesinger to Meggers, 17 Jan 1920 (MC).

28. Meggers to Burns, 22 Jan 1920, and Burns to Meggers, 6 Feb 1920 (MC).

29. Merrill to Meggers, 1 Mar 1920 (MC).

30. Meggers to O. H. Truman, 8 Feb 1921 (MC). Truman apparently did not follow up the idea, which had been suggested by Lindemann during the war. A. F. and F. A. Lindemann (1916).

31. Wright to Hale, 27 Feb 1920 (LO).

32. Hale to Wright, 11 Mar 1920, and Hale to Campbell, 20 May 1920 (LO); G. E. Hale (1921a).

33. A. Pais (1982), 508.

34. *London Times* (1919b). See J. Crelinsten (1980), 116.

35. H. H. Turner (1919), 445.

36. Ibid.

37. *MNRAS* (1919), 96–103.

38. "Astronomers on Einstein," *London Times*, 13 Dec 1919, 9.

39. Ludwig Silberstein had some facility with the theory, but did not like it. See Chapter Nine.

40. J. Crelinsten (1980b), 187–189. For more on Lodge's contributions to physics and his attitudes toward relativity before the general theory was developed, see S. Goldberg (1968), 303–370, esp. 306–315, and (1970), esp. 100–113.

41. *New York Times*, 11 Nov 1919, 17.

42. Ibid.

43. *New York Times*, 16 Nov 1919. See J. Crelinsten (1980a), 121.

44. *New York Times*, 18 Nov 1919. See J. Crelinsten (1980a), 122.

45. Merrill to Curtis, 24 Feb 1922 (CP).

46. Eddington to Einstein, 1 Dec 1919, in A. Hentschel and K. Hentschel (2004), 159.

47. A. S. Eddington (1920); Eddington's notebook, "Books: Sales and receipts." I am indebted to the late A. V. Douglas for letting me look at this and other material that she had collected in writing her biography of Eddington.

48. A. S. Eddington (1923). By contrast, his widely acclaimed book, *The Internal Constitution of the Stars*, published in 1926, sold 1,417 copies in the first five years. During the 1930s the sales of Eddington's *Mathematical Theory of Relativity* were almost four times those of *Internal Constitution of the Stars*. Cf. Eddington's notebook.

49. A. S. Eddington (1915), 97.

50. A. S. Eddington (1916–17), 377.

51. Ibid., 379–380.

52. A. S. Eddington (1918), 54–56.

53. A. S. Eddington (1919). Cf. J. Crelinsten (1981), 172–174.

54. A. S. Eddington (1919), 121. Underlining in original. At this time, Eddington was willing to entertain any amount of light deflection, as determined by the ratio of weight to mass for light. He later focused on only three possible values: zero, half, or full Einstein deflection, because each of these values was predicted by a specific theory. Historians John Earman and Clark Glymour have suggested that if Eddington and Frank Dyson had not "repeatedly claimed" that there were only three possible results, then the role of the British eclipse results as evidence for Einstein's prediction would have "been greatly weakened." At the time these historians made this suggestion, they were unaware that Campbell's negative results from 1918 were widely known, having been presented at a meeting in Pasadena as well as in England by Campbell. Astronomers had long been wrestling with the possibility that the law of gravitation might have to be modified, and Einstein's new theory was a specific alternative that had remarkably far-reaching implications. Many astronomers nonetheless remained skeptical after the British results were announced. Astronomers agreed that the result should be retested at the next eclipse. The enormous public attention that the British announcement galvanized was largely due to nonscientific issues and would likely have occurred with any positive announcement from the British, even one qualified by scientific subtleties. Cf. J. Earman and C. Glymour (1980a), 58, 79–80; J. Crelinsten (1980), 167–174.

55. A. S. Eddington (1919), 122.

56. Ibid.

57. W. S. Adams (1924), 5–6.

58. Brown to Schlesinger, 7 Nov 1920 (SM).

59. Eddington and Synge quoted in J. Eisenstaedt (1989), 285.

60. Quoted in A. Fölsing (1997), 457.

61. Einstein to Haenisch, 6 Dec 1919, quoted in A. Fölsing (1997), 472.

62. K. Hentschel (1997), 49–50, 58–61, quote on 49. Freundlich was referring to Commission 1 of the newly created International Astronomical Union, which Germans were not invited to join. See Epilogue.

63. Quoted in ibid., 51.

64. A. Fölsing (1997), 473; K. Hentschel (1997), 52 and 170, note 32. The inflation rate in 1919 was 58.1%; in 1920: 113.1%; in 1921: 28.1%; in 1922: 1,024.6%, and in 1923: 105.8 million percent.

65. A. Fölsing (1997), 474.

66. Einstein to Kneser, 7 Jun 1918, CPAE, vol. 8, 791, note 3; and Kneser to Einstein, 7 Jul 1918, ibid., 829.

67. Einstein to Maja Winteler-Einstein, May or Jun 1919, quoted in A. Fölsing (1997), 460.

68. London Times (1919d). See J. Crelinsten (1980a), 117–118.

69. A. Fölsing (1997), 460–465; A. Hermann (1979), 53–61. See also H. Goenner (1993).

70. Quoted in A. Hermann, (1979), 57.

71. A. Fölsing (1997), 464.

72. Quoted in A. Hermann (1979), 58.

73. Ibid., 59–61.
74. J. R. Goodstein (1991), 64–87.
75. Hale diary for Jan and Mar 1921 (HM); R. Kargon (1977a), 12–15.
76. J. R. Goodstein (1991), 97–99.
77. Hale to Merriam, 28 Jul 1921 (HM).
78. Hale to Larmor, 21 Dec 1921 (HM).
79. G. E. Hale (1921b), 291.
80. Ibid., 294.
81. Ibid., 297. See also R. Kargon (1977a).

CHAPTER SEVEN. TACKLING THE SOLAR REDSHIFT PROBLEM

1. J. Evershed (1920a), 155.
2. Ibid., 156.
3. Ibid.
4. Ibid., 157.
5. Ibid.
6. C. E. St. John (1920a), 159.
7. Ibid., 161.
8. Ibid., 162.
9. See J. Earman and C. Glymour (1980b), esp. 197–204. J. Crelinsten, (1981), 185–189.
10. L. Grebe and A. Bachem (1919), also in *Berichte der Deutschen Phys. Ges., 13/14* (1919); L. Grebe and A. Bachem (1920). See also Grebe and Bachem to Einstein, 23 Dec 1919, in A. Hentschel and K. Hentschel (2004), 196–197.
11. J. Earman and C. Glymour (1980b), 194–195. See also Einstein to Meyer, 28 Dec 1919, in A. Hentschel and K. Hentschel (2004), 200.
12. Einstein to Besso, 6 Jan 1920; Einstein to Eddington, 2 Feb 1920, and Eddington to Einstein, 15 Mar 1920, in A. Hentschel and K. Hentschel (2004), 206, 245, 296, respectively.
13. H.S.U. (1920).
14. *Nature* (1920).
15. R. D. Carmichael (1920).
16. If the slit was not parallel to the axis, then the two ends of the lines would be affected by the Sun's rotation in opposite directions, rendering measurment much more difficult.
17. C. E. St. John (1920b), 261.
18. For instance, the first few lines of any Rowland grating were irregularly ruled, since Rowland used to begin the ruling as he started up the ruling machine. These irregularities might produce asymmetries in the arc lines for long exposures.
19. C. E. St. John (1920b), 262.
20. Others criticized the Bonn results as well. See W. G. Duffield (1920); L. C. Glaser (1922). The latter was derisive of the Bonn experiments and Einstein's theory. See J. Earman and C. Glymour (1980b), 195–196.
21. Cf. A. Pais (1982), 508–510.
22. *Observatory* (1920b).

23. *MNRAS* (1920).

24. C. E. St. John and S. B. Nicholson (1920a), 227.

25. C. E. St. John and S. B. Nicholson (1920b); J. Evershed (1923a), 302.

26. MNRAS (1920), 413.

27. J. Evershed (1920b), 301.

28. Ibid., 302.

29. *Observatory* (1921), 159.

30. J. Evershed (1921), 244.

31. Ibid., 245. See C. E. St. John and S. B. Nicholson (1921).

32. C. E. St. John (1920c), 242.

33. Ibid., 243.

34. Ibid., 244.

35. G. E. Hale (1921a).

36. R. H. Fowler and E. A. Milne (1923).

37. J. Evershed (1923a), 300–301.

38. Ibid., 301.

39. Ibid., 302.

40. Ibid., 304.

41. Ibid.

42. C. E. St. John and S. B. Nicholson (1921). After the Earth effect issue was resolved, the Mount Wilson researchers went on to use the extensive Venus data for a determination of the solar parallax and a study of the rotation period of Venus. St. John and Nicholson (1922).

43. C. E. St. John (1922).

44. C. E. St. John and H. D. Babcock (1922–23).

45. Heiliger (St. John) to Hale, 13 May 1923 (HM).

46. Hale to Heiliger (St. John), 10 Jul 1923 (HM).

Chapter Eight. More Eclipse Testing

1. Correspondence of Curtis and Schlesinger, 15 Jan–20 Mar 1920 (SM); Campbell to Merriam, 23 Jan and 23 Mar 1920, Curtis to Campbell, 16 Apr 1920, Campbell to David P. Barrows, 20 Apr 1920 (LO); Curtis to Family, 1 Feb and 15 Mar 1920 (LO copy from MA); W. W. Campbell (1920).

2. Adams to Campbell, 29 Apr 1920, and Campbell to Adams, 26 May 1920 (LO).

3. Trumpler, biographical sketch (LO).

4. A. G. Marshall to Trumpler, 18 May 1919; Trumpler to "Herr Kollege," 28 Apr 1919; and Sproul to Trumpler, 9 Jun 1920 (LO).

5. Biefeld to Trümmler [*sic*], 13 Dec 1919, and W. H. Wright to Tracey R. Kelley, 28 May 1924 (LO).

6. Campbell to Schlesinger, 9 Jun 1922 (SM).

7. *PASP, 32* (1920), 191.

8. Moore to Campbell, 18 Jun 1920 (LO).

9. Campbell to Curtis, 15 Jul 1920 (LO).

10. Curtis to Campbell, 22 Jul 1920, and 2 August 1920 (LO).

11. Campbell to Curtis, 26 Jul 1920 (LO).

12. Curtis to Campbell, 2 and 18 Aug 1920 (LO).

13. Campbell to Curtis, 4 Oct 1920 (LO).

14. Curtis to Campbell, 11 Oct 1920 (LO).

15. Campbell to Curtis, 20 Dec 1920 (CP). Campbell and Moore had determined the three-minute exposure from consideration of the faintness of the Goldendale images. Several months later they realized that the eclipsed star images had been doubled by an unwanted movement of the apparatus. Their judgment of faintness, based upon only part of the image, had led to overcompensation.

16. Curtis to Campbell, 29 Dec 1920 (CP).

17. Campbell to Curtis, 17 Jan 1921 (CP).

18. Campbell to Curtis, 2 Mar 1921 (LO).

19. Ibid.

20. Campbell to Curtis, 31 Mar 1921 (LO).

21. Ibid.

22. Campbell, correspondence with Curtis and Burckhalter, 31 Mar–6 Apr 1921 (LO).

23. Campbell to Hale, 13 Apr 1921 (HM); Campbell to Stebbins, 12 Apr 1921 (AAS).

24. Campbell to Curtis, 19 May 1921 (LO).

25. Curtis to Campbell, 24 May 1921 (LO).

26. E. L. Larkin (1921) in Lick news clippings, 197 (LO).

27. Campbell to Dyson, 23 Nov 1921 (RGO).

28. Slosson to Campbell, 3 Dec 1921 (LO).

29. Bulletin prepared for display at Lick, 17 Dec 1921 (LO).

30. Campbell to Slosson, 12 Dec 1921, and Slosson to Campbell, 19 Dec 1921 (LO).

31. *New York Times*, 19 Dec 1921.

32. J. von Soldner (1804). Lenard's reprint appeared under the same title in *Annalen der Physik*, *65* (1921), 593–604. In a footnote he described von Soldner as a farmer's son who had enjoyed "the advantage of not having attended too much school." See R. Clark (1971), 254–264; A. Hermann (1977); J. Bernstein (1973), 123.

33. *New York Times*, 19 Dec 1921.

34. McSorley to Campbell, 1 Jan 1922, and Campbell to McSorley, 3 Jan 1922 (LO).

35. Campbell to Curtis, 23 Feb 1922 (LO).

36. Slosson to Campbell, 17 Mar 1922 (LO).

37. Campbell to Curtis, 3 Jul 1922 (LO).

38. Curtis to Campbell, 11 Jul 1922 (LO).

39. T.J.J. See (1922).

40. Shepard to Slipher, 1 May 1922, and Slipher to Welch, 11 May 1922 (SP). John Earman and Clark Glymour (1980a) studied a part of the story of the Lick attempt to measure the light bending at the 1918 eclipse, but they relied solely on archival material from 1920 to 1922, primarily located in the Curtis collection in the University of Pittsburgh archives. Unfortunately, this incomplete study of the existing material led them to the erroneous conclusion that the negative results

were not made public. They claimed that since the results were never published, they "had virtually no effect on the reception of general relativity" (64). This conclusion overlooks the importance of scientific meetings and informal communication among scientists. The account presented here shows that as a result of the Lick work, which was announced at several meetings and discussed in correspondence, skepticism about the light-bending test persisted for a number of years until it could be repeated reliably.

41. Cf. A. Pais (1982), 510.

42. *Observatory* (1920a), 143–144.

43. Campbell to Adams, 8 Jun 1920, and Curtis to Campbell, 24 Aug 1920 (LO).

44. Campbell to Adams, 17 Nov 1920 (LO); W. W. Campbell (1921).

45. Campbell to Curtis, 29 Nov 1920 (LO).

46. Campbell to Curtis, 12 Sep 1920 (LO).

47. Curtis to Campbell, 19 Sep 1920, Campbell to Curtis, 4 Oct 1920, Curtis to Campbell, 23 Dec 1920 (quote) (LO); Campbell to Curtis, 20 Dec 1920, and Curtis to Campbell, 29 Dec 1920 (quote) (LO); Curtis to Eddington, 23 Dec 1920 (CP).

48. Campbell to Chant, 3 Dec 1920, and Chant to Campbell, 26 Nov 1920 (LO). See also the Chant-Curtis correspondence in the Robarts Library, University of Toronto.

49. Campbell to Curtis, 17 Jan 1921 (CP).

50. Curtis to Campbell, 7 Feb 1921, and Campbell to Curtis, 2 Mar 1921 (LO).

51. Curtis to Campbell, 1 Jun 1921 (LO).

52. Campbell to Curtis, 4 May 1921 (LO).

53. L. Courvoisier (1913), cited in W. W. Campbell and R. J. Trumpler (1928), 155; L. Courvoisier (1920).

54. W. W. Campbell (1923a), on 15.

55. Campbell to Chant, 11 Oct 1921 (LO); Dyson to Campbell, 8 Nov 1921 (RGO).

56. Campbell to Chant, 11 Oct 1921 (LO).

57. Dyson, Eddington, and Davidson (1920), 317–319, 330.

58. Curtis to Eddington, 23 Dec 1920 (CP).

59. C. Davidson (1922).

60. Dyson to Campbell, 12 Jul 1922 (LO), and Jones to Dyson, 28 Aug 1922, published in *Observatory* (1922).

61. Campbell to Mitchell, 14 Feb 1922 (LO).

62. Campbell, "Notes on the Crocker Eclipse Expedition to Australia," undated typewritten ms. (LO).

63. Campbell to Chant, 2 Jun 1922, Campbell to Trumpler, 4 Jul 1922, and other documents (LO, 1922 eclipse box). See also Campbell (1923a), 21–22.

64. Campbell to Hale, 26 April 1922 (LO).

65. Dyson to Campbell, 8 Nov 1921 (LO).

66. J. Evershed (1923b).

67. Campbell to Mitchell, 28 Mar 1922 (LO).

68. Campbell to Trumpler, 31 Mar 1922 (LO). See also G. F. Dodwell and C. R. Davidson (1924), 150–151.

69. Campbell to Trumpler, 23 Jan and 25 Feb 1922, Trumpler to Campbell, 25 Jan 1922, Chant to Campbell, 16 Feb 1922; quotation in Curtis to Campbell, 1 Jun 1922 (all in LO).

70. W. W. Campbell and R. Trumpler (1923a), 42–45.

71. Campbell to Schlesinger, 9 Jun 1922 (SM); Curtis to Campbell, 11 Jul 1922 (LO).

72. *Observatory, 45* (1922), 61, 142–144, 317–318; Trumpler to Campbell, 31 Oct 1922 (LO).

73. The same coelostat mirror had been used by the British at Sobral in 1919 and had caused bad images then, but these had been attributed to temperature effects. Evershed noted the same problem but judged the cause to be faulty surfacing of the mirror. Contrary to those who believed that heating of coelostat mirrors made the technology unsuitable for the Einstein problem, Evershed believed it was actually well suited to the problem because "only with a coelostat is it practically possible to get an adequate scale."

74. J. Evershed (1923b).

75. *Sydney Morning Herald*, 22 Mar 1923; *Sydney Daily Telegraph*, 3 Mar 1923, Lick news clippings, 69 (LO), quote in former.

76. W. W. Campbell (1923a), 23–26.

77. Campbell to Trumpler, 5 Aug 1922 (LO).

78. W. W. Campbell (1923a), 40.

79. Trumpler to Campbell, undated (LO).

80. Ibid.

81. Campbell to Mitchell, 31 Mar 1923 (LO).

82. Science Service to Campbell, 22 Nov 1922, and Campbell to Science Service, 25 Nov 1922 (LO).

83. Campbell to Trumpler, 24 Nov 1922, and Trumpler to Campbell, 29 Dec 1922 (LO).

84. Campbell to Trumpler, 29 Dec 1922 (LO).

85. Mitchell to Campbell, 30 Nov 1922, Dyson to Campbell, 9 Jan 1923, Campbell to Dyson, 2 Feb 1923 (LO).

86. Campbell to Mr. E. C. McCobb (*Michigan Chimes*), 16 Jan 1923, quote in Campbell to Mr. J. B. Gum, International Newsreel Corp., 5 Jan 1923 (LO).

87. Russell to Campbell, 6 Dec 1922 (LO).

88. See to Campbell, 28 Dec 1922 (underlining in original), and Campbell to See, 29 Dec 1922 (LO).

89. Campbell to Dyson, 2 Feb 1922 (LO). See also C. J. Struble to Campbell, 29 Jan 1923, and Campbell to Board of Regents, 2 Feb 1923 (LO).

90. Trumpler to Campbell, 2 Feb 1923 (LO).

91. E. McConn (1923), Lick news clippings (LO).

92. Correspondence with Russell and Abbot, 6 Dec 1922–9 Mar 1923 (LO).

93. Correspondence with Science Service, 13–31 Mar 1923, Campbell to Slosson, 7 Apr 1923 (LO).

94. J. R. Brokenshire to Campbell, 9 Apr 1923, Turner to Campbell, 11 Apr 1923, Secretary to San Jose *Mercury*, 11 Apr 1923, giving itinerary of Campbell's trip to the Atlantic Coast (LO), Campbell to Einstein, 12 Apr 1923.

95. Campbell to Dyson, 12 Apr 1923 (RGO); W. W. Campbell (1923b).

96. R. Trumpler, "A test of Einstein's predicted light deflection near the Sun at the total solar eclipse of Sept. 21, 1922. Report to W. W. Campbell for Washington lecture, April 1923 (not published)" (LO).

97. Ibid.

98. W. W. Campbell and R. Trumpler (1923a).

Chapter Nine. Emergence of the Critics

1. Dyson to Campbell, 15 Apr 1923 (LO).

2. *Observatory* (1923a).

3. H. H. Turner (1923). The Australians took comparison plates in March 1923, and, because of faulty measuring equipment at Adelaide, did the reductions at Greenwich. Their results, published in January 1924, yielded a limb deflection of 1.77″ with a rather high probable error of 0.″3. G. F. Dodwell and C. R. Davidson (1924), 156, 162.

4. Russell to Aitken, 11 Apr 1923 (Center for History of Physics, AIP. Microfilm copy of Henry Norris Russell Collection, Princeton University archives).

5. St. John to Hale, 13 May 1923 (HM); see also Chapter Seven.

6. Mitchell to Campbell, 7 Apr 1923 (LO).

7. S. A. Mitchell (1923), 370–419, 392.

8. Ibid., 417–418. See pp. 231ff. and Chapter Ten, 236–240.

9. S. A. Mitchell (1923), 418.

10. Perrine to Campbell, 30 Jan and 2 Jun 1923 (LO).

11. Chant to Curtis, 15 Apr 1923, and Curtis to Chant, 18 Apr 1923 (CP).

12. Paddock to Curtis, 17 Aug 1923, and Curtis to Paddock, 29 Sep 1923 (CP). For Curtis's work at the eclipse of September 1923 and after, see Chapter Ten.

13. Curtis to Dr. Vogtherr, 19 Sep 1923 (CP).

14. T.J.J. See (1922). See also M. Lecat (1924), 115, item 3.

15. *Philadelphia Journal* (12 Apr 1923) and *San Francisco Chronicle* (12 Apr 1923), Lick news clippings, 131, 174 (LO).

16. See Chapter Six, 164–165. See also R. W. Clark (1971), 254–264.

17. *San Francisco Chronicle* (12 April 1923).

18. Ibid.

19. Cf. H. A. Bumstead to Hale, 23 Jul 1918 (HM); D. J. Kevles (1970); B. Schroeder-Gudehus (1973).

20. B. Hoffmann and H. Dukas (1972), 135; W. H. McCrea (1979); Newall to Dyson, 4 Jan 1920 (RGO).

21. "Professor See again attacks Einstein test," unidentified clipping, Lick news clippings, 135 (LO).

22. A. Einstein (1923); see also *Nature, 112* (22 Sep 1923), 448.

23. "Professor See again attacks Einstein test," Lick news clippings.

24. T.J.J. See, "Einstein a second Dr. Cook?" 7 Jun 1923 (LO); "Einstein a trickster. Noted European scientists, charging plagiarism, reject theory of relativity," 15 Jun 1923 (LO).

25. T.J.J. See, "Einstein a second Dr. Cook?" 7 Jun 1923 (LO).

26. Campbell to Trumpler, 22 Jun 1923.

27. Campbell to Aitken, 29 Jun 1923; Aitken to Campbell, 3 Jul 1923 (LO).

28. R. Trumpler (1923a); also in *Science, 58* (1923), 161–163.

29. Trumpler, "Einstein's theory of relativity, 1922–24," handwritten notes (LO).

30. "Father Ricard brands Einstein's theory as insult to common sense," 27 Jul 1923, Lick news clippings, 130 (LO).

31. *AMS, 4* (1927), 812.

32. See to Aitken, 6 and 8 Nov 1923; Aitken to See, 12 Nov 1923 (LO).

33. Campbell to Curtis, 21 Sep and 4 Oct 1921, and Curtis to Campbell, 28 Sep 1921 (LO).

34. Hale to Campbell, 26 Nov 1926 (HM).

35. "Prof. Campbell endorses Einstein," 6 Dec 1923, "Campbell sanctions theory of relativity," Lick news clippings, 123, 135 (LO).

36. *San Francisco Journal*, 9 Dec 1923, in ibid., 136.

37. C. L. Poor (1905), (1908). Poor to Curtis, 10 Mar 1925 (CP). The physicist R. H. Dicke proposed a similar idea in the 1960s. Cf. R. H. Dicke (1964), 28–29; R. H. Dicke and H. M. Goldenberg (1967); P. G. Bergmann (1968), 188–190.

38. C. L. Poor (1921).

39. C. L. Poor (1922), 48.

40. Poor to Curtis, 14 May 1923 (CP).

41. Poor to Curtis, 30 May 1923 (CP).

42. C. L. Poor (1922), 60.

43. Ibid., 67.

44. Ibid., 189.

45. Ibid., 196.

46. Ibid., 222.

47. Ibid., 240.

48. Ibid., 254–255.

49. Ibid., 257.

50. Poor to Campbell, Dec 1922, and 18 Dec 1922 (LO).

51. Campbell to Poor, 5 Apr 1923 (LO).

52. Poor to Campbell, 10 May 1923, Campbell to Poor, 21 May 1923 (LO).

53. Curtis to Poor, 7 May 1923, and Poor to Miller, 9 May 1923 (CP).

54. Curtis to Poor, 7 May 1923 (CP).

55. Poor to Curtis, 8 May 1923 (LO).

56. Poor to Miller, 9 May 1923 (CP).

57. Curtis to Poor, 15 May 1923 (CP); F. Slocum (1921).

58. Poor to Miller, 9 May 1923 (CP).

59. Poor to Curtis, 14 May 1923 (CP).

60. Curtis was referring to an effect called "halation," a halo effect caused by reflected light from the back of the photographic plate onto the back of the emulsion.

61. Curtis to Poor, 15 May 1923 (CP).

62. Poor to Curtis, 30 May 1923, and Curtis to Alter, 29 Jun 1923 (CP); J. A. Miller and R. W. Marriott (1925), 88.

63. E. W. Morley and D. C. Miller (1905). Cf. Loyd Swenson (1972), 48–52, 141–154, 190–212; R. S. Shankland (1941); H. Fletcher (1945).

64. Swenson (1972), 192–193.

65. K. Hentschel (1992), 603, note 32; A. Pais (1982), 113–114.

66. Merritt to Hale, 22 Nov 1920 (HM).

67. L. Silberstein (1920), 164.

68. Silberstein to Louis King, 14 Dec 1923 (MA); L. Silberstein (1922), 2.

69. Millikan to Hale, 16 Jul 1921; Hale to Millikan, 26 Jul 1921; and Millikan to Hale, 6 Aug 1921 (HM). See J. Crelinsten (1981), 307.

70. L. Swenson (1972), 196–198.

71. *New York Times*, 13 May 1921, 7; Michelson to Hale, 7 Jun 1921 (HM).

72. D. C. Miller (1925a); Hale to Miller, 28 Jul 1921 (HM).

73. Michelson to Silberstein, 28 Jul 1921; cited in L. Swenson, (1972), 198.

74. Hale to Miller, 28 Jul 1921 (HM). See also Hale to Merriam, 28 Jul 1921 (HM).

75. Michelson to Hale, 31 Oct 1922 (HM).

76. L. Swenson, (1972), 207.

77. D. C. Miller (1922). Abstract of presentation to National Academy of Sciences in April 1922.

Chapter Ten. The Debate Intensifies

1. Campbell to Curtis, 21 Sep 1921, Miller to Campbell, 8 Feb 1923, and Campbell to Miller, 16 Feb 1923 (LO).

2. Larmor to Hale, 9 Jan 1922 (HM).

3. J. Larmor (1919).

4. Larmor to Hale, 2 Jan 1923 (HM); J. L. Le Roux (1921), (1922).

5. Hale diary for 1922; Hale to Adams, 11 Jan 1923.

6. Larmor to Hale, 26 Jan 1923 (HM).

7. Adams to Hale, 19 Feb 1923, Millikan to Hale, 17 Feb 1923 (HM).

8. Larmor to Hale, 28 Feb 1923 (HM).

9. Adams to Hale, 17 Apr 1923 (HM); *Observatory* (1923b).

10. Page (McClanahan) to Campbell, 11 Jun 1923 (LO). Emerson Page was the nom de plume of A. C. McClanahan.

11. Emerson Page, "A New Fact with Old Distortions," *Delta County Tribune*, undated news clipping (LO).

12. Trumpler to McClanahan, 16 Jun 1923 (LO).

13. W. H. Wright (1923); R. Trumpler (1923b). Leo Courvoisier of Berlin had reported displacements of stars away from the Sun in all directions, based on meridian observations, which he attributed to refraction in the Earth's atmosphere. Near the Sun, the stellar displacement was about 1/2″ and it diminished very gradually, being about 0.″1 at 90 degrees and zero at 180 degrees. Einstein predicted a 1.″7 displacement near the Sun and a much more rapid decrease with increasing distance from the Sun.

14. *PASP* (1923a); Curtis to Poor, 29 Sept 1923 (CP).

15. *PASP* (1923b); C. E. St. John (1927). (The American Astronomical Society brought out their publications every several years, hence the 1927 date instead of 1923.) R. Trumpler (1927a); R. Trumpler, "Deflection of light in the Sun's gravitational field," 6-page ms. (LO).

16. *Observatory* (1923c), 344.

17. Evershed to Turner, 22 Dec 1923; Turner to Evershed, 23 Dec 1923 (RAS).

18. E.g., J. Earman and C. Glymour (1980b), 196, argue that "the eclipse measurements may loom larger in the confirmation of Einstein's red shift formula than the red shift measurements themselves." For the opposite view, see K. Hentschel (1993).

19. C. L. Poor (1927a). Eddington had derived an expression for the velocity of light passing near the Sun and found that it was a function of direction. As this variation was "inconvenient," he altered the coordinates slightly, setting $r = r_1 + m$, where m was a constant of integration derived earlier in finding the coefficients of the line element in a gravitational field. This substitution allowed Eddington to obtain an expression for the velocity of $1 - 2m/r \simeq 1 - 2m/r_1$, which was independent of the direction. Second-order terms of m/r_1 were neglected. Poor convinced himself that such a procedure prejudiced the rest of the calculation.

20. Curtis to Poor, 29 Sep 1923 (CP).

21. Ibid.

22. Miller to Aitken, 22 Oct 1923 (LO).

23. Miller and Marriott (1925), 87.

24. Curtis to Miller, 2 Dec 1923 (CP).

25. J. A. Miller and R. W. Marriott (1925), 88–89.

26. Poor to Curtis, 9 Jan 1924 (CP).

27. J. A. Miller and R. W. Marriott (1925), 84.

28. Ibid., 92.

29. Ibid.

30. Trumpler notes on "Einstein's theory . . ." (LO). Miller submitted his paper for publication in July 1924 and it appeared in March 1925. Trumpler must have seen an advance copy as his notes were written before 1925.

31. Poor to Curtis, 5 Oct 1923, and Curtis to Poor, 26 Oct 1923 (CP).

32. Poor to Curtis, 5 Oct 1923 (CP).

33. Curtis to Poor, 29 Sep 1923 (CP).

34. Curtis to Poor, 26 Oct 1923 (CP).

35. Curtis to Stratton, 13 Nov 1922 (CP); Meggers to Curtis, 21 May 1923 (MC).

36. C. L. Poor, "The deflection of light," 19 Nov 1923, enclosed with Poor to Curtis, 9 Jan 1924 (CP); *New York Times*, 20 Nov 1923.

37. Poor to Campbell, 10 Jun 1927, and Campbell to Poor, 18 Jun 1927 (LO).

38. Curtis to Poor, 24 Jan 1924 (CP).

39. Curtis to Miller, 7 Jan 1923 [*sic*] (CP). From the context, it is obvious that Curtis wrote this letter in 1924, and being only seven days into the New Year, had forgotten to use the new date.

40. Meggers to Burns, 18 Jan 1924 (MC).

41. W. W. Campbell and R. Trumpler (1923a), 54, (1923b), 163.

42. H.S.J. [Jeffreys] (1923), 282. Jeffreys was obviously referring to the Sobral value of 1."98 and not Eddington's Principe value of 1."61.

43. C. L. Poor (1924).

44. Miller to Campbell, 15 Feb 1924 (LO).

45. Miller to Aitken, 25 Feb 1924, and Aitken to Miller, 28 Feb 1924 (LO).

46. Curtis to David White, 19 Mar 1924, and White to Curtis, 24 May 1924 (CP).

47. Stevens to Curtis, 26 Mar 1924 (CP).

48. Curtis to Merrill, 4 Apr 1924 (CP).

49. Poor to Curtis, 12 Apr 1924, and Curtis to Poor, 24 Apr 1924 (CP).

50. "Report for the year July 1st, 1923 to June 30, 1924" (LO); R. Trumpler (1924a), (1924b).

51. R. Trumpler (1924b), 223.

52. J. Hopmann (1923).

53. R. Trumpler (1924b), 223.

54. Courvoisier had initiated a correspondence with Trumpler in the fall of 1923 after the 15-foot camera results had appeared. See Courvoisier to Trumpler, 4 Oct 1923, 9 Apr 1924, and 31 May 1924, and Trumpler to Courvoisier, 8 Feb 1924 and 9 May 1924 (LO). Courvoisier met Einstein in January 1924 and corresponded with him until October 1928, with no agreement being reached. See K. Hentschel (1992), 613.

55. R. Trumpler (1924b), 224.

56. *APS, 63* (1924), xii. The published proceedings do not indicate what was said, although they note that Burns's paper was discussed by St. John, Arthur Edwin Kennelly (an electrical engineer and member of the National Academy), and Burns; and St. John's paper was discussed by Burns.

57. H. D. Curtis (1924a); C. E. St. John (1924a).

58. H. D. Curtis (1924a), 442.

59. Ibid., 443.

60. W. S. Adams (1924), 314.

61. C. E. St. John (1924b), 91.

62. C. E. St. John (1910), (1914).

63. Julius had suggested his anomalous dispersion to explain the general redshift (a suggestion rejected by Evershed and St. John), but St. John appealed only to the mechanism to explain the excess redshift over the relativity shift at the solar limb.

64. S. A. Mitchell (1923), 404–416.

65. Almost forty years after the Allegheny—Mount Wilson controversy began, Eric Forbes judged from extensive research conducted by solar spectroscopists over the intervening years that the Allegheny data "should not be used to test St. John's hypothesis, nor do they constitute any valid evidence against it"; E. Forbes (1961), 148. In discussing the excess limb shift, Forbes did not like St. John's explanation. From research by Evershed and M. S. Adam of Oxford, he concluded that a real limb excess existed and could not be explained by St. John's method; ibid., 157. Nonetheless, he concluded that relativity and Doppler convection currents seem "capable of providing a satisfactory quantitative description of existing solar red shift data, thereby settling the controversial problem of the origin of this

phenomenon," despite some residual difficulties that required further research; ibid. 163–164.

66. Curtis to Poor, 3 May 1924, and Poor to Curtis, 16 Jun 1924 (CP).

67. Curtis to Poor, 3 May 1924.

68. Babcock to Curtis, 2 Dec 1924, and Curtis to Babcock, 8 Dec 1924 (CP).

69. H. D. Curtis (1924a); Poor to Curtis, 7 Feb 1925, and Curtis to Poor, 9 Feb 1925 (CP).

70. C. L. Poor (1926); Cattel to Robertson, 26 Mar 1940 (CA). See also R. C. Tobey (1977), 111.

71. These stars are reddish in color, indicating low temperatures. They are also highly luminous. Since their low temperature rules out a high energy output per unit area, they must be very large to appear so luminous. Thus the name "red giant."

72. See A. V. Douglas (1956), 58–87, on 69–70.

73. Hale to Michelson, 4 Oct 1920, in D. DeVorkin (1975), 10, 11–12.

74. Eddington to Hale, 17 Jan 1921, in A. V. Douglas (1956), 70.

75. G. E. Hale (1921b), 293.

76. A. V. Douglas (1956), 72–73; Eddington to Adams, 22 Mar 1924 (MW).

77. Quoted in A. V. Douglas (1956), 73. See A. S. Eddington (1924).

78. Eddington to Adams, 13 Jan 1924 (MW). See also A. V. Douglas (1956), 75.

79. Astronomers often measure redshifts as equivalent Doppler shifts, or "radial velocities" rather than angstroms or wavelength.

80. Eddington to Adams, 13 Jan 1924 (MW). In January 1924, Eddington believed that Sirius B had an A-type spectrum indicating a temperature of about 8500 degrees. By March, he realized it had an F-type spectrum with a lower temperature of about 8000 degrees.

81. Adams to Aitken, 9 Jan 1924 (LO).

82. W. S. Adams (1924a), 3.

83. Ibid., 7–8.

84. W. S. Adams (1915).

85. Adams to Eddington, 12 Feb 1924 (MW).

86. Adams to Eddington, 3 Mar 1924 (MW).

87. Ibid.

88. Eddington to Adams, 22 Mar 1924 (MW).

89. Adams to Eddington, 24 Apr 1924 (MW).

CHAPTER ELEVEN. RELATIVITY TRIUMPHS

1. Miller to Curtis, 25 Jan 1924, and Curtis to Miller, 28 Jan 1924 (CP).

2. These were at Vassar, Cornell, Wesleyan, Brown, Yale, and Columbia, though the latter would be just on the edge of the shadow; H. Benioff (1923a).

3. Mitchell to Aitken, 8 Dec 1924 (LO). Mitchell listed over a dozen different programs being planned for the eclipse. Except for Mount Wilson, all of them were from the East. See H. Benioff (1923b).

4. Aitken to Miller, 5 Jan 1925 (LO).

5. Miller to Aitken, 13 Jan 1925 (LO).

6. Mitchell to Aitken, 8 Dec 1924 (LO).

7. Curtis to Merrill, 8 Jan 1925 (CP).

8. *New York Times*, 23 Jan 1925, and 24 Jan 1925.

9. Miller to Aitken, 2 Feb 1925 (LO); J. A. Miller and R. W. Marriott (1928a); *New York Times*, 26 Jan 1925.

10. Curtis to Poor, 9 Feb 1925, and Poor to Curtis, 23 Feb 1925 (CP).

11. Curtis to Poor, 7 Mar 1925 (CP).

12. Poor to Curtis, 10 Mar 1925, and Curtis to Poor, 13 Mar 1925 (CP).

13. Poor to Curtis, 15 May 1925, and Curtis to Poor, 20 May 1925 (CP).

14. *New York Times*, 14 Oct 1924, and 16 Oct 1924a,b,c.

15. H. D. Curtis (1924b).

16. Meggers to Stebbins, 12 Nov 1921 (MC).

17. Meggers to Burns, 3 May 1921 (MC).

18. Curtis to Campbell, 11 May 1921 (LO).

19. Meggers to Burns, 18 Sep 1925, and Meggers to Einstein, 13 Feb 1925 (MC); K. Burns (1925).

20. Meggers to Raymond T. Birge, 29 Sep 1926 (MC).

21. G. Strömberg (1924), 473.

22. Adams to Hale, 6 Nov 1922 (HM).

23. G. Strömberg (1924), 477.

24. Besso to Einstein, 24 Dec 1920, quoted in A. Fölsing (1997), 481; A. Einstein, "Ether and the theory of relativity," lecture delivered in Leiden, 27 Aug 1920, published in English as Einstein (1922). See L. Dostro (1988).

25. D. C. Miller (1926), 437.

26. Ibid., 439.

27. Right ascension 17–1/2 hours or 262° and declination +65°.

28. D. C. Miller (1926), 442.

29. *New York Times*, 5 Jan 1925.

30. *New York Times*, 28 Apr 1925. Loyd Swenson has pointed out that the situation was ambiguous. Various scientists, including Michelson, later judged the experiment as not providing a crucial test for general relativity as they had previously thought. See Swenson (1972), 207–208. Nonetheless, the perception was that Michelson had again proved relativity to be correct via another ether-drift experiment.

31. D. C. Miller (1925a), (1925b); *New York Times*, 29 Apr 1925.

32. Curtis to Poor, 20 May 1925 (CP); Silberstein to King, 12–13 May 1925 (MA); L. Silberstein (1925a); A. S. Eddington (1925); L. Silberstein (1925b); O. Lodge (1926); Adams to Miller, 5 Jun 1925. See also Swenson (1972), 210.

33. Aitken to Eddington, 15 Jan 1924, and Eddington to Aitken, 6 Feb 1924 (LO); "Lectures by Professor Eddington at the University of California," *PASP*, 36 (1924), 230–231; Adams to Eddington, 9 Jun 1924 (MW).

34. K. C. Babcock to H. D. Curtis, 2 Dec 1924, re series of lectures at the University of Illinois.

35. W. S. Adams (1925); reprinted in *Communications from the Mount Wilson Observatory, to the N.A.S., 2* (1918–27), 205–210, quote on 206.

36. David White to Curtis, 3 Apr 1925, and Curtis to White, 4 Apr 1925 (CP).

37. Adams to Hale, 26 Apr 1925 (HM). "Beta" and "Gamma" refer to lines of hydrogen which Adams measured.

38. F.J.M. Stratton (1956b), 5.

39. *New York Times*, 28 Apr 1925, 20, col. 8, and 29 Apr 1925, 35, col. 1.

40. Slosson to Adams, 29 Apr 1925 (MW).

41. R. W. Smith (1982), chaps. 3 and 5. See Epilogue.

42. Adams to Eddington, 14 May 1925 (MW); W. S. Adams (1925).

43. W. S. Adams (1925), 208–209.

44. Early in 1926 Gustav Strömberg did an independent calculation of the orbit of Sirius and its companion, using updated values from work by Robert Aitken of Lick. The revised value for the redshift of the companion was +0.29 angstrom or +19 km/sec. See G. Strömberg (1926); and W. S. Adams (1926).

45. W. S. Adams (1925), 209–210.

46. Adams to Eddington, 14 May 1925, and Eddington to Adams, 4 Jun 1925 (MW).

47. *New York Times*, 22 Jul 1925, 21, col. 1.

48. Eddington's Notebook; A. S. Eddington (1926), 172–173.

49. J. A. Miller and R. W. Marriott (1928a), 101; *PASP* (1925); Aitken to Miller, 26 Mar 1925 (LO); *Memoirs RAS, 64* (1927), 105.

50. *PASP* (1926); S. A. Mitchell (1932), 215.

51. Miller to Aitken, 26 Mar 1926 (LO).

52. Curtis to Arnold, 20 Jun 1926 (CP).

53. Trumpler to Aitken, 20 Mar 1926, Aitken to Trumpler, 22 Mar 1926, and Aitken to Miller, 24 Apr 1926 (LO); S. A Mitchell (1932), 216.

54. Miller to Curtis, 14 Sep 1926 (CP).

55. Curtis to Miller, 16 May 1927, and Miller to Curtis, 30 May 1927 (CP). Quotation in former.

56. Miller to Curtis, 24 Jun 1927 and 7 Dec 1927 (CP).

57. S. A. Mitchell (1932), 216–219.

58. Miller to Curtis, 7 Dec 1927 (CP).

59. Ibid.

60. Curtis to Miller, 8 Dec 1927 (CP).

61. Curtis to Miller, 22 Mar 1928 (CP).

62. J. A. Miller and R. W. Marriott (1928a), 105.

63. J. A. Miller and R. W. Marriott (1928b).

64. Curtis to Miller, 15 Nov 1928 (CP).

65. R. Trumpler (1927b). At the 1919 eclipse the star observed nearest to the Sun from Sobral had a theoretical displacement of 0.″88 and at the 1922 eclipse the nearest star as seen from Wallal would be deflected 0.″85 away from the Sun.

66. D. H. Menzel (1929a), 122.

67. K. Hentschel (1997), 103–116.

68. Adams to Aitken, 5 Oct 1929, and Aitken to Adams, 28 Oct 1929 (LO).

69. L. Swenson (1972), 212.

70. A. Einstein (1926), quoted and discussed in K. Hentschel (1992), 604–606. Hentschel's translation, Einstein's emphasis.

71. Quoted in K. Hentschel (1992), 604, in response to a letter from Ralph E. Zuar, correspondent for *Popular Wireless and Wireless Review*, 18 Apr 1926.

72. R. J. Kennedy (1926); Michelson et al. (1928), 373.

73. J. C. Duncan to C. O. Lampland, 14 Jul 1926 (LP).

74. Adams to Michelson, 24 Nov 1926, cf. L. Swenson (1972), 214.

75. Hale to Einstein, 2 Dec 1926, and Einstein to Hale, 25 Dec 1926 (HM).

76. St. John to Leuschner, 25 Jan 1927 (LO).

77. Michelson et al. (1928), 401–402.

78. Ibid., 382, 385, 399–401.

79. Ibid., 364.

80. Ibid., 395.

81. Ibid., 345.

82. Ibid., 373.

83. Ibid., 386–388.

84. Ibid., 389.

85. F. G. Pease (1930), 197.

86. A. A. Michelson, F. G. Pease, and F. Pearson (1929).

87. *New York Times*, 3 Nov 1928, 21.

88. J. H. Moore (1928), 231.

89. Adams to Aitken, 25 Oct 1926 (LO).

90. Aitken to Adams, 3 Nov 1926 (LO).

91. Moore (1928), 231.

92. Ibid., 233.

93. C. E. St. John (1928).

94. Heiliger (St. John) to Hale, 14 Sep 1927 (HM). "Heiliger" was a nickname, meaning "saint" in German.

95. Back in 1926, Meggers was hoping to explain the variation of redshift with intensity using anomalous dispersion. Anomalous dispersion should vary with wavelength, being strongest for blue lines. Since all lines of high intensity showed the excess redshift, he had asked Burns whether it might be "probable that the gas absorbing strong red lines descends faster than the gas absorbing strong blue lines?" St. John's interpretation that the lines of high intensity are high in the solar atmosphere allowed him to apply Milne and Merfield's mechanism and provide a better explanation of the larger redshifts for broad lines of high intensity. This general picture is now the accepted explanation of the solar redshifts.

96. C. E. St. John (1928), 195.

97. E. G. Forbes (1961), 148–164.

98. C. E. St. John (1928), 196.

99. Ibid., 236.

100. Ibid.

101. Ibid.

102. St. John remeasured these lines, and the shifts were essentially the same as before.

103. C. E. St. John (1928), 238.

104. Ibid., 239.

105. R. Trumpler (1928), 131.

106. W. W. Campbell and R. Trumpler (1928), 152.

107. R. Trumpler (1928), 132–133.

108. Ibid., 133.

109. Ibid., 134. Trumpler gave no details of this study in his article, but Campbell and Trumpler dealt at length with Courvoisier and abnormal refraction studies in their more detailed version. Cf. W. W. Campbell and R. Trumpler (1928), 155–160.

110. R. Trumpler (1928), 134.

111. Merrill to Aitken, 7 Feb 1927, and Aitken to Merrill, 10 Feb 1927 (LO).

112. R. G. Aitken (1927).

113. Aitken to Poor, 31 May 1927 (quote), 3 May 1928, Poor to Aitken, 8 Jun 1927, Poor to Redman, 29 Aug 1927, and Aitken to W. H. Pickering, 31 Jan 1927 (LO).

114. Aitken to Russell, 9 Mar 1926 (LO).

115. Russell to Aitken, 15 Mar 1926 (LO).

116. R. G. Aitken (1928).

CHAPTER TWELVE. SILENCING THE CRITICS

1. J. G. Porter (1927).

2. Curtis to Porter, 13 May 1927, and Porter to Curtis, 19 May 1927 (CP).

3. Chant to Trumpler, 28 Feb 1929 (LO).

4. C. L. Poor (1927b), 227.

5. A. Einstein (1911). See A. Einstein et al. (1952), 106.

6. C. L. Poor (1927b), 228–229.

7. Ibid., 229, 232. Poor's notation.

8. Ibid., 234.

9. Aitken to Poor, 19 Apr 1927 (LO).

10. Poor to Aitken, 30 Apr 1927 (LO).

11. R. D. Carmichael et. al. (1927).

12. J. G. Porter (1927).

13. P. Fox (1927–28).

14. Poor to Aitken, 15 May , and Aitken to Poor, 25 May 1928 (LO).

15. Poor to Aitken, 11 Jun 1928 (LO).

16. Aitken to Poor, 28 Jun 1928 (LO).

17. R. Trumpler (1929a), quotation on 23.

18. Chant to Trumpler, 28 Feb 1929 (LO).

19. C. L. Poor (1929); R. Trumpler (1929b).

20. Campbell and Trumpler (1928), 137; italics added.

21. Ibid., 154–157.

22. C. L. Poor (1929), 357.

23. R. Trumpler (1929b), 362.

24. Chant to Trumpler, 12 Jun 1929 (LO).

25. Trumpler (1929a).

26. J. G. Porter (1929).

27. R. G. Aitken (1929).

28. Meggers to Burns, 10 May 1929 (emphasis in the original), Richtmyer to Meggers, 17 May 1929, and Meggers to Richtmyer, 23 May 1929 (MC).

29. *PASP* (1929).

30. Curtis to Miller, 4 Oct 1929 (CP).

31. Miller to Curtis, 28 Sept 1929 (CP); Miller to Lampland, 2 Oct 1929 (LP).

32. Curtis to Miller, 2 Oct 1929 (Takengon is on the west coast of Sumatra), and Miller to Curtis, 7 Oct 1929 (CP).

33. *JOSA* (1930); "The red-shift of solar and stellar lines," 8-page ms. (MC).

34. C. L. Poor (1930); K. Burns (1930); H. R. Morgan (1930).

35. F. K. Richtmyer (1928). A second edition was published the year Richtmyer died, and three subsequent editions came out in 1942, 1947, and 1955. The later editions after Richtmyer's death had coauthors beside Richtmyer's name. See S. Goldberg (1968), 419.

36. D. C. Miller (1929–30); *Science, 70* (1929), 560–561; *JRASC, 24* (1930), 82–84. Ironically, on the next day Howard Percy Robertson, a physicist on the Princeton University staff, presented a paper on "Foundations of relativistic cosmology," auguring the beginning of a new era in astronomy for general relativity. See H. P. Robertson (1929–30). See also Epilogue.

37. E. Freundlich et al. (1931a,b).

38. R. Trumpler (1932a), reprinted in (1932b); also (1932c). For a reply to the latter, see E. Freundlich et al. (1932).

39. St. John obviously meant to write "stars" here, but wrote "lines" by force of habit. He was used to arguing that his own solar spectrum results benefited from the large number of "lines" in his sample.

40. St. John to Campbell, 24 Apr 1930 (LO).

41. St. John to Campbell, 2 May 1930 (LO).

42. Campbell to St. John, 6 May 1930 (LO).

43. D. C. Miller (1931), (1933a,b,c).

44. Aitken to Russell, 25 Nov 1930, and Russell to Aitken, 19 Dec 1930 (LO).

45. C. E. St. John (1932).

46. D. C. Miller (1926), 437–438.

47. C. E. St. John (1932), 278.

48. Ibid., 276.

49. Ibid., 284. In 1945 H.R. Morgan found that observations of the Sun show a secular advance in the perihelion of the Earth's orbit in good agreement with the value predicted by general relativity. See G. de Vaucouleurs (1957), 198.

50. E. Freundlich et al. (1931b). A *réseau* is a network of fine lines scratched onto a glass plate. Two sets of equally spaced parallel lines perpendicular to each other are ruled onto the glass, forming a grid of squares of standard size.

51. He included another remeasurement of the Potsdam plates by the British astronomer J. Jackson. Jackson had found that a different method of reduction brought the limb deflection down to 1."98 with a higher probable error than the Germans had obtained.

52. C. E. St. John (1932), 295. Quoted from J. H. Jeans (1929), 72–73.

Epilogue. The Emergence of Relativistic Cosmology

1. *Trans. IAU* (1922), 7.

2. Eddington to Curtis, 20 Oct 1920 (CP).

3. *Trans. IAU* (1922), 11, 157.

4. Ibid., 19.

5. *Trans. IAU* (1925), 177.

6. *Trans. IAU* (1928), 64; (1932), 47.

7. For details on the history of relativistic cosmology, see J. D. North (1965); R. Berendzen et al. (1976); N. S. Hetherington (1970); R. W. Smith (1982).

8. V. M. Slipher (1913), W. G. Hoyt (1980).

9. Hale to Adams, 27 Jul 1916 (HM).

10. E. Hubble (1929).

11. W. de Sitter (1917–18). Static solutions were ones in which the coefficients of the metric were time independent. The general consensus at the time was that time-dependent, or nonstatic solutions were ruled out because there was no indication from astronomical observations of any systematic motion of the stellar system.

12. E. Hubble (1929), 173.

13. A. Friedmann (1922); G. Lemaître (1927a,b); A. Eddington (1930); W. de Sitter (1930). De Sitter discussed his static solution and then moved to Lemaître's non-static one with the remark: "A dynamical solution . . . is given by Dr. G. Lemaître in a paper published in 1927, which had unfortunately escaped my notice until my attention was called to it by Professor Eddington a few weeks ago" (482). See also J. D. North (1990).

14. D. H. Menzel (1929b), 225–226.

15. Ibid., 228. De Sitter's description called for a slowing down of the time for distant objects (technically due to the curvature of time). Menzel pointed out that though the distant nebula would not be moving at the implied high recessional velocity when the light was emitted, by the time the light is observed, the nebula would have acquired that speed. "The observed shift in colour of the spectral lines," he explained, "though due to space-time curvature, is a sort of anticipation of the real velocity the nebula will have, by the time its light reaches our system"; Menzel (1929b), 228–229. The picture was somewhat paradoxical since de Sitter's solution was derived for a static universe. Once the nonstatic solutions to Einstein's equations were discovered, the strict interpretation of Hubble's observations as a true expansion of the universe would replace the de Sitter picture.

16. Ibid., 229.

17. Ibid.

18. *Science, 73* (1931).

19. Ibid., 381.

20. *PASP, 40* (1928). Rockefeller had a policy to support science in a university or educational context, hence there had to be an educational thrust to the entire enterprise. Caltech was the natural institution to flow the funds and manage the project.

21. D. Menzel (1929b), 225.

FINAL REFLECTIONS

1. E.g., R. W. Clark (1971), chaps. 8–9; B. Hoffmann and H. Dukas (1972), chap. 8; J. Bernstein (1973), chap. 11; A. Pais (1982), chap. 16; A. Fölsing (1997),

chaps. 18, 22; D. Overbye (2000), chaps. 18, 19, 25. Fölsing and Overbye deal with earlier attempts but none after 1919.

2. S. Chandrasekhar (1975); W. H. McCrea (1979); D. F. Moyer (1979); J. Earman and C. Glymour (1980a). Earman and Glymour devote more attention to attempts before the British to measure the bending of light at an eclipse, but they discount the role these attempts played in the reception of Einstein's theory. K. Hentschel (1994) deals with other attempts before and after 1919 by the German astronomer Erwin Freundlich.

3. For a useful review of the literature on the acceptance of relativity, see Stephen G. Brush, "Why was relativity accepted?" *Phys. Perspect.*, *1* (1999), 184–214.

4. Jean Eisenstaedt has characterized the period 1925–55 as the "low water mark" of general relativity. He notes that physicists, preoccupied with quantum mechanics, were not interested in general relativity and considered its empirical foundations to be weak and limited. General relativity's nonlinear equations intimidated most physicists, who stayed away from working in the field because the links to experience were minimal. It seemed not to be worth the effort to master the theory. Only in cosmology did physical situations arise where Einstein's theory was more useful than Newton's. By focusing on the physics literature, and neglecting astronomy, Eisenstaedt may be overstating his case, especially for the period 1925–35 or even later. An interesting study would be to follow the careers of some of Eddington's, de Sitter's, and H. P. Robertson's students to identify emerging "schools" or centers of research that elaborated general relativity within the field of cosmology. See Eisenstaedt (1986) and (1989).

Abbot, Charles Greeley. 1935. "Charles Edward St. John" *Ap. J., 82*: 273–283.

Adams, Walter Sydney. 1910. "An investigation of the displacements of the spectrum lines at the Sun's limb," *Ap. J., 31*: 30–61.

———. 1915. "The spectrum of the companion of Sirius." *PASP*, 27: 236–237.

———. 1924a. "Summary of the year's work at Mt. Wilson," *PASP, 36*: 313–319.

———. 1924b. "Address of the retiring president of the society in awarding the Bruce Medal to Professor A. S. Eddington," *PASP, 36*: 2–9.

———. 1925. "The relativity displacement of the spectral lines in the companion of Sirius," *Proc. NAS, 11*: 382–387.

———. 1926. "The radial velocity of the companion of Sirius," *Observatory*, 49: 88.

Aitken, Robert Grant. 1917. "A total eclipse of the sun, Third Adolfo Stahl Lecture, delivered in San Francisco on January 12, 1917," *PASP, 29*: 25–40.

———. 1920. "The Einstein theory of relativity," *PASP, 32*: 71–73.

———. 1927. *PASP, 39*: 65–66.

———. 1928. "Progress in astronomical research at Pacific Coast observatories in the year 1927–1928," *PASP, 40*: 239–247.

———. 1929. "Editor's note," *PASP, 41*: 172.

Beck, Anna (trans.), and Don Howard (consultant). 1995. *The Collected Papers of Albert Einstein*, Vol. 5, *The Swiss Years: Correspondence, 1902–1914*, English translation. Princeton, N.J.: Princeton University Press.

Benioff, Hugo. 1923a. "The total eclipse of the sun January 24, 1925," *PASP, 35*: 337–338.

———. 1923b. "A list of total eclipses of the sun to 1952," *PASP, 35*: 338.

Berendzen, R., R. Hart, and D. Seeley. 1976. *Man Discovers the Galaxies*. New York: Neale Watson.

Bergmann, Peter G. 1968. *The Riddle of Gravitation*. New York: Scribner's.

Bernstein, Jeremy. 1973. *Einstein*. New York: Viking.

Bertotti, B., R. Balbinot, S. Bergia, and A. Messina. 1990. *Modern Cosmology in Retrospect*. Cambridge, U.K.: Cambridge University Press.

Born, Max. 1916. "Einsteins Theorie der Gravitation und der allgemeinen Relativität," *Phys. Z., 17*: 51–59.

Bracher, Katherine. 1979. "The famous eclipse of June 8, 1918," *Sky and Telescope, 58*: 411–413.

Brown, Ernest W. 1910. "The secular motions of the elements of the inner planets," *MNRAS, 70*: 342–344.

Brush, Stephen G. 1999. "Why was relativity accepted?" *Phys. Perspect., 1*: 184–214.

Bucherer, Alfred S. 1908. "Messungen an Becquerelstrahlen. Die experimentelle Bestätigung der Lorentz-Einsteinschen Theorie," *Phys. Z., 9*: 755–762.

Buchwald, Diana Kormos, et al., eds. 2004. *The Collected Papers of Albert Einstein*, vol. 9, *The Berlin Years: Correspondence, January 1919–April 1920*. Princeton, N.J.: Princeton University Press. Referred to in the Notes as *CPAE*, vol. 9; English translation: A. Hentschel and K. Hentschel (2004).

Bumstead, H. A. 1909. *Amer. J. Sci.*, 26: 493.

Burns, Gavin J. 1910. "The principle of relativity," *JBAA*, 21: 110–113.

Burns, Keivin. 1925. "Standard solar wavelengths," *Pub. Allegh. Obs.*, 6: 105.

———. 1930. "A comparison of laboratory and solar wave lengths," *JOSA*, 20: 212–224.

Burton, C. V. 1910. "The Sun's motion with respect to the Aether," *Phil. Mag.*, 19: 417–423.

Campbell, William Wallace. 1901. "The D. O. Mills expedition to the southern hemisphere," *PASP*, 13: 28–29.

———. 1903. "A brief account of the D. O. Mills expedition to Chile," *PASP*, 15: 70–75.

———. 1907. "Organization and history of the D. O. Mills expedition to the southern hemisphere," Lick Observatory, *Publications*: 5–12.

———. 1911a. "On the motions of the brighter class B stars," *LOB*, 6: 101–124.

———. 1911b. "Some peculiarities in the motions of the stars," *LOB*, 6: 125–135.

———. 1913a. *Stellar Motions: With Special Reference to Motions Determined by Means of the Spectrograph*. New Haven: Yale University Press.

———. 1913b. "Concerning some forces affecting cosmical motion," *PASP*, 25: 164–166.

———. 1913c. "International meetings of astronomers in Germany," *PASP*, 25: 244–255.

———. 1918a. "Return of eclipse instruments from Russia," *PASP*, 30: 312.

———. 1918b. "The Crocker eclipse expedition from the Lick Observatory, University of California, June 8, 1918," *PASP*, 30: 220–240.

———. 1918c. "The total solar eclipse of June 8, 1918," *LOB*, 10, no. 318: 1–3.

———. 1918d. "Clouds fall away for solar eclipse," *New York Times*, 10 Jun 1918.

———. 1918e. "Total solar eclipses of the near future," *PASP*, 30: 256–257.

———. 1920. "Resignation of Dr. Curtis," *PASP*, 32: 201–202.

———. 1921. "The total eclipse of September 21, 1922," *PASP*, 33: 254–255.

———. 1923a. "The total eclipse of the sun, September 21, 1922," *PASP*, 35: 10–44.

———. 1923b. "Sun eclipse pictures prove Einstein theory; Lick Observatory finds star light is bent," *New York Times*, April 12.

Campbell, William Wallace, and Robert Trumpler. 1923a. "Observations on the deflection of light in passing through the sun's gravitational field, made during the total solar eclipse of Sept. 21, 1922," *LOB*, 346: 41–54.

———. 1923b. "Observations on the deflection of light in passing through the sun's gravitational field, made during the total solar eclipse of Sept. 21, 1922," *PASP*, 35: 158–163.

————. 1928. "Observations made with a pair of five-foot cameras on the light-deflection in the sun's gravitational field at the total solar eclipse of September 21, 1922," *LOB, 397*: 130–160.

Carmichael, R. D. 1920. "Einstein's third victory," *New York Times*, 28 Mar, sec. 3, 11.

Carmichael, R. D., et. al. 1927. *A Debate on the Theory of Relativity*. Chicago: Open Court.

Chandrasekhar, Subramanyan. 1975. "Verifying the theory of relativity," *Bulletin of the atomic scientists, 31:6*: 17–22.

Clark, Ronald W. 1971. *Einstein: The Life and Times*. New York: World Publishing.

Clarke, Agnes M. 1903. *Problems in Astrophysics*. London: Adams and Charles Black.

Comstock, Daniel F. 1910. "The principle of relativity," *Science, 31*: 767.

Courvoisier, Leo. 1913. "Systematische Abwegungen der Sternpositionen im Sinne einer jährlichen Refraktion," Königliche Sternwarte, Berlin, *Beobachtungs-Ergebnisse, 15*.

————. 1920. "Jährliche Refraktion und Sonnenfinsternisaufnahmen 1919," *AN, 211*: 305–312.

Crelinsten, Jeffrey. 1980a. "Einstein, relativity, and the press: The myth of incomprehensibility," *Physics Teacher, 18*: 115–122.

————. 1980b. "Physicists receive relativity: Revolution and reaction," *Physics Teacher, 18*, 187–193.

————. 1981. "The reception of Einstein's general theory of relativity among American astronomers: 1910–1930," Ph.D. diss., University of Montreal.

————. 1983. "William Wallace Campbell and the 'Einstein Problem,' *HSPS, 14*: 1–91.

Curtis, Heber Doust. 1911. "The theory of relativity," *PASP, 23*: 219–229.

————. 1913. "The influence of gravitation on light," *PASP, 25*: 77–80.

————. 1917. "Space, time, and gravitation," *PASP, 19*: 63–64.

————. 1919. "The Einstein effect: Eclipse of June 8, 1918," *PASP, 31*: 197.

————. 1924a. "Allegheny results on the shift of the solar lines predicted by the theory of relativity," *Science, 59*: 442–443.

————. 1924b. "Two laboratory arcs," *JOSA, 8*: 697–700.

Danby, J.M.A. 1975. "Henry Crozier Plummer," *DSB, 11*: 49.

Davidson, C. 1922. "Observations of the Einstein displacement in eclipses of the sun," *Observatory, 45*: 224–225.

de Sitter, Willem. 1911. "On the bearing of the principle of relativity on gravitational astronomy," *MNRAS, 77*: 388–415.

————. 1913. "The secular variations of the elements of the four inner planets," *Observatory, 36*: 296–303.

————. 1915–16. "On Einstein's theory of gravitation and its astronomical consequences. First Paper," *MNRAS, 76*: 699–728.

————. 1916. "Space, time, and gravitation," *Observatory, 39*: 412–419.

————. 1916–17a. "On Einstein's theory of gravitation and its astronomical consequences. Second Paper," *MNRAS, 77*: 155–184.

de Sitter, Willem. 1916–17b. "Planetary motion and the motion of the moon according to Einstein's theory," *Koninklijke Akademie van Wetenschappen te Amsterdam, Section of Sciences. Proceedings, 19*: 367–381.

———. 1917–18. "On Einstein's theory of gravitation and its astronomical consequences. Third Paper," *MNRAS, 78*: 3–28.

———. 1930. "On the distances and radial velocities of extra-galactic nebulae, and the explanation of the latter by the relativity theory of inertia," *Proc. NAS, 16*: 474–488.

de Vaucouleurs, Gérard. 1957. *Discovery of the Universe*. London: Faber and Faber.

DeVorkin, David H. 1975. "Michelson and the problem of stellar diameters," *JHA, 6*: 1–18.

———. 1981. "Community and spectral classification in astrophysics: The acceptance of E. C. Pickering's system in 1910," *Isis, 72* (Mar): 29–49.

Dicke, Robert H. 1964. *The Theoretical Significance of Experimental Relativity*. New York: Gordon and Breach.

Dicke, R. H., and H. M. Goldenberg. 1967. "Solar oblateness and general relativity," *Phys. Rev. Lett., 18* (Feb): 313–316.

Dingle, Herbert. 1972. "William Huggins," *DSB, 6*: 540–543.

———. 1973. "Joseph Norman Lockyer," *DSB, 8*: 440–443.

Dodwell, G. F., and C. R. Davidson. 1924. "Determination of the deflection of light by the sun's gravitational field from observations made at Cordillo Downs, South Australia, during the total eclipse of 1922 September 21," *MNRAS, 84*: 150–162.

Dostro, Ludwik. 1988. "An outline of the history of Einstein's relativistic ether concept," in Eisenstaedt and Kox (1988), 260–280.

Douglas, Allie Viebert. 1956. *Arthur Stanley Eddington*. Toronto: Thomas Nelson.

Duffield, W. G. 1920. "The displacements of the spectrum lines and the equivalence hypothesis," *MNRAS, 80*: 262–272

Dyson, Frank W. 1916–17. "On the opportunity afforded by the eclipse of 1919, May 29, of verifying Einstein's theory of gravitation," *MNRAS, 77*: 445–447.

Dyson, F. W., A. S. Eddington, and C. Davidson. 1920. "A determination of the deflection of light by the sun's gravitational field, from observations made at the total eclipse of May 29, 1919," Royal Society, *Transactions, 220*: 291–333, reprinted in 1920, Royal Astronomical Society, *Memoirs, Supplement*: 1–43.

Earman, John, and Clark Glymour. 1980a. "Relativity and eclipses: The British eclipse expeditions and their predecessors," *HSPS, 11*: 49–85.

———. 1980b. "The gravitational red shift as a test of general relativity: History and analysis," *Studies in History and Philosophy of Science, 11*: 175–214.

Eddington, Arthur Stanley. 1915. "Gravitation," *Observatory, 38*: 97–98.

———. 1916. "Gravitation and the principle of relativity," *Nature, 98*: 328–330.

———. 1916–17. "Einstein's theory of gravitation," *MNRAS, 77*: 377–383.

———. 1917a. "Einstein's theory of gravitation," *Observatory, 40*: 93–95.

———. 1917b. "Stellar distribution and motions," *MNRAS, 77*: 370–374.

———. 1918. *Report on the Relativity Theory of Gravitation for the Physical Society of London*. London: Royal Society of London.

————. 1919. "The total eclipse of 1919, May 29, and the influence of gravitation on light," *Observatory, 42*: 119–122.

————. 1920. *Space, Time and Gravitation.* Cambridge, U.K.: Cambridge University Press.

————. 1923. *Mathematical Theory of Relativity.* Cambridge, U.K.: Cambridge University Press.

————. 1924. "Relation between the masses and luminosities of the stars," *MNRAS, 84*: 308.

————. 1925. "Ether-drift and the relativity theory," *Nature, 115*: 870.

————. 1926. *The Internal Constitution of the Stars.* Cambridge, U.K.: Cambridge University Press.

————. 1930. "On the instability of Einstein's spherical world," *MNRAS, 90*: 688.

Einstein, Albert. 1905a. "Über einen die Erzeugung und Umwandlung des Lichtes betreffenden heuristischen Standpunkt [On a heuristic viewpoint concerning the generation and transformation of light], *Ann. der Phys., 17*: 132–84.

————. 1905b. "Über die von der molekulartheoretischen Theorie der Wärme geforderte Bewegung von in ruhenden Flüssigkeiten suspendierten Teilchen [On the movement of particles suspended in fluids at rest, as postulated by the molecular theory of heat], *Ann. der Phys., 17*: 549–560.

————. 1905c. "Zur Elektrodynamik bewegter Körper," *Ann. der Phys., 17*, 891–921; "On the electrodynamics of moving bodies," in A. Einstein et al., *The Principle of Relativity.* New York: Dover, 1952, sec. 6: 51–55 (English translation of Einstein's original 1905 paper).

————1905d. "Ist die Trägheit eines Körpers von seinem Energieinhalt abhängig?" [Does the inertia of a body depend on its energy content?], *Ann. der Phys., 17*: 639–641.

————. 1907. "Über das Relativitätsprinzip und die aus demselben gezogenen Folgerungen [On the relativity principle and the conclusions drawn from it], *Jahrbuch für Radioaktivität und Elektronik, 4*: 411–462.

————. 1911. "Ueber den Einfluss der Schwerkraft auf die Ausbreitung des Lichtes," *Ann. der Phys., 35*: 898–908.

————. 1916a. "Die Grundlage der allgemeinen Relativitätstheorie," *Ann. der Phys, 49*: 769–822. English translation in A. Einstein et al. (1952), 111–164.

————. 1916b. *Die Grundlage der allgemeinen Relativitätstheorie.* Leipzig: Barth.

————. 1916c. "Näherungsweise Integration de Feldgleichungen der Gravitation," *Königlich Preussische Akademie der Wissenschaften* (Berlin), *Sitzungsberichte*, 768–770.

————. 1917a. *Über die spezielle und die allgemeine Relativitätstheorie (Gemeinverständlich).* Braunschweig: Vieweg.

————. 1917b. "Kosmologische Betrachtungen zur allgemeinen Relativitätstheorie," *Königlich Preussische Akademie der Wissenschaften, Sitzungsberichte*, 142–152. English translation in A. Einstein et al. (1952), 177–188.

————. 1919. "Prüfung der allgemeinen Relativitätstheorie," *NW, 7*: 776.

————. 1922. *Sidelights on Relativity.* London: Methuen; New York: E. P. Dutton.

Einstein, Albert. 1923. "Zur affinen Feldtheorie," *Preussiche Akademie der Wissenschaften, Sitzungsberichte*, 137–140.

———. 1926. "Meine Theorie und Millers Versuche," *Vossische Zeitung*,19 Jan.

———. 1934. *Mein Weltbild*. Amsterdam: Querido.

———. 1965. "On the method of theoretical physics," in *Ideas and Opinions*. New York: Crown, 263–270.

Einstein, Albert, et al. 1952. *The Principle of Relativity*. New York: Dover.

Eisenstaedt, Jean. 1982. "Histoire et singularités de la solution de Schwarzschild (1915–1923)," *Archive for History of Exact Sciences, 27*, no. 2: 157–198.

———. 1986. "La relativité générale à l'étiage: 1925–1955," *Archive for History of Exact Sciences, 35*, no. 2: 115–185.

———. 1989. "The low water mark of general relativity: 1925–1955," in Howard and Stachel (1989), 277–292.

Eisenstaedt, Jean, and A. J. Kox. 1988. *Studies in the History of General Relativity*. Boston: Birkhäuser.

Evershed, John. 1909a. "Radial motion in sunspots," *Kod. Bull., 15*: 291.

———. 1909b. "Pressure in the reversing layer," *Kod. Bull., 18*: 131–134.

———. 1913. "A new interpretation of the general displacement of the lines of the solar spectrum towards the red," *Kod. Bull., 36*: 45–52.

———. 1916a. *Observatory, 39*: 35.

———. 1916b. *Observatory, 39*: 432.

———. 1918a. "The displacement of the cyanogen bands in the solar spectrum," *Observatory, 41*: 371–375.

———. 1918b. "Kodaikanal and Madras Observatories," *MNRAS, 78*: 278–281.

———. 1919. "The displacement of the solar lines reflected by Venus," *Observatory, 42*: 51–52.

———. 1920a. "Displacement of the lines in the solar spectrum and Einstein's prediction," *Observatory, 43*: 153–157.

———. 1920b. "On the displacements of the triplet bands near $\lambda3883$ in the solar spectrum," *Kod. Bull., 64*: 297–302.

———. 1921. "The relativity shift in the solar spectrum," *Observatory, 44*: 243–245.

———. 1923a. "The Einstein effect in the solar spectrum," *Observatory, 46*: 299–304.

———. 1923b. "Report of the Indian eclipse expedition to Wallal, West Australia," *Kod. Bull., 72*: 45.

Evershed, J., and T. Royds. 1914. "On the displacements of the spectrum lines at the Sun's limb," *Kod. Bull., 39*: 71–81.

Fabry, Charles, and Henri Buisson. 1910. "Mesure des petites variations de longueur d'onde par la méthode interférentielle: Application à différents problèmes de spectroscopie solaire," *J. de Phys., 9*: 298.

Fletcher, Harvey. 1945. "Dayton Clarence Miller," *BMRAS, 23*: 61–74.

Fölsing, Albrecht. 1997. *Albert Einstein: A Biography*. New York: Viking.

Forbes, Eric Gray. 1961. "A history of the solar red shift problem," *Annals of Science, 17*: 129–164.

Forbes, Eric G., ed. 1975. *Greenwich Observatory . . . the Story of Britain's Oldest Scientific Institution, the Royal Observatory at Greenwich and Herstmonceux, 1836–1975*. London: Taylor and Francis.

Forman, Paul, John L. Heilbron, and Spencer Weart. 1975. "Physics *circa* 1900: Personnel, funding, and productivity of the academic establishments," *HSPS*, 5: 1–185.

Fowler, R. H., and E. A. Milne. 1923. "The intensities of absorption lines in stellar spectra," *MNRAS, 83*: 419.

Fox, Philip. 1927–28. " 'An attempt to detect the Einstein displacement at the limb of Jupiter,' address at the fall meeting of the National Academy of Sciences, 28 October 1927, held at the University of Illinois, Urbana IL," *Ann. Report NAS*: 16.

Freundlich, Erwin Finlay. 1913. "Ueber einen Versuch, die von A. Einstein vermutete Ablenkung des Lichtes in Gravitationsfeldern zu prüfen," *AN, 193*: 369–372.

———. 1914. "Ueber die Verschiebung der Sonnenlinien nach dem roten Ende auf Grund der Hypothesen von Einstein und Nordström," *Phys. Zs., 15*: 369.

———. 1915/16a. "Ueber die Gravitationsverschiebung der Spektrallinien bei Fixsternen," *Phys. Zs., 16*: 115–117.

———. 1915/16b. "Ueber die Gravitationsverschiebung der Spektrallinien bei Fixsternen," *AN, 202*: 17–24.

———. 1916. *Die Grundlagen der Einsteinschen Gravitationstheorie*. Berlin: Springer.

Freundlich, Erwin, Harold von Klüber and A. von Brunn. 1931a. "Die Ablenkung des Lichtes im Schwerefeld der Sonne," *Abhandl. Preuss. Ak. W., Math.-Phys. Kl., No. 1*: 1–61.

———. 1931b. "Ergebnisse der Potsdamer Expedition zur Beobachtung der Sonnenfinsternis von 1929, Mai 9, in Takengon (Nordsumatra). 5. Mitteilung: Ueber die Ablenkung des Lichtes im Schwerefeld der Sonne," *Zs. Astrophys., 3*: 171–198.

———. 1932. "Bemerkung zu Herrn Trümplers Kritik," *Zs. Astrophys, 4*: 221–223.

Friedmann, A. 1922. "Über die Krümmung des Raumes," *Zs. f. Phys., 10*: 377–386.

Galison, Peter L. 1979. "Minkowski's space-time: From visual thinking to the absolute world," *HSPS, 10*: 85–121.

Gehrcke, Ernst. 1916. "Zur Kritik und Geschichte der neueren Gravitationstheorien," *Ann. der Phys., 51*: 119–124.

Gerber, P. (1898). "Die räumliche und zeitliche Ausbreitung der Gravitation," *Zeitschrift für Mathematik und Physik, 43*: 93–104, reprinted in 1917 as "Die Fortpflanzungsgeschwindigkeit der Gravitation," *Ann. der Phys., 52*: 415–441.

Glaser, L. C. 1922. "Über die Gravitationsverschiebung der Fraunhoferschen Linien," *Phys. Z., 23*: 100–102.

Goenner, Hubert. 1993. "The reaction to relativity theory I: The anti-Einstein campaign in Germany in 1920," *Science in Context, 6*: 107–133.

Goldberg, Stanley. 1968. "Early responses to Einstein's theory of relativity, 1905–1911: A case study in national differences," Ph.D. diss., Harvard University.

Goldberg, Stanley. 1970. "In defense of ether: The British response to Einstein's special theory of relativity: 1905–1911," *HSPS*, 2: 89–126.

———. 1984. *Understanding Relativity: Origin and Impact of a Scientific Revolution*. Boston: Birkhäuser.

Goodstein, Judith R. 1991. *Millikan's School: A History of the California Institute of Technology*. New York: W. W. Norton.

Grebe, L., and A. Bachem. 1919. "Ueber den Einsteineffekt im Gravitationsfeld der Sonne," *Verh. Deutsch. Phys. Ges.*, 21: 454–464; also in *Berichte der Deutschen Phys. Ges., 13/14* (1919).

———. 1920. "Ueber die Einsteinverschiebung im Gravitationsfeld der Sonne," *Zs. f. Phys.*, 1: 51–4.

Hale, George Ellery. 1903. "The Snow Horizontal Telescope," *Ap. J.*, 17: 14.

———. 1908a. "The Tower Telescope of the Mt. Wilson Solar Observatory," *Ap. J.*, 27: 204–212.

———. 1908b. "Report of the Director of the Solar Observatory, Mount Wilson, California," *CIW Yearbook, No. 7*: 146–157.

———. 1915. "Annual Report of the Director of the Mt. Wilson Solar Observatory," *CIW Yearbook, 14*: 251–293.

———. 1917. "Annual Report of the Director of the Mount Wilson Solar Observatory," *CIW Yearbook, 16*: 199–235.

———. 1921a. "A summary of the year's work at Mt. Wilson," *PASP, 33*: 18–30.

———. 1921b. "A summary of the year's work at Mount Wilson," *PASP, 33*: 291–312.

Hartmann, Eduard. 1917. "Einsteins allgemeine Relativitätstheorie," *Philosophisches Jahrbuch der Görresgesellschaft* [Fulda], 30: 363–387.

Harvey, G. M. 1979. "Gravitational deflection of light. A re-examination of the observations of the solar eclipse of 1919," *Observatory*, 99: 195–198.

Hentschel, Ann (trans.), and Klaus Hentschel (consultant). 1998. *The Collected Papers of Albert Einstein*, Vol. 8, *The Berlin Years: Correspondence, 1914–1918*. English translation. Princeton, N.J.: Princeton University Press.

———. 2004. *The Collected Papers of Albert Einstein*, Vol. 9, *The Berlin years: Correspondence, January 1919–April 1920*. English translation. Princeton, N.J.: Princeton University Press.

Hentschel, Klaus. 1992. "Einstein's attitude towards experiments," *Studies in History and Philosophy of Science*, 23: 593–624.

———. 1993. "The conversion of St. John: A case study on the interplay of theory and experiment," *Science in Context*, 6: 137–194.

———. 1994. "Erwin Finlay Freundlich and testing Einstein's theory of relativity," *Archive for the History of the Exact Sciences*, 47: 143–201.

———. 1997. *The Einstein Tower: An Intertexture of Dynamic Construction, Relativity Theory, and Astronomy*. Stanford, Calif.: Stanford University Press.

Hermann, Armin. 1977. "Der Kampf um die Relativitäts theorie," *Bild der Wissenschaft*, 14: 108–116.

———. 1979. *The New Physics: The Route into the Atomic Age*. Munich: Heinz Moos Verlag.

Hetherington, Norris S. 1970. "The development and early application of the velocity-distance relation." Ph.D. dissertation, Indiana University.

Hirosige, Tetu. 1976. "The ether problem, the mechanistic worldview, and the origins of the theory of relativity," *HSPS*, 7: 3–82.

Hoffmann, Banesh, and Helen Dukas. 1972. *Albert Einstein: Creator and Rebel*. New York: New American Library.

Holton, Gerald. 1960. "On the origins of the special theory of relativity," *Amer. J. Phys.*, 28: 627–636, reprinted in Holton (1973), 165–183.

———. 1969. "Einstein, Michelson, and the 'crucial experiment,'" *Isis*, 60: 133–197, reprinted in Holton (1973), 261–352.

———. 1973. *Thematic Origins of Scientific Thought: Kepler to Einstein*. Cambridge, Mass.: Harvard University Press.

Hopmann, J. 1923. "Die Deutung der Ergebnissen der amerikanischen Einstein-expedition," *Phys. Zs.*, 24: 476–486.

Hoskin, Michael A. 1971. "Heber Doust Curtis," *DSB*, 3: 508–509.

Howard, D., and John Stachel, eds. 1989. *Einstein and the History of General Relativity*. Boston: Birkhäuser.

Hoyt, William Graves. 1976. *Lowell and Mars*. Tucson: University of Arizona Press.

———. 1980. "Vesto Melvin Slipher, November 11, 1875–November 8, 1969," *BMNAS*, 52: 410–419.

Hubble, Edwin. 1929. "A relation between distance and radial velocity among extra-galactic nebulae," *Proc. NAS*, 15: 168.

Huggins, W. 1897. "The new astronomy: A personal retrospect," *The Nineteenth Century*, June.

Jeans, James Hopwood. 1917. "Einstein's theory of gravitation," *Observatory*, 40: 57–58.

———. 1919. *Problems of Cosmogony and Stellar Dynamics*. Cambridge, U.K.: Cambridge University Press.

———. 1929. *The Universe around Us*. Cambridge, U.K.: Cambridge University Press.

Jeffreys, Harold S. 1923. "The Lick Observatory eclipse expedition results," *Observatory*, 46: 280–282.

JOSA. 1930. "Session devoted to experimental data bearing on the theory of relativity," 20: 142.

Joy, Alfred H. 1958. "Walter Sydney Adams," *BMNAS*, 31: 1–31.

Jungnickel, Christa, and Russell McCormmach. 1986. *Intellectual Mastery of Nature: Theoretical Physics from Ohm to Einstein*. Chicago: University of Chicago Press.

Kahn, Karla and Franz. 1975. "Letters from Einstein to de Sitter on the nature of the universe," *Nature*, 257: 451–454.

Kargon, Robert. 1977a. "Temple to science: Cooperative research and the birth of the California Institute of Technology," *HSPS*, 8: 3–31.

———. 1977b. "The conservative mode: Robert A. Millikan and the twentieth-century revolution in physics," *Isis*, 68: 509–526.

Kennedy, Roy J. 1926. "A refinement of the Michelson-Morley experiment," *Proc. NAS*, 12: 628.

Kevles, Daniel J. 1970. "Into hostile political camps. The reorganization of inter-national science in World War I," *Isis, 62*: 47–60.

King, Henry C. 1955. *The History of the Telescope*. London: C. Griffin.

Klein, Martin J. 1970. *Paul Ehrenfest*. Amsterdam: North Holland.

———. 1972. "Mechanical explanation at the end of the 19th century," *Centaurus, 17*: 58–82.

Klein, Martin, A. J. Kox, and Robert Schulmann, eds. 1993. *The Collected Papers of Albert Einstein*, vol. 5, *The Swiss Years: Correspondence, 1902–1914*. Princeton, N.J.: Princeton University Press. Referred to in the Notes as *CPAE*, vol. 5; English translation: A. Beck and D. Howard (1995).

Konen, Heinrich. 1912. *Reisebilder von einer Studienreise durch Sternwarten und Laboratorien der Vereinigten Staaten*. Görres-Gesellschaft, II, Vereinsschrift für 1912. Cologne: Bachem.

Langley, Samuel P. 1884. *The New Astronomy*. Boston: Houghton.

Lankford, John. 1980. "A note on T.J.J. See's observations of craters on Mercury," *JHA, 11*: 129–132.

Larkin, E. L. 1921. "Einstein's theory is not proved," *N.Y. American*.

Larmor, Joseph. 1919. "The relativity of the forces of nature," MNRAS, 80: 109–111, 118–138.

Lecat, Maurice. 1924. *Bibliographie de la relativité*. Brussels: Lametin.

Lemaître, Georges. 1927a. "Un univers homogène de masse constante et de rayon croissant, rendant compte de la vitesse radiale des nebuleuses extra-galactiques," *Ann. Soc. Sci. Bruxelles, 47*: 49–56; translated in *MNRAS, 91*: 483–490, as: "A homogeneous universe of constant mass and increasing radius accounting for the radial velocity of extra-galactic nebulae."

———. 1927b. "The expanding universe," *MNRAS, 91*: 490–501.

Le Roux, J. L. 1921. "Théorie de la relativité et mouvement séculaire du perihelia de Mercure," *Comptes Rendus, 173*.

———. 1922. "Gravitation dans la mécanique classique et dans la théorie d'Einstein," *Comptes Rendus, 175*.

Lewis, Gilbert N., and Richard Chase Tolman. 1909. "The principle of relativity and non-Newtonian mechanics," *Phil. Mag., 17*: 510.

Lindemann, A. F. and F. A. 1917. "Daylight photography of stars as a means of testing the equivalence postulate in the theory of relativity," *MNRAS, 77*: 140–151.

Lodge, O. 1926. "On Prof. Miller's ether drift experiment," *Nature, 117*: 854.

London Times. 1919a. "Revolution in science," 7 Nov, 12.

———. 1919b. "The revolution in science," 8 Nov.

———. 1919c. "The revolution in science, astronomers' discussion," 15 Nov, 14.

———. 1919d. "Einstein on his theory," 28 Nov, 12–13.

———. 1919e. "Astronomers on Einstein," 13 Dec, 9.

Martin, Daniel. 1976. "Edmund Taylor Whittaker," *DSB, 14*: 316–318.

Maunder, Walter. 1913. *Sir William Huggins and Spectroscopic Astronomy*. London: T. C. and E. C. Jack.

McConn, Edith. 1923. "Lick astronomer tells of eclipse expediton," *Evening News* (San Jose, Calif.), 12 Feb.

McCormmach, Russell. 1970a. "H. A. Lorentz and the electromagnetic view of nature," *Isis, 61*: 459–497.

———. 1970b. "Einstein, Lorentz, and the electron theory," *HSPS, 2*: 41–87.

———. 1975. "Editor's foreword," *HSPS, 6*: xi–xiv.

McCrea, William H. 1957. "Edmund Taylor Whittaker," *Journal of the London Mathematical Society, 32*: 234–256.

———. 1979. "Einstein's relationship with the Royal Astronomical Society," *Observatory, 99*: 105–107.

Menzel, Donald H. 1929a. "Expeditions to the total solar eclipse of May 9, 1929," *PASP, 41*: 120–122.

———. 1929b. "Progress in astronomy," *PASP, 41*: 224–231.

Michelson, A. A., H. A. Lorentz, D. C. Miller, R. J. Kennedy, E. R. Hedrick, and P. S. Epstein. 1928. "Conference on the Michelson-Morley experiment," *Ap.J., 68*: 341–402.

Michelson, A. A., F. G. Pease, and F. Pearson. 1929. "Repetition of the Michelson-Morley experiment," *JOSA, 18*: 181–182.

Miller, Dayton Clarence. 1922. "Ether-drift experiments at Mount Wilson in 1921 and at Cleveland in 1922," *Science, 55*: 496.

———. 1925a. "Ether drift experiments at Mount Wilson," *Science, 61*: 617–621.

———. 1925b. "Report on ether drift experiments," *Ann. Report NAS*: 48–49.

———. 1926. "Significance of the ether-drift experiments of 1925 at Mount Wilson," *Science, 68* (Dec): 433–443.

———. 1929–30. "Report on aether drift experiments of 1929 and considerations of other experimental evidence indicating a motion of the solar system," *Ann. Report NAS*: 14.

———. 1931. "Ether-drift experiments at Cleveland in 1930," *Proc. BAAS*.

———. 1933a. "Absolute motion of the solar system and the orbital motion of the earth determined by the ether-drift experiment," *Science, 77*: 587–588.

———. 1933b. *Phys. Rev., 43*: 1054.

———. 1933c. *Rev. Mod. Phys., 5*: 203–242.

Miller, John A. 1935. "A photometric study of the photographs of five solar eclipses," *Sproul Observatory Publications, 13*: 3–36.

Miller, J. A., and R. W. Marriott. 1925. "Observations of the total solar eclipse of 10 September, 1923, by the Sproul Observatory," *Ap. J., 61*: 73–96.

———. 1928a. "A diameter of the *moon* determined from photographs of a total solar eclipse," *AJ, 38*: 101–105.

———. 1928b. "Report of the Swarthmore College expedition to observe the total solar eclipse of January 14, 1926," *PASP, 40*: 83 et seq.

Millikan, Robert A. 1910. "A new modification of the cloud method of determining the elementary electrical charge and the most probable value of that charge," *Phil. Mag., 19*: 209.

Millikan, Robert A. 1916. "A direct photoelectric determination of Planck's 'h'," *Phys. Rev., 7*: 355.

Minkowski, H. 1908. "Die Grundgleichungen für die elektromagnetischen Vorgänge in bewegten Körpern," *Nachrichten der königlichen Gesellschaft der*

Wissenschaften und Georg-August Universität zu Göttingen, Math.-phys. Klasse: 53–111.

Mitchell, Samuel Alfred. 1923. *Eclipses of The Sun*. New York: Columbia University Press.

——. 1924. *Eclipses of The Sun*, 2nd ed. New York: Columbia University Press.

——. 1932. *Eclipses of The Sun*, 3rd ed. New York: Columbia University Press.

MNRAS. 1919. "Discussion on the theory of relativity," *80*: 96–118.

——. 1920. "Proceedings at meeting of Friday, November 12, 1920," *43*: 411–413.

Moore, Joseph Haines. 1928. "Recent spectrographic observations of the companion of Sirius," *PASP, 40*: 229–233.

——. 1939. "William Wallace Campbell," *Ap. J., 89*: 143–151.

Morgan, H. R. 1930. "The observed motion of the perihelion of Mercury," *JOSA, 20*: 225–229.

Morley, E. W., and D. C. Miller. 1905. "Report of an experiment to detect the Fitzgerald-Lorentz effect," *Phil. Mag., 9*: 680–685.

Moszkowski, A. 1919. "Die Sonne bracht' es an den Tag," *Berliner Tageblatt*, 8 Oct.

Moyer, Don F. "Samuel Pierpont Langley," *DSB, 8*: 1921.

——. 1979. "Revolution in science: The 1919 eclipse test of general relativity," in B. Dursonogly, A. Perlmutter, and L. F. Scott, eds., *On the Path of Albert Einstein*. New York: Plenum, 55–101.

New York Times. 1919a. "Eclipse showed gravity variation," 9 Nov, 4.

——. 1919b. "Accepts Einstein gravitation theory," 11 Nov, 17.

——. 1919c. "Amateurs will be resentful," 11 Nov.

——. 1919d. "Noguchi tells discovery," 12 Nov, 16.

——. 1919e. "Jazz in scientific world: Prof. Charles Lane Poor of Columbia explains Prof. Einstein's astronomical theories," 16 Nov, sec. 3: 8.

——. 1919f. "Nobody need be offended," 18 Nov.

——. 1921. "Einstein theory again is verified," 19 Dec.

——. 1923. "Einstein theory assailed," 20 Nov 1923, 3.

——. 1924. "Prof. See declares Einstein in error," 14 Oct, 14.

——. 1924a. "Denies See proved Einstein wrong: Professor Eisenhart takes issue with astronomer's views on starlight deflection," 16 Oct, 12.

——. 1924b. "Defends Einstein against Capt. See: Sir Frank Dyson contradicts American's charges sent to Royal Society," 16 Oct, 12.

——. 1924c. "Denies error in relativity: Prof. Eddington declares Captain See's disproof 'all bosh,' " 16 Oct, 12.

——. 1925a. "Michelson proves Einstein theory. Experiments conducted with 5,200-foot vacuum tube show light displacement. Ether drift is confirmed. Rays found to travel at different speeds when sent in opposite directions," 9 Jan, 2, col. 5.

——. 1925b. "13 observatories to go into action," 23 Jan, 10.

——. 1925c. "New Haven buzzes with eclipse talk," 24 Jan, 2.

——. 1925d. "Einstein tests hindered," 26 Jan, 19.

———. 1925e. "New Einstein data given by Michelson. Chicago physicist's paper before National Academy of Science offer proofs by light," 28 April, 20, col. 8.

———. 1925f. "Strikes a blow at Einstein theory. Professor Miller tells Academy of Sciences there is a drift of ether," 29 April, 35.

Newall, Hugh Frank. 1919. "Note on the physical aspect of the Einstein prediction," *MNRAS, 80*: 22–25.

Nature. 1920. *104*: 565.

North, J. D. *The Measure of the Universe.* Oxford: Clarendon Press.

———. 1990. "The early years," in Bertotti et al. (1990), 12–30.

Observatory. 1919a. "Meeting of the Royal Astronomical Society, Friday, 1919 July 11," *42*: 297–306.

———. 1919b. "Joint eclipse meeting of the Royal Society and the Royal Astronomical Society 1919, November 6," *42*: 389–398.

———. 1919c. "Meeting of the Royal Astronomical Society, Friday, 1919 November 14," *42*: 421–431.

———. 1920a. "Meeting of the Royal Astronomical Society, Friday, 1920 March 12," *43*: 137–150.

———. 1920b. "Meeting of the British Astronomical Association," *43*: 277–278.

———. 1921. "The relativity shift in the solar spectrum," *44*: 159–160.

———. 1922. "Christmas Island eclipse expedition," *45*: 317–319.

———. 1923a. "The total solar eclipse of 1922 September 21," *46*: 164–165.

———. 1923b. "American expeditions to observe the eclipse of 1923 Sept. 10," *46*: 256–257.

———. 1923c. "The theory of relativity and the displacement of spectral lines," *46*: 343–344.

Osterbrock, Donald E. 1979–80. "Lick Observatory solar eclipse expeditions," *Astronomical Quarterly, 3*: 67–79.

———. 1988. *Eye on the Sky: Lick Observatory's First Century.* Berkeley: University of California Press.

Overbye, Dennis. 2000. *Einstein in Love: A Scientific Romance.* New York: Viking.

Pais, Abraham. 1982. *'Subtle is the Lord' . . . The Science and the Life of Albert Einstein.* New York: Oxford University Press.

Pannekoek, Antonie. 1961. *A History of Astronomy.* London: Allen and Unwin.

PASP. 1919. "The Einstein Effect," *PASP, 31*: 320.

———. 1923a. "The eclipse of the sun, September 10, 1923," *PASP, 35*: 267–268.

———. 1923b. "The Los Angeles-Pasadena astronomical meetings," *PASP, 35*: 268–269.

———. 1925. "The total eclipse of the sun of January 14, 1926," *PASP, 37*: 287–289.

———. 1926. "The total eclipse of the sun, January 14, 1926," *PASP, 38*: 55.

———. 1928. "The astrophysical observatory of the California Institute of Technology," *PASP, 40*: 363–368.

———. 1929. "The eclipse of May 9, 1929," *PASP, 41*: 288.

Pease, Francis G. 1930. "Ether drift data," *PASP, 42*: 197–202.

Perrine, Charles D. 1916. "The total eclipse of February 3, 1916," *PASP, 28*: 247.

———. 1923. "Contribution to the history of attempts to test the theory of relativity by means of astronomical observations," *AN, 219*: 281–284.

Pickering, E. C. 1914. "The study of the stars," *Pop. Astr., 22*: 65–74.

Plummer, Henry Crozier. 1909. "On the theory of aberration," *MNRAS, 69*: 496–508.

———. 1910. "On the theory of aberration and the principle of relativity," *MNRAS, 70*: 252–266.

Poincaré, Henri. 1905. "Sur la dynamique de l'électron," *Rendiconti del circolo matematico di Palermo, 21*: 129.

Poor, Charles Lane. 1905. "The figure of the sun," *Ap. J., 22*.

———. 1908. "Figure of the sun, possible variations in size and shape," *New York Academy Annals, 18*.

———. 1921. "Motion of the planets and the relativity theory," *Science, 54*: 30–34.

———. 1922. *Gravitation versus Relativity: A Non-technical Explanation of the Fundamental Principles of Gravitational Astronomy and a Critical Examination of the Astronomical Evidence Cited as a Proof of the Generalized Theory of Relativity.* London: G. P. Putmann's Sons Ltd.

———. 1924. "Eclipse casts shadow on relativity. Astronomers say photograph measurements disagree with Einstein's figures on the deflection of star rays passing the sun," *New York Times*, 6 Jan, sec. 8: 4.

———. 1926. Letter to the editor. *New York Times*, 5 Sep, sec. 7: 14.

———. 1927a. "Relativity: An approximation (Abstract)," *PAAS, 5*: 80.

———. 1927b. "The relativity deflection of light," *JRASC, 21*: 225–238.

———. 1929. "Lick Observatory Bulletin, No. 397," *Pop. Astr., 37*: 355–359.

———. 1930. "The deflection of light as observed at total solar eclipses," *JOSA, 20*: 173–211.

Porter, J. G. 1927. "Recent textbooks and relativity," *Pop. Astr., 35*: 193–194.

———. 1929. "The relativity deflection of light: Facts versus theory," *PASP, 41*: 171–172.

Pyenson, Lewis. 1974. "The Göttingen reception of Einstein's general theory of relativity." Ph.D. diss. Johns Hopkins University, Baltimore.

———. 1976. "Einstein's early scientific collaboration," *HSPS, 7*: 83–124; reprinted in Pyenson (1985), 215–246.

———. 1977. "Hermann Minkowski and Einstein's special theory of relativity," *Archives for History of Exact Sciences, 17*: 70–95.

———. 1985. *The Young Einstein: The Advent of Relativity.* Bristol and Boston: Adam Hilger.

Richtmyer, F. K. 1928. *Introduction to Modern Physics*, 1st ed. New York: McGraw-Hill.

Robertson, Howard Percy. 1929–30. "Foundations of relativistic cosmology," *Ann. Report NAS*: 14.

Rosenfeld, L. 1973. "Gustav Robert Kirchoff," *DSB, 7*: 379–383.

Rothenberg, Mark. 1974. "The educational and intellectual background of American astronomers, 1825–1875." Ph.D. diss. Bryn Mawr College, Pennsylvania.

Russell, Henry Norris. 1920. "Note on the Sobral eclipse photographs," *MNRAS, 81*: 154–164.

———. 1921. "Modifying our ideas of nature: The Einstein theory of relativity," *Annual Report, Smithsonian Institution*, 197–211.

San Francisco Journal. 1923. "Einstein again supported," 9 Dec.

Schroeder-Gudehus, Brigitte. 1973. "Challenge to transnational loyalties: International scientific organizations after the First World War," *Science Studies, 3*: 93–118.

———. 1978. *Les scientifiques et la paix. La communauté scientifique internationale au cours des années vingt.* Montreal: Les Presses de l'Université de Montréal.

Schwarzschild, Karl. 1914. "Ueber die Verschiebungen der Bande bei 3,883A im Sonnenspektrum," *Königlich Preussische Akademie der Wissenschaften* (Berlin) *Sitzungsberichte*, 1201.

———. 1916a. "Ueber das Gravitationsfeld eines Massenpunktes nach der Einsteinschen Theorie," *Königlich Preussische Akademie der Wissenschaften* (Berlin) *Sitzungsberichte*, 189–196.

———. 1916b. "Über das Gravitationsfeld einer Kugel aus inkompressibler Flüssigkeit nach der Einsteinschen Theorie," *Königlich Preussische Akademie der Wissenschaften* (Berlin), *Sitzungsberichte*, 424–434.

Science. 1931. "Professor Einstein at the California Institute of Technology: Address at the dinner in his honor," *73*: 375–381.

See, Thomas Jefferrson Jackson. 1916. "Einstein's theory of gravitation," *Observatory, 39*: 511–512.

———. 1922. "New theory of the aether," *AN, 217*: 193–284.

Shankland, R. S. 1941. "Dayton Clarence Miller: Physics across fifty years," *Amer. J. Phys., 9* (October): 273–283.

Schulmann, Robert, et al., eds. 1998. *The Collected Papers of Albert Einstein*, vol. 8 in two vols, *The Berlin Years: Correspondence, 1914–1918.* Princeton, N.J.: Princeton University Press. Referred to in the Notes as *CPAE*, vol. 8; English translation: A. Hentschel and K. Hentschel (1998).

Silberstein, Ludwig. 1920. "The recent eclipse results and Stokes-Planck's aether," *Phil. Mag., 39*: 161–164.

———. 1922. *General Relativity and Gravitation.* Toronto: University of Toronto Press.

———. 1925a. "D. C. Miller's recent experiments, and the relativity theory," *Nature, 115*: 798–799.

———. 1925b. "Ether drift and the relativity theory," *Nature, 116*: 98.

Slipher, Vesto Melvin. 1913. "The radial velocity of the Andromeda nebula," *Lowell Obs. Bull. No. 58, 2*: 26–27.

Slocum, F. 1921. "Photographic distortion on eclipse plates and the Einstein effect," *Pop. Astr., 29*: 273–274.

Smart, W. M. 1945–46. "Arthur Robert Hinks," *Observatory, 66*: 89.

Smith, Robert W. 1982. *The Expanding Universe: Astronomy's "Great Debate," 1900–1931*: London: Cambridge University Press.

———. 1979. "The origins of the velocity-distance relation," *JHA, 10*: 133–165.

St. John, Charles E. 1910. "The general circulation of the mean and high-level calcium vapor in the solar atmosphere," *Ap. J., 32*: 36–82.

———. 1913a. "Radial motion in sun-spots. I. The distribution of velocities in the solar vortex," *Ap. J., 37*: 322–353.

———. 1913b. "Radial motion in sun-spots. II. The distribution of the elements in the solar atmosphere," *Ap. J., 38*: 341–391.

———. 1913c. "Radial motion in sun-spots," read at 83rd meeting BAAS, *Observatory, 36*: 395–397.

———. 1913d. "Sondage de l'atmosphère solaire par les mésures de vitesses radials dans les taches," *Compte rendus, 157*: 428–430.

———. 1914. "On the distribution of the elements in the solar atmosphere as given by flash spectra," *Ap. J., 40*: 356–376.

———. 1915a. "Anomalous dispersion in the sun in light of observations," *Ap. J., 41*: 28–41.

———. 1915b. "Critique of the hypothesis of anomalous dispersion in certain solar phenomena," *Proc. NAS, 1*: 21–25.

———. 1917a. "A search for an Einstein relativity-gravitational effect in the sun," *Proc. NAS, 3*: 450–452.

———. 1917b. "The principle of generalized relativity and the displacement of Fraunhofer lines toward the red," *Ap.J., 46*: 249–263. Reprinted in *Contributions from the Mount Wilson Observatory, 138*: 261–277.

———. 1918. "On pressure in the solar atmosphere," *PAAS, 3*: 95.

———. 1920a. "Displacement of solar lines and the Einstein effect," *Observatory, 43*: 158–162.

———. 1920b. "The displacement of solar spectral lines," *Observatory, 43*: 260–262.

———. 1920c. "Displacement of solar lines and relativity," *CIW Yearbook, 20*: 242–244.

———. 1922. "The displacement of solar lines," *CIW Yearbook, 21*: 214–216 in G. E. Hale and W. S. Adams, "Mount Wilson Summary of the Year's Work, 198–255.

———. 1927. "The atmosphere of the sun and relativity (abstract)," *PAAS, 5*: 84–86.

———. 1928. "Evidence for the gravitational displacements of lines in the solar spectrum predicted by Einstein's theory," *Ap. J., 67*: 195–239.

———. 1932. "Observational basis of relativity," *PASP, 44*: 277–295.

St. John, Charles E., and Harold D. Babcock. 1914. "A displacement of arc lines not due to pressure," *Phys. Rev., 3*: 487–488.

———. 1915a. "A study of the pole effect in the iron arc," *Ap. J., 42*: 231–262.

———, 1915b. "On the pole effect in the iron arc," *Proc. NAS, 1*: 295–298.

———. 1922–23. "Center and limb displacements," *CIW Yearbook, 22*: 189.

———. 1924a. "Gravitational influence of spectral lines," *Science, 59*: 442–443.

———. 1924b. "General relativity and the displacement of solar lines," *CIW Yearbook, 23*: 90–92.

St. John, C. E., and S. B. Nicholson. 1920a. "Relative wave-lengths of skylight and sunlight reflected from Venus," *CIW Yearbook, 19*: 227–228.

———. 1920b. "Relative wave-lengths of skylight and Venus-reflected sunlight (Abstract)," *PASP, 32*: 194.

———. 1921. "On systematic displacements of lines in spectra of Venus," *Ap. J., 53*: 380–391.

———. 1922. "Wave-lengths in skylight and in the spectrum of Venus," *CIW Yearbook, 21*: 218.

Stratton, F.J.M. 1945–46. "James Hopwood Jeans," *Observatory, 66*: 392.

———. 1956a. "John Evershed, 1864–1956," *BMRAS, 2*: 41–51.

———. 1956b. "Walter Sydney Adams," *BMRS, 2*: 1–18.

Strömberg, Gustav. 1924. "The motions of the stars and the existence of a velocity-restriction in a universal world-frame," *Sci. Monthly, 19*: 465–478.

———. 1926. "Note concerning the radial velocity of the companion of Sirius," *PASP, 38*: 44.

Swenson, Loyd. 1972. *The Ethereal Aether: A History of the Michelson-Morley-Miller Aether-Drift Experiments, 1880–1930*. Austin: University of Texas Press.

Sydney Daily Telegraph. 1923. "Einstein theory. Negative results locally," 3 March.

Sydney Morning Herald. 1923. "Relativity. Eclipse observations. Unsatisfactory results," 22 Mar.

Temple, George. 1956. "Edmund Taylor Whittaker," *BMRS, 2*: 299–325.

Tobey, Ronald C. 1977. *The American Ideology of National Science*. Pittsburgh: University of Pittsburgh Press.

Tolman, Richard Chase. 1910. "The second postulate of relativity," *Phil. Rev., 31*: 26.

Trans. IAU. 1922. *1*.

Trans. IAU. 1925. *2*.

Trans. IAU. 1928. *3*.

Trans. IAU. 1932. *4*.

Trumpler, Robert. 1923a. "Historical note on the problem of light deflection in the sun's gravitational field," *PASP, 35*: 185–188.

———. 1923b. "Note on the program of photography of precision," *PASP, 35*: 288–289.

———. 1924a. "A method for differential measurement of stellar photographs," *PASP, 36*: 9.

———. 1924b. "Preliminary results on the Einstein eclipse test from observations with the five-foot camera (abstract)," *PASP, 36*: 221–224.

———. 1927a. "Relativity as represented by the Einstein Eclipse Problem (Abstract)," *PAAS, 5*: 90–92.

———. 1927b. "The star field surrounding the eclipsed sun at the total solar eclipse of May 9, 1929," *PASP, 39*: 377–379.

———. 1928. "Final results on the light deflections in the sun's gravitational field from observations made at the total solar eclipse of September 21, 1922," *PASP, 40*: 130–135.

———. 1929a. "The relativity deflection of light," *PASP, 41*: 23–35.

———. 1929b. "Reply to Professor Poor's review of Lick Observatory Bulletin No. 397," *Pop. Astr., 37*: 359–362.

Trumpler, Robert. 1932a. "The deflection of light in the sun's gravitational field," *Science,* 75 (20 May): 538–540.

———. 1932b. "The deflection of light in the sun's gravitational field," *PASP,* 44: 167–173. Reprint of above.

———. 1932c. "Die Ablenkung des Lichtes im Schwerefeld der Sonne," *Zs. Astrophys.* 4: 208–220.

———. 1935. "Observational evidence of a relativity red shift in class O stars," *PASP,* 47: 249–256.

———. 1938. "William Wallace Campbell 1862–1938," *The Sky,* December, 18.

Turner, Herbert Hall. 1899. "On the fundamental law of increase of gaseous reputation," *Observatory,* 22: 292.

———. 1912. *The Great Star Map, Being a Brief General Account of the International Project Known as the Astrographic Chart.* London: J. Murray.

———. 1913a. "Oxford notebook," *Observatory,* 36: 356–366.

———. 1913b. "Oxford notebook," *Observatory,* 36: 382–386.

———. 1913c. "Oxford notebook," *Observatory,* 36: 413–417.

———. 1919. "From an Oxford note-book," *Observatory,* 42: 442–445.

———. 1923. "From an Oxford note-book," *Observatory,* 46: 171.

U., H. S. 1920. "The Einstein displacement of solar lines," *Amer. J. Sci.,* 200: 394.

von Laue, Max. 1917. "Die Fortpflanzungsgeschwindigkeit der Gravitation. Bemerkungen zur gleichnamigen Abhandlung von P. Gerber," *Ann. der Phys.,* 52: 19–21.

von Seeliger, Hugo. 1901. "Kosmische Staubmassen und das Zodiakallicht," *Akademie der Wissenschaften, Berlin, Math.-phys. Klasse, Sitzungsberichte,* 26: 265–292.

———. 1906a. "Das Zodiakallicht und die empirischen Glieder in der Bewegung der inner Planeten," *Sitzgsb. Ak. W. München,* 36: 595–622.

———. 1906b. "Die empirischen Glieder in der Theorie der Bewegungen der Planeten Merkur, Venus, Erde und Mars," *Vierteljahrschrift,* 41: 234.

von Soldner, J. 1804. "Ueber die Ablenkung eines Lichtstrals von seiner geradlinigen Bewegung, durch die Attraktion eines Weltkörpers, an welchem er nahe vorbei geht," *Berliner Astronomisches Jahrbuch,* 161–604.

Wacker, Fritz. 1906. "Ueber Gravitation und Elektromagnetismus," *Phys. Zs.,* 7: 300–302.

Waterfield, Reginald L. 1938. *A Hundred Years of Astronomy.* London: Duckworth.

Watson, Warren Zachary. 1969. "A historical analysis of the theoretical solution to the problem of the advance of the perihelion of Mercury." Ph.D. diss., University of Wisconsin, Madison.

Weart, Spencer. 1979. "The physics business in America, 1919–1940: A statistical reconnaissance," in Nathan Reingold, ed., *The Sciences in the American Context: New Perspectives,* 295–348. Washington, D.C.: Smithsonian Institution Press.

Whitehead, Alfred North. 1926. *Science and the Modern World.* London.

Whitney, Charles A. 1976. "Henry Draper," *DSB,* 13: 474–475.

Whittaker, Edmund Taylor. 1910a. "Recent researches of space, time, and force," *MNRAS,* 70: 363–366.

————. 1910b. *A History of the Theories of Aether and Electricity from the Age of Descartes to the Close of the Nineteenth Century.* London: Longmans, Green.

————. 1953. *A History of the Theories of Aether and Electricity,* Vol. 2, *The Modern Theories, 1900–1926.* London: Nelson and Sons.

————. 1955. "Albert Einstein," *BMRS,* 37–67.

Wiechert, Emil. 1911a. "Relativitätsprinzip und Aether," *Phys. Z.,* 12: 689–707.

————. 1911b. "Relativitätsprinzip und Aether," *Phys. Z.,* 12: 737–758.

Wilkens, Alexander. 1927. *Hugo von Seeligers wissenschaftliches Werk.* Munich.

Will, Clifford. 1986. *Was Einstein Right? Putting General Relativity to the Test.* New York: Basic.

Wilson, E. B. 1917. "Generalized co-ordinates, relativity, and gravitation," *Ap. J.,* 45: 244–253.

Wright, Helen. 1966. *Explorer of the Universe. A Biography of George Ellery Hale.* New York: E. P. Dutton.

Wright, Helen, Joan Warnow, and Charles Weiner, eds. 1972. *The Legacy of George Ellery Hale.* Cambridge, Mass.: MIT Press.

Wright, William Hammond. 1923. "The Lick Observatory-Crocker expedition to Ensenada, Lower California, September 10, 1923," *PASP,* 35: 275–280.

————. 1941. "William Wallace Campbell," *BMNAS,* 25: 35–74.

INDEX

Abbot, Charles Greeley, 207
aberration of starlight, 28–29
Abraham, Max, 73, 78
absolute motion, 31–32, 39
absolute space, xxi; cosmological model of universe and, 106; relativity and, 5, 38–39; Whittaker on, 31–32
absolute time, xxi, 38–39
acceleration, and gravitation, 47–48
Adams, Charles Edward, 183, 195, 287
Adams, Walter Sydney, 319; 1923 eclipse expedition, 237–38; on D. C. Miller's work, 277; limb-center shift studies, 66–68, 73, 74–75; magnesium redshift smaller than Einstein shift, 172; pressure theory of solar line shifts, 66, 68; Sirius B measurements, 259–62, 278–82; spectroscopic research on solar rotation, 17; speech during Einstein's 1931 visit, 318–19; on Strömberg's work, 274
Adler, Friedrich, 6–7
Africa/Brazil eclipse (1919), 129–31, 138–40, 143–44, 346n54. *See also* British eclipse expedition (1919)
Aitken, Robert Grant: acceptance of relativity by, 322; on advances in astronomy, 296–99; on British eclipse results, 149, 151; Curtis asks to withhold Goldendale paper, 141; on modifying equipment for Einstein test, 113; on Poor's relativity critique, 303, 304; replies to See's letter, 223–24; on See, 250
Aiyar, Narayana, 126, 171
Allegheny Observatory: 1925 eclipse, 265–66; Curtis becomes director, 8, 183; parallax program at, 8; solar redshift experiments at, 246, 247–48, 251, 254–56; viewing conditions at, 8, 25–26. *See also* California/Mexico eclipse (1923)
Alter, Dinsmore, 231
Ambronn, Leopold, 183
America. *See* United States
American eclipse expeditions. *See specific astronomers, eclipses, and observatories*
American eclipses: California/Mexico (1923), 228–31, 236–41, 249–50, 252–54; New Haven (1925), 265–66. *See also* Goldendale eclipse (1918)
American Philosophical Society, 206, 207, 248, 250, 251, 253–54, 265, 303
Anderson, John, 258
angular deviation of light, gravitational, 49–50. *See also* light bending
Annalen der Physik, 5, 47, 95
anomalous dispersion, 72–73, 105, 109, 225, 356n63, 360n95
antirelativity campaign: Curtis and, 242, 245–46, 247–48, 251–52, 257, 270, 308; difficulties encountered by, 256–57; effect on evaluation of theory, 322; ether drift advocates and, 231–35; in Germany, 102–3, 153, 164–65, 191, 216–18, 219; group effort organized by Poor, 246, 251–52; Lick/Mount Wilson response to, 239, 252–56, 269, 273, 310–14; Miller (D. C.) and, 309, 311; Optical Society symposium and, 307–9; Porter and, 300–301, 304, 307; proponents of, 269–73, 303–4; See and, 98–99, 216–25, 269–70. *See also* Poor, Charles Lane
anti-Semitism, 6–7, 164–65, 192
Arrhenius, Svante, 153, 175
astronomers: attitudes toward relativity, 132–33, 147–52, 213–15, 228–29, 243, 270, 296–97, 299; duties of, 8–9
astronomers, first publications on relativity by: Burns, 36–38; Curtis, 40–43; de Sitter, 38–40; Plummer, 28–31; Whittaker, 31–36
astronomical consequences of relativity, 38–40, 43, 48–50, 54, 88, 91. *See also* light bending; perihelia of planets, residual motions; redshift, gravitational
astronomical photography: and astrophysics, 9–10; development of, xxii, 50–51; and solar eclipse studies, 50–51, 54–55. *See also* eclipses; photographic plates of eclipses
Astronomical Society of the Pacific, 20, 51, 133–34, 138, 297
astronomical time, definition of, 40